THE END OF SEX AND THE FUTURE OF HUMAN REPRODUCTION

THE END OF SEX

and the Future of Human Reproduction

HENRY T. GREELY

Harvard University Press

Cambridge, Massachusetts, & London, England

Printed in the United States of America

First Harvard University Press paperback edition, 2018
First printing

Library of Congress Cataloging-in-Publication Data

Names: Greely, Henry T., author.
Title: The end of sex and the future of human reproduction / Henry T. Greely.
Description: Cambridge, Massachusetts: Harvard University Press, 2016. |
Includes bibliographical references and index.
Identifiers: LCCN 2015043931| ISBN 9780674728967 (hc : alk. paper) |
ISBN 9780674984011 (pbk.)
Subjects: LCSH: Preimplantation genetic diagnosis. | Preimplantation genetic
diagnosis—Moral and ethical aspects. | Human reproduction.
Classification: LCC RG628.3.P74 G74 2016 | DDC 618.3/042—dc23

LC record available at http://lccn.loc.gov/2015043931

FOR LAURA.

Of course.

CONTENTS

THE END OF SEX AND THE FUTURE OF HUMAN REPRODUCTION

INTRODUCTION

CHANGES

This is a book about the future of our species, about the likely development of revolutionary biological technologies, and about the deep ethical and legal challenges our societies will face as a result. But the best way to sum it up, I think, is to say that it is about the coming obsolescence of sex.

It is not about the disappearance of all the things we mean by the word "sex." Humans will still (usually) appear at birth having physical attributes of one sex or the other and will be loudly pronounced as either baby girls or baby boys, with the appropriately colored, and gendered, accessories. Our descendants will still (almost all the time) have genetic contributions from both an egg and a sperm, thereby achieving the mixing of parental genes that is also sex or, at least, sexual reproduction. And, I am confident, people will continue to practice sexual intercourse in myriad different ways and for almost all of the current varying, complicated (and uncomplicated) reasons. Except one.

I expect that, sometime in the next twenty to forty years, among humans with good health coverage, sex, in one sense, will largely disappear, or at least decrease markedly. Most of those people will no longer use sexual intercourse to conceive their children. Instead of being conceived in a bed, in the backseat of a car, or under a "Keep off the Grass" sign, children will be conceived in clinics. Eggs and sperm will be united through in vitro fertilization (IVF). The DNA of the resulting embryos will then be sequenced and carefully analyzed before decisions are made (passive voice intentional) about which embryo or embryos to transfer to a womb for possible development into one or more living, breathing babies.

Prospective parents will be told as much as they want to know about the DNA of, say, 100 embryos and the implications of that DNA for the diseases, looks, behaviors, and other traits of the child each of those embryos might become. Then they will be asked to pick one or two to be transferred into a womb for possible gestation and birth. And it will all be safe, legal, and, to the prospective parents, free.

In short, we humans will begin, very broadly, to select consciously and knowingly the genetic variations and thus at least some of the traits and characteristics of our children. This idea is not new. It has been a subject of hundreds, probably thousands, of stories and novels—*Brave New World* by Aldous Huxley being, if not the first, certainly the first truly memorable example.[1] It has been the subject of other forms of fiction, notably the 1997 movie *Gattaca*.[2] And it has been the subject of tens of thousands of books, articles, sermons, and other nonfiction analyses—usually viewed with alarm, but occasionally with (prospective) pride.

This book is different. Not, at its heart, a discussion of the consequences of such a world (although Part III does try to analyze them to some extent), it is a description of precisely how and why that world is going to arrive. Two insights drive the book. The first is the way new techniques, drawn from several different areas of modern bioscience research, will combine to make this future not just possible but cheap and easy. The second is the way economic, social, legal, and political forces will combine to make this future not just achievable but, as I believe, inevitable, in the United States and in at least some other countries. Those insights turn these questions from interesting, goosebump-inducing speculation to real problems that will confront real people—ourselves, our children, and our grandchildren—in the next few decades.

The technical innovations will come from two worlds: genetics and stem cell research. We can already do preimplantation genetic diagnosis (PGD) on embryos. We can take away a few cells from an early "test tube" embryo, test them for a genetic trait or two, and use that information to decide whether to give the embryo a chance to become a baby. PGD sounds like science fiction to many people but it has been used for over a quarter century—the first child born after PGD is now over twenty-five years old. And every year now, around the world, thousands of new children are born after being subjected to PGD as embryos.

But today PGD is only weakly informative, as well as expensive, unpleasant, and even dangerous, thanks both to the limitations of genetic testing and to the necessity of using IVF as part of PGD. These constraints will change. Genetics will allow us to do cheap, accurate, and fast sequencing of the entire 6.4 billion base pair genome of an embryo and will give us an increasingly deep understanding of what that sequence means for disease risks, physical characteristics, behaviors, and other traits of the child that embryo would become. And stem cell research will allow couples to avoid the expensive and (for the women involved) unpleasant and physically risky process of maturing and retrieving human eggs by allowing us to make eggs (and sperm) from stem cells. The result will be a cheap, effective, and painless process I call "Easy PGD."

Of course, just because technological innovations are possible does not mean they will be adopted. The supersonic commercial jetliner came and went; the flying car and the rocket backpack were never really launched, though both are technically feasible. But unlike those technologies, Easy PGD has a clear path to acceptance in the United States and likely paths to adoption in many other countries. It may not be approved everywhere, but in an increasingly global world, that could well be irrelevant.

The ideas in the last few paragraphs are the core of this book. I will also discuss some of the potential consequences that widespread adoption of DNA-based embryo selection using Easy PGD will have for individuals, for families, for societies, and for humanity. The fields of genetic selection have been frequently plowed before; I hope the specificity of Easy PGD as the method of choice for parents to select their children's traits, as well as the near immediacy of the questions it raises, will add some value to my analysis over those that have come before.

Concretely, the book is divided into three parts. Part I provides background information on the science and technology involved in Easy PGD. It gives a nonscientist a guide to the varied ways living things reproduce; to the specifics of how humans reproduce, naturally and by IVF; to DNA, genes, chromosomes, and genetic testing; and to stem cell research. Much of it will be helpful in understanding what follows; I must confess that some of it is here in the hope that you will come to share the excitement and fascination of biology with me, a person whose last biology class was in tenth grade. Part II explains how and why Easy PGD will happen, looking first at the technical developments in genetics (or genomics) and in stem cell science and then at the medical, economic, legal, and political factors that will make it not just acceptable, but widely adopted. Part III examines the broader implications of Easy PGD. It looks at issues of safety, family, equality, coercion, and nature, along with some other more practical consequences of the technology.

I've gotten lots of good advice in writing this book, but I haven't taken all of it. Although IVF, the fountainhead of modern assisted reproductive technologies, is less than forty years old, it has already spawned a vast literature on a wide range of issues, including many fascinating and important matters for which Easy PGD would be relevant, such as surrogacy, parental status, gamete donor rights (and duties), and the positions and roles of religious beliefs, among others. This book could and perhaps should be longer; however, practical considerations mean that the likely interactions between Easy PGD and other issues I do not analyze must await future treatments.

More fundamentally, some people have told me to make an argument—to take a position and fight for it, guns blazing. But I'm a law professor, trained as a lawyer. Lawyers do many things. Sometimes they argue zealously in court for their clients' positions, whether they believe them or not. But sometimes they lay out all the facts and implications, as they see them, to help clients make their own decisions. I have some views about ways we might want to regulate Easy PGD, but they are tentative, based on glimpses and guesses of the future and on my own preferences and principles. I will share them, but I do not insist on them. But I will ask you to develop opinions. Easy PGD will give prospective parents—including perhaps some who are reading these words—more choices but it will also set some hard questions for all of us. My goals are, first, to get you interested in those questions—as parents, as grandparents, as citizens, as *humans*—and second, to give you information to help you come to your own conclusions.

Aldous Huxley's famous novel takes its title from one of Shakespeare's last plays, *The Tempest*. Years before the play starts, plotters abandon Prospero, who is both the Duke of Milan and a magician, at sea with his infant daughter, Miranda. They survive on an island with only non-human company. The years go by—Miranda grows up, and fate, working through Shakespeare, delivers the plotters to the island and into Prospero's hands. Miranda sees them, almost the very first humans she ever remembers seeing, and, not knowing that some of them had long ago plotted her death along with her father's, she famously exclaims:

O, wonder!
How many goodly creatures are there here!
How beauteous mankind is! O brave new world,
That has such people in't![3]

That is often remembered. What few recall (though I am sure Huxley did) was Prospero's immediate reply: "'Tis new to thee." My hope is that when Easy PGD opens the prospects of some kind of brave new world, you will be more knowledgeable, and more sophisticated, than Miranda. (And that things will work out as well for you as, happily, they do for her in the end.)

PART I

THE SCIENCE

This part of the book sets out, in six chapters, the scientific background that I think is useful for understanding Easy PGD and its implications. The chapters cover basic information about cells, DNA, and genes; "normal" reproduction among living things, including humans; assisted reproduction in humans, genetics, genetic testing, and stem cells. I have tried to write about them to make the information understandable to anyone interested, even those of you who, like me, last took a biology course at the age of fifteen.

Some of you will have educational and professional backgrounds that give you far more knowledge of the areas than I can convey, or know (although, given the increasing specialization of science and medicine, I suspect very few of you will be expert in all of these fields). I will not be offended if you skip some or all of these chapters. Others of you, without a background in these sciences, will be determined to stay that way and will not want to read these chapters. I hope you change your minds. I came to biology late in life, as an amateur, and I fell in love with it—with its breadth, its combination of deep unities and myriad complexities, its many rules—each with exceptions and every exception with provisos—and its infinite surprises. In many ways it reminds me of my professional field, the law. One of my goals for this book is to bring to some of you a love of biology. For that, I need you to read the next six chapters. And I think even those of you who plan to read the next six chapters may want to read the rest of this introduction—it should be useful to guide your deeper reading, though it does mean that some parts of the next chapters will be familiar.

Part I of the book is a fairly shallow look at the science; before my last edits it was twice as long and still shallow. But I know that for some of you even those shortened chapters will be far too long, so the rest of this introduction is for you: it is the relevant biology in a nutshell—and not a big one (pistachio, maybe?). Please remember that everything that follows is incomplete and much of it is, at least in some particular and unusual applications, wrong—or at least not quite right. (And if you want references, read the endnotes to the following chapters.)

Living organisms are made out of cells, sacks of materials held together like water balloons by membranes or walls. Most living things have only one cell; the vast majority of them are bacteria or archaebacteria, which have only very simple cells. Some one-celled organisms and all multicelled organisms, from plants to ants to us, have more complicated types of cells, which have distinct different "organelles" inside them. One of those "little organs" is the nucleus of the cell. The nucleus contains (almost) the cell's entire DNA, a molecule known more fully as deoxyribonucleic acid. The DNA in the nucleus is organized into distinct bodies called chromosomes, which come (mainly) in pairs. Humans normally have 46 chromosomes, one pair each of chromosomes 1 through 22 (the autosomes) and two sex chromosomes, either two X chromosomes (in women) or an X and a Y chromosome (in men).

Cells normally reproduce by doubling their chromosomes and splitting in half, sending the right number of pairs of chromosomes to each daughter cell. Each of the two daughter cells is a "clone" of the parent cell—they are genetically identical. Most life on this planet reproduces by cloning, but most of the life visible to our naked eye does not. Instead, it reproduces sexually. Sexual reproduction around the biosphere is much more varied and complicated than it is in humans, but, at its core, it ensures that instead of being genetically identical copies, an organism's offspring are a new combination of the chromosomes from two different "gametes," sperm and eggs.

Human sexual reproduction is so complicated that it is amazing any of us gets born. But, basically, sperm from a man makes a long and arduous journey to meet with a woman's ripe egg, which has made its own shorter but difficult trip. The sperm and the egg each carry 23 chromosomes from the man or woman, half the usual number. The sperm merges into the much larger egg (think of a small pea going into a basketball). After fertilization the egg is renamed a zygote and chromosomes from the egg and sperm eventually merge to form a new nucleus, which begins to divide. After four or five days of dividing, the resulting embryo is a hollow ball, about five one-thousandths of an inch wide, perhaps visible to someone with good eyes in good light. Shortly thereafter, it needs to be in the womb, attaching to its lining and becoming implanted, if it is to have any chance to be born.

Some couples cannot have babies the usual way. Sometimes the problem is with the woman's eggs getting to or implanting in the womb, sometimes it is with the man's sperm getting to and fertilizing the eggs, and sometimes the cause is unknown. In many cases assisted reproduction can help, often through IVF.

In IVF, the woman's ovaries are artificially forced to ripen extra eggs, which are then surgically extracted. This process is expensive, unpleasant, and somewhat risky for the woman involved. The eggs are usually mixed with sperm and become fertilized, although often a procedure called intracytoplasmic sperm injection (ICSI) is used. (Of course, if one of the would-be parents has no eggs or sperm, IVF alone is insufficient and the couple will need "donated" eggs or sperm—often sold.) Either way, some of the eggs will be fertilized successfully, and the resulting zygotes will begin to divide in containers in the clinic. If the zygotes divide successfully for a long enough time, they will be transferred into a woman's womb sometime between the third and sixth day after fertilization, in the hope that they will implant and eventually become babies.

Now we need to go back to the chromosomes and the DNA they contain. DNA is famously called "the double helix." For those of you who, like me, don't have a good mental image of a double helix, think of a very long ladder that has been twisted into a spiral. The sides of the DNA ladder are unimportant; the rungs are crucial. Each rung is made up of two out of four molecules: adenosine, cytosine, guanine, and thymine—widely known as A, C, G, and T. But A will only combine with T to make a rung and C will only combine with G. The rungs, therefore, are made up of "base pairs" consisting of either A-T, C-G, G-C, or T-A. By reading the bases attached to one side of the ladder, you get the DNA's "sequence"—for example, AGCGAGTTTTCG. (The "other" sequence, attached to the other side of the ladder, *must* read TCGCTCAAAAGC.) But instead of just the twelve bases in that example, the sequence of a whole human chromosome is between 50 and 250 million bases long.

Humans normally have 46 chromosomes, one copy of chromosomes 1 to 22 plus a sex chromosome (either an X or a Y) from their fathers and another copy of those autosomes plus, necessarily, an X chromosome from their mothers (the mothers only have X chromosomes to give—if they had a Y chromosome, they would be male). The sequences of all chromosomes from one parent make up "the human genome" and are about 3.2 billion bases long. Each of us has two copies of the human genome, one from each parent. These copies are very similar to each other (except for men, whose Y chromosome is quite different—much smaller and less important—from the X chromosome), but they do differ in about one base, or "letter," in one thousand. Your complete genome sequence then, is about 6.4 billion bases long. If you think of each base as a character in the English language—a letter, punctuation mark, or space—your genome is about as long as 700 copies of the King James Version of the Bible.

Most of the bases in the human genome have no known (and quite possibly no unknown) meaning, but about 1.5 percent of them spell out instructions ("code") to the cell on how to make particular proteins, the molecules that make up most of the substance of our bodies. Another chunk of the DNA letters—whether it is 5 percent, 10 percent, or more is controversial—control when and how much those protein-coding regions will be turned on or off, up or down, as well as making other useful molecules of a type called RNA. The exact meaning of the term "gene" is surprisingly unclear, but the human genome contains about 23,000 protein-coding regions, which can make over 100,000 different human proteins. By reading the genetic code of the sequence, we can know what those proteins are made of and whether they are normal, dangerously abnormal, or abnormal in ways that might or might not be important.

Human genetic testing has taken place for about fifty years, using many different methods. Today (and increasingly in the future) it involves looking at DNA sequences in regions of chromosomes that are known to be important and trying to figure out whether a person's sequence is normal or dangerous. For example, the famous breast and ovarian cancer gene, *BRCA1* (by convention, gene names should be italicized), is made up of about 80,000 bases near the end of the long arm of chromosome 17. It can be sequenced to see if a woman has a normal version (the vast majority of people), a version known to be dangerous (less than 1 percent of people), or a version that is not normal but may or may not be dangerous (another roughly 5 percent).

It has only been possible to sequence a person's entire genome for less than fifteen years. The first whole genome sequence cost about $500 million and took years. Today you can get your genome sequenced in a few days for about $1,500. Observers expect this price to continue to fall, very soon to about $1,000 and eventually much further. Most people expect whole genome sequencing to be widely used for genetic tests in a few years as the price drops.

Genetic testing can be used in many different contexts. Adults or children can be tested to diagnose, or predict, diseases or traits. Fetuses can be tested before birth, through three different technologies, starting between the tenth and the eighteenth weeks of pregnancy. And embryos created through IVF can be tested before they are transferred for possible implantation and pregnancy, usually about five days after fertilization. The later process, PGD (short, recall, for preimplantation genetic diagnosis), involves taking a few cells from the embryo and then testing those cells. The results of those tests are then used to decide whether to transfer an embryo. In the past twenty-six plus years of PGD's use, it could only be used to test any particular embryo for one or a handful of genes. PGD has been used to look for DNA associated with a genetic disease found in the family, for DNA that would allow an embryo to become a baby that could be a cord blood donor to a family member, or for the embryo's future sex.

What all can genetic testing tell us? It depends. For some things, genetic tests reveal destiny. Anyone whose DNA has the version of the *Huntingtin* gene associated with Huntington disease can only avoid dying of that disease by dying first of something else. But a woman with a genetic variation of the *BRCA1* gene has only about a 60 to 85 percent chance of being diagnosed with breast cancer during her lifetime and only about a 30 percent chance of an ovarian cancer diagnosis. A man with a dangerous variation in the *BRCA2* gene has about 100 times the normal man's risk of breast cancer, but his risk is still only a few percent. The percentage of people with a particular DNA variation who will get a disease or a trait associated with that variation is called the variation's "penetrance."

Today, genetic testing can give us strong information about a few thousand genetic diseases, almost all of them rare, as well as some nondisease traits, like ABO blood type. It can give us weaker information about other diseases or traits and very weak to no information about others. In the long run, though, DNA sequences should be able to reveal much, though not everything, about disease and trait "risks" that can be lumped into five categories: highly penetrant, serious, early-onset diseases; other diseases; cosmetic traits (hair color, eye color, and so on); behavioral traits (math ability, sports ability, personality type); and sex—boy or girl.

Whole genome sequencing has now been used experimentally on early embryos. Today it is too expensive and inaccurate to be widely used but that will change and when it does, PGD should become more popular because it will be able to make far more predictions about an embryo's possible future. But there is still one more barrier—PGD requires IVF and IVF is difficult. The answer is our last area of science: stem cells.

Most human cells have limited lifespans. After a certain number of divisions, usually about forty to eighty depending on the cell type, they stop dividing and die. Stem cells don't—they just keep dividing, perhaps indefinitely. Furthermore, some stem cells divide into different kinds of cells. So blood-forming stem cells can eventually make all the scores of different kinds of blood cells in our bodies.

Human embryonic stem cells (hESCs) are created by taking the cells inside the hollow ball that is a five-day-old embryo and growing them in a laboratory. They can become any cell type in the human body. We know that because those cells on the inside of the embryo go on to become every cell type in your body and mine. Extraction of hESCs has been extremely controversial because it requires the destruction of a human embryo. In 2007, Shinya Yamanaka in Japan produced the first human "induced pluripotent stem cells" (iPSCs). These are cells from normal body tissue (usually from the skin) that he treated in a way that made them act like hESCs. They, too, are expected to be able to become every human cell type, including eggs and sperm. And, in fact, baby mice have already been created from mouse eggs and mouse sperm derived from both mouse embryonic stem cells and mouse iPSCs.

If ripe human eggs could be derived from a person's skin cells, it would avoid most of the cost, almost all of the discomfort, and all of the risk of IVF. It should also provide an unlimited supply of eggs, from women at any age. Along with accurate, inexpensive whole genome sequencing of early embryos, that should make PGD much easier and more attractive, leading to what this book calls Easy PGD.

For more information on any or all of the science discussed above, please read some or (better) all of the next six chapters. But if you have had enough, proceed to the First Interlude, between the end of Part I and the beginning of Part II, to pick up the story of *The End of Sex*.

1

CELLS, CHROMOSOMES, DNA, GENOMES, AND GENES

Nineteenth-century author Samuel Butler wrote, "It has, I believe, been often remarked, that a hen is only an egg's way of making another egg."[1] Butler could not have known it, but his statement could have been made more foundational by saying "a hen is only chicken DNA's way of making more chicken DNA." Richard Dawkins's famous term "selfish DNA" encapsulates that idea.[2] Deoxyribonucleic acid—DNA—is the thread that connects generation with generation.

This book will have quite a lot to say about DNA, but some basic knowledge of cells, chromosomes, DNA, genomes, and genes is essential at the very beginning. This chapter provides that first, shallow background, starting with cells.[3]

Cells

Life as we know it is made up of cells that contain DNA. Cells make up tiny bacteria, enormous whales, and everything in between, including us. These cells are living containers of proteins, fats, sugars, and other molecules, held together within an external membrane. They are often busy things, taking in molecules, giving off molecules, expanding, contracting, moving, and otherwise interacting with their environments. And, from time to time, they split into two identical copies of themselves, using division to multiply.

Of course, in biology few if any bald statements are without exceptions. Viruses, which some consider living (though not me) are not made up of cells.[4] Sometimes apparently multiple cells are actually contained within one vast membrane, along with multiple copies of DNA; examples include not only slime molds[5] but also human skeletal muscles.[6] Some unquestionable cells, like our red blood cells, contain no DNA. And some cells, like many neurons in our brains, after being created by cell division, will never divide again. Nonetheless, the DNA-containing cell is truly the building block of life.

Most biologists now divide the world of living things into three great branches, the "domains" of bacteria, archaea, and eukarya. Bacteria and archaea are single-celled organisms that are dramatically smaller that eukaryotic cells and lack much of the internal specialization found in those cells. In particular, they lack a cell nucleus. Bacteria and archaebacteria are collectively called prokaryotes, meaning, from Greek, *pro* (before) *karyon* (kernel or nut).

This chapter will focus on eukaryotes ("good kernels"), as all regularly multicellular organisms and, hence, (almost) all organisms visible to the naked eye are eukaryotes. Though remember that many eukaryotes are neither multicellular nor visible. Lots of single-celled microscopic beasties, from the malaria plasmodium to high school's familiar paramecium, are eukaryotes.

Eukaryotes are organisms whose cells contain nuclei, an area inside the cell that is set off from the rest of the cell by its own "nuclear membrane." All (well, almost all) eukaryotic cells have nuclei, along with various other specialized bodies known as organelles. Almost all the DNA in eukaryotic cells is normally contained in the nucleus, organized onto chromosomes.

Chromosomes

Chromosome just means "colored body" in Greek; the name comes from the fact that some dyes strongly stain these parts of the cell—they are not colorful without the dyes. Each chromosome is basically one very long molecule of DNA, wrapped around a backbone of protein. In eukaryotic cells chromosomes usually come in pairs. Thus, humans have 46 chromosomes. Forty-four of them are always paired and are named chromosomes 1 through 22. Geneticists refer to these chromosomes as the "autosomes." Geneticists named the human autosomes in order from the largest, chromosome 1, which is made up of about 250 million base pairs of DNA, to the shortest, chromosome 22, with about 50 million base pairs. In two places, though, they goofed—chromosome 21 is actually shorter than chromosome 22 and chromosome 19 is shorter than chromosome 20.

In addition to these 22 pairs of autosomes, humans have two more chromosomes, the X and the Y chromosomes. These are called the sex chromosomes for the good reason that they (largely) determine a person's sex. Someone whose cells have two copies of the X chromosome (and hence a grand total of 23 pairs of chromosomes) is (almost always) female; someone who has one X chromosome and one Y chromosome is (almost always) male. The X chromosome is moderately large, about as big as chromosome 7, and contains many genes. The Y chromosome is the third-smallest human chromosome and contains the fewest genes.

You may have seen pictures of a human cell's chromosomes, neatly stained into a striped pattern and laid out in pairs, looking a bit like misshapen barber poles. These kinds of images of chromosomes are called karyotypes and have long been important in some forms of genetic testing. These pictures also nicely show the centromere, a special part of the chromosome that is usually near the middle and that separates each chromosome into two arms, a short one (called p for "petit") and a long one (called q for "the letter after p"). Less clearly visible are the special structures at each end of the chromosomes, the telomeres, long stretches of repetitive and probably "meaningless" DNA.

But chromosomes in cells almost never look like karyotypes. Not only are they largely invisible unless stained, but they are also rarely neat and compact. Most of the time, the DNA on a chromosome sprawls wildly through the nucleus, which will be like a sphere filled with 46 long strands of very, very thin pasta. Just how filled and just how thin is impressive. The chromosomes in any one human cell, if straightened out, would stretch for about 2 meters—roughly six and a half feet. But they all fit into a cell nucleus that is about six millionths of a meter in diameter. By comparison, that is as though a basketball held forty-six pieces of string, ranging from about 250 meters (810 feet) to 55 meters (180 feet) long. The string clearly would have to be very thin indeed and so are chromosomes.

The chromosomes cannot be condensed during most of the cell's life because the cell cannot use the DNA in the chromosomes unless it is "unwound." Only during the process of cell division do the chromosomes condense from this massive and tangled ball of angel hair pasta into the discrete, rodlike chromosomes we see in karyotypes.

Obviously, it is much easier to see, and to think about, the chromosomes in their condensed forms and so that is how they are imagined. Thanks to the bands made by the stains, and the more subtle subbands and sub-subbands within the bands, these condensed chromosomes can then be divided into particular parts and given "addresses." For example, 5q32 means a location in the second subband on the third band of the long arm (q) of chromosome 5. 5q35.2 would be the second sub-subband on the fifth subband on the third band on the long arm of chromosome 5.

Deoxyribonucleic Acid—DNA

From chromosomes we now need to jump down to DNA before coming back to genes. And here we come to DNA's famous "double helix," the discovery of which made James Watson and Francis Crick immortal (and left Rosalind Franklin, whose work was essential to their discovery, largely out in the cold).[7]

When I first heard the term, double helix did not make sense to me, in large part because I did not have a good sense for what a helix was. It helped me to think of the double helix as a ladder twisted into spirals. Each side of the ladder is one helix, twisting around the other but connected to it by the rungs.

The sides of the ladder, which provide the structure of the DNA molecule, are made up of unvarying, and uninteresting, components. These backbones are made up of molecules of a particular kind of sugar ("deoxyribose," which means a "ribose" sugar that is missing one oxygen molecule) connected to each other by phosphate molecules. It is just an unvarying sugar-phosphate-sugar-phosphate-sugar-phosphate combination, over and over for tens of millions of sugars and phosphates.

DNA's power is in the twisted ladder's rungs. The rungs of the ladder are made up of two other molecules, each attached to a deoxyribose sugar molecule on the side. These rungs are the famous "base pairs," base because they are basic, not acidic, and pairs because, in DNA, they always come in pairs. Four kinds of bases make up the base pairs of DNA: adenine, cytosine, guanine, and thymine (A, C, G, and T). Collectively, along with another molecule, uracil, or U, which is not found in DNA but in its important cousin, RNA, they are called the nucleotides.

The deoxyribose sugars on the sides of the twisted ladder are happy to bond with any of the four DNA nucleotides and, in the DNA molecule, each of them will, in fact, be joined to an A, C, G, or T. But a nucleotide on one side of the ladder will connect in the middle with one and only one other kind of nucleotide. Adenine bonds only with thymine in DNA; cytosine bonds only with guanine: A with T, C with G. In normal DNA every rung is complete, so in every place where one side of the DNA molecule has an A, the other side *must* have a T, and so on. Every rung is either AT, CG, GC, or TA.

The "sequence" of the DNA is the order of these nucleotides, as attached to one side of the ladder. For example, ATTCGATAGACT would be the sequence for one stretch of a dozen nucleotides. Of course, that is the sequence on only one side of the DNA molecule. But once we know that sequence, we know that the sequence on the other side *must* be TAAGCTATCTGA, because A and T always bind to each other, as do C and G. (The two sides are identified as 5′—"five prime"—and 3′—"three prime"—and the sequence is, by convention, read from the 5′ side.)

This is the great secret of DNA because it provides a way for one cell to become two copies of itself. If the DNA is split down the middle—if the twisted ladder is, in what I hope will not be a confusing mixed metaphor, "unzipped"—each side of the ladder will be floating free with half rungs (unattached nucleotides) sticking out into the now-unconnected middle. Everywhere there is an A, a T will be attached; every unpaired G will match up with a C. One molecule of DNA, split down the middle, can become—in fact, normally *does* become—two molecules of DNA, identical to the first molecule. Here is the way to turn one twisted ladder into two twisted ladders, each identical to the other and to the ladder that split to produce them. And so, if the DNA contains the instructions for the cell, this is how it can become two identical copies of one set of instructions.

Watson and Crick acknowledge this in a famous understatement near the end of their very short first publication on the structure of DNA: "It has not escaped our notice that the specific pairing we have postulated immediately suggests a possible copying mechanism for the genetic material."[8] The need to copy, precisely, the genetic material that passes from one cell, and one organism, to another is crucial. The mechanisms by which DNA is copied turn out to be quite complex—Nobel prizes have been and continue to be won through clarifying them—but Watson and Crick saw the basic story—and transformed biology.

Looked at in gross, DNA is dull, a huge molecule made up of deoxyribose, phosphate, and, among its nucleotides, roughly 21 percent cytosine, 21 percent guanine, 29 percent adenine, and 29 percent thymine. It is at the level of detail, in the sequences of the millions of bases, that DNA becomes impressively complex. With four choices for every position and roughly 6.4 billion positions in a full human genome, the theoretical number of different genomes—of different sequences of the entire 6.4 billion base pairs of the genome—is four times four 6.4 billion times. It would take only about 130 base pairs to offer as many combinations as there are estimated elementary particles in the observable universe. The amount of information that a DNA sequence can carry is, quite literally, beyond astronomical.

Genomes

The sequence of the entire DNA in an organism's chromosomes (and hence in its cells' nuclei and thus its "nuclear DNA") is called its genome. That is almost, but not quite, all the DNA in the cell. Some of the organelles, the "little organs," inside eukaryotic cells have their own small bits of DNA, organized in circles. The mitochondria, the "energy powerhouses of the cell," have their own genome; in humans it is made up of 16,569 base pairs, about one four-millionth the size of the human nuclear DNA. Green plants have, in addition to mitochondria, organelles called chloroplasts, necessary to photosynthesis, that have their own DNA. The human mitochondrial genome is important but we generally talk of it as separate from "the" human genome.

The human genome, then, is the sequence spelled out on the 46 chromosomes, the 22 pairs of autosomes and the individual's two sex chromosomes. One member of each chromosome pair, as well as one of each of the sex chromosomes, came from each parent. In each pair of chromosomes, the paternal and maternal copies will be very similar. They normally will be the same length, have the same banding, and carry almost exactly the same sequence.

This leads to another tricky issue of vocabulary. Does a human genome have about 3.2 billion base pairs or about 6.4 billion? That depends on whether you are talking about the "haploid genome," the genome on the chromosomes derived from just one parent, or the "diploid genome," the (doubled) sequence that is the actual sequence of all the DNA in a person's cell. Of course, if those two sequences, from the mother and the father, were absolutely identical, it would not matter. The diploid genome would be just the haploid genome "printed" twice. In fact, in each human, the two haploid genomes are *almost* identical—almost, but not quite.

On average, two diploid human genomes differ at about one base pair in a thousand. That may not sound like much, but, remember, each genome has over 3.2 billion base pairs. That means each of the two genomes inside any one person will differ about three million times; when two people compare their diploid genomes, they will vary about six million times.

These variations come in several different forms. Let's pretend we are looking at one small length of DNA and that on the 5′ side of the maternal chromosome a nine-base stretch of DNA in that area reads CTTAGACTA while the corresponding stretch of the paternal chromosome reads CCTAGACTA. In this kind of change, the identity of just one of the bases in a stretch of DNA sequence is different. This is called a SNP (pronounced "snip"), a "single nucleotide polymorphism," where "polymorphism" is just a fancy way of saying "difference."

Now assume that, instead of a SNP, that maternal chromosome has three extra bases inserted—CAGATTAGACTA instead of CTTAGACTA—or is missing two of the bases, let's say the first two Ts—CAGACTA instead of CTTAGACTA. When base pairs are added, it is called an "insertion"; when they are missing, it is a "deletion." Insertions and deletions are collectively referred to as indels. SNPs and indels are among the most common variations found in human genomes.

Of course, in any particular pair of chromosomes, if one of them has two more base pairs in a particular location than the other one does, how do you know whether it is an insertion (two extra were added to the longer strand) or a deletion (two are missing from the shorter strand)? To do that you need to know something about the usual sequence in humans in that location. There is no one human genome sequence; there are currently over fourteen billion—two each for over seven billion people, minus a bit less than 1 percent for those of identical twins. But we could invent a so-called reference sequence by taking the most common sequence at each location. The current human reference sequence, compiled and maintained by the Genome Reference Consortium, is a more complicated effort to agree upon commonly found variations of the human genome.[9]

Genes

You may have noticed that I have scarcely mentioned genes, even though I have talked about genetics and genomes. Although knowledge that offspring tend to inherit traits from their parents is ancient, the modern idea that there are discrete units of heredity dates from Gregor Mendel's discoveries in the early 1860s and, more importantly, their simultaneous rediscovery by three different European researchers in 1900. The name "genes" was given to the units that were responsible for inheritance, but at first they were largely abstract units. It was not until the 1910s that scientists realized the genes had a physical presence on the chromosomes. Even then, the conventional wisdom was that it was the proteins of the chromosome that contained hereditary information. Proteins are complicated molecules made up of twenty different units; DNA, by comparison, was boring.

The first solid evidence that DNA carried the genes came in 1944 in a famous experiment by Oswald Avery at Rockefeller University.[10] He used a solution made from chromosomes to change an inherited characteristic of one strain of pneumococcus bacteria into an inherited characteristic of another. What made the experiment special was that he had treated the solution in a way that removed all the protein. Many scientists resisted the Avery finding that something other than protein in the chromosomes must be the basis for inheritance. Watson and Crick's discovery greatly boosted the idea that genes were made of DNA by providing a plausible physical explanation for how genetic information could be passed down between generations.

So what are genes? Genes are stretches of DNA that carry inheritable information. At first, people thought genes were only stretches of DNA that, through the intermediation of a closely related molecule called ribonucleic acid (RNA), defined the structure of proteins. The so-called Central Dogma of Molecular Biology, first set out by Francis Crick in 1958 (and somewhat qualified by the Nobel Prize–winning discovery of retroviruses by David Baltimore and Howard Temin in 1970), is that "DNA makes RNA makes protein." Most of what we think of as genes are still, in fact, stretches of DNA that tell the cell how to make proteins. Proteins are made of twenty different units, each one variety of a class of molecules called amino acids. The same twenty amino acids make up proteins in all humans, and in nearly all living things. The bases in DNA spell out the identities of those twenty amino acids, plus "start" and "stop," using the base pairs as the letters to form three-letter words, called codons.

Four letters can make sixty-four three-letter combinations: four possible first letters (A, C, G, T) times four possible second letters times four possible third letters. In the genome each of those possible sixty-four words "spells" something. The DNA sequence TGG "spells" the amino acid tryptophan. The sequence TAA spells "stop"—but so do TAG and TGA. The sequence ATG spells "start"—but it also spells the amino acid methionine. It should not take sixty-four codons to spell out twenty amino acids, plus start and stop; twenty-two would do. But most of the amino acids can be spelled in different ways. Serine, for example, can be spelled in six ways: TCT, TCC, TCA, TCG, AGT, and AGC. Although most of the amino acids (and "stop") can be spelled in different ways, every three-letter word (codon) spells only one thing (except for ATG—which is both "start" and "methionine"). The association between the codons and the amino acids, start, and stop is called the genetic code, and all known living things use the same code or something very close to it.

Each cell, from a bacterium's to our own, uses machinery to turn the genetic code of DNA into proteins. The DNA in the nucleus is "unzipped" and "transcribed" into a kind of RNA called messenger RNA. That messenger RNA is then moved outside the nucleus and "translated," using another form of RNA called transfer RNA that puts the correct amino acid in place for each codon in the messenger RNA.

But an average chromosome has 150 million base pairs. Where does a gene that codes for protein start—and what defines how you count off the three bases in a codon? The answer is ATG, the codon that spells both "methionine" and "start." When the cell's transcription machinery sees an ATG, it starts transcribing DNA into messenger RNA. After those first three letters, it keeps going, turning DNA into messenger RNA three bases at a time, until it runs across a codon that says TAA, TAG, or TGA: Stop! That stretch of DNA, from a start codon to a stop codon, is an "open reading frame" (ORF) and usually codes for protein. There are about 23,000 ORFs in the human genome; these are our protein-coding genes.

But genes, at least in eukaryotes, are more complicated. Not all of the codons in an ORF become protein. Almost all human ORFs contain some stretches that "code" for protein and some that do not. The parts that code for protein are called "exons"; the parts that do not are called "introns." (This has always confused me—the "in's" seem like the important ones, not the "ex's"—I remember which is which by recalling that it is counterintuitive.) The entire ORF, exons and introns, is transcribed into messenger RNA, but the messenger RNA is then edited by "splicing" out the introns before the remainder is translated into a protein.

Human genes often have many exons and many introns, with the introns usually accounting for more base pairs than the exons. *BRCA1*, for example, has twenty-two exons (and hence twenty-one introns—remember, the introns are always between exons). The whole open reading frame contains 81,188 base pairs of DNA from start to finish, located at 17q21 (the first subband of the second band of the long arm of chromosome 17). The exons, though, contain only 5,592 base pairs, which leads to an amino acid transcript of 1,863 amino acids—5,592 divided by three (three base pairs per codon) minus one (for the stop codon).

The move from DNA to protein is often still more complicated. The messenger RNA can be spliced together in various different ways. Sometimes exons are excluded; sometimes their borders are changed. And sometimes the proteins will be changed by subsequent modifications. The result is that the roughly 23,000 human protein-coding genes can make over 100,000 human proteins.

And the move from DNA to gene is also more complicated than just ORFs. Other stretches of DNA, outside these open reading frames, also provide crucial inherited information. Some of these are regulatory regions, stretches that help determine when particular genes are turned on or turned off. It is not clear whether to think of these regulatory regions as being themselves genes, or as being parts of the genes they regulate, or as being something else entirely. Researchers have identified hundreds of regulatory sequences.

Other stretches of DNA provide the code for many different kinds of small RNA molecules, which can serve a variety of functions within cells. These are genes even though they do not result in proteins; we know about several thousand of them.

So how many human genes are there? Well, at least the roughly 23,000 protein-coding ORFs plus several thousand RNA-coding genes and possibly, depending on your definition, hundreds or thousands of regulatory regions.

Genomes Again

And now we come to perhaps the strangest fact about the human genome and the genomes of other complicated living things. The vast majority of each genome does not appear to do anything. DNA may seem like a finely engineered machine, but, in fact, it is more like your grandmother's attic, with occasional treasures half-hidden in the mass of useless clutter.[11]

Only about 1.5 percent of the human genome is made up of the exons of protein-coding genes. RNA-coding genes and regulatory sequences make another few percent. And DNA stretches that have structural value—the chromosomes' centromeres and telomeres—add a few more percent. Although this remains somewhat controversial, the rest of the human genome, probably 80 to 90 percent of it, seems to have no known information content—or other use—and so has been called, by careless researchers and commentators, junk DNA. Careful researchers and commentators, on the other hand, might call it "junk" DNA.

DNA science has provided enough surprises that one would be reckless to omit the scare quotes. After all, shortly before the human genome was first sequenced, the estimates of the number of protein-coding human genes ranged from about 150,000 to about 30,000—almost no one expected as few as were discovered.

Some 5 to 10 percent of the "non-gene" genome is strongly conserved—many species have the same sequence, or close to the same sequence, even though they have been separated from each other by scores or hundreds of millions of years. This implies that those sequences are being preserved—being selected—by evolution, from which one can infer that those sequences do something important. We just don't usually know *what*.

In sum, it seems highly likely that some of those highly conserved "junk" DNA sequences do have functions that we just don't understand yet. But it also seems highly likely that most "junk" DNA does not. This "other" DNA falls into several large categories, including "pseudogenes," transposons and retrotransposons, LINEs, and SINEs.

Over time, stretches of DNA, including those with genes, may get copied. Pseudogenes are sequences that look an awful lot like other genes that we have but that have lost their ability to make protein. Somewhere along the way, an ancestor passed on a copy that contained a SNP, an indel, or some other mutation that blocked the gene from functioning and now it sits useless in the genome, like a ghost at a party.

The largest single chunk of the human genome, though, around half the total, is the result of transposons and retrotransposons. These are stretches of DNA sequences that can become copied and have their copies incorporated into other parts of the genome. These could be viewed as the ultimate in DNA parasites—they "live," "grow," and "reproduce" inside our genomes.

The classic retrotransposon is actually a bit of nonhuman DNA (or RNA) contained in a retrovirus. A retrovirus, like HIV, invades a human cell and uses a protein called reverse transcriptase, made from its own RNA genome, to turn its RNA into DNA, which is then incorporated into the genome of the infected human cell. Should that cell, or one of its descendants, ever give rise to an egg or a sperm that in turn gives rise to a baby, that baby will have the "stitched in" part of the HIV genome as part of its own genome, which it in turn will pass on to its descendants. About 8 percent of the human genome can be traced to specific ancestral retroviruses; the remaining retrotransposon portion is probably descended from retroviruses so ancient that the actual virus can no longer be discerned. Other copies come from transposons (not *retro*transposons) that incorporate themselves directly into other regions of the genome without first going through an RNA stage.

The most common genome products of these transposons and retrotransposons are LINEs and SINEs. "LINEs" is short for "long interspersed elements" and "SINEs" stands for "short interspersed elements." Researchers have identified about 500,000 copies of LINEs in the human genome, making up about 17 percent of the total genome. SINEs are short, 500 base pairs or fewer. There are about 1.5 million copies of SINEs and they account for about 11 percent of the genome.

It is possible that pseudogenes, retrotransposons, and transposons are doing some good in our genome, but, for the most part it is hard for anyone to see much immediate use for these repeated sequences. And this leads to the mystery. Copying DNA is expensive to cells. Presumably, a mutant cell that did not have to copy as much useless or potentially harmful DNA would have a competitive advantage over a cell that did. One would expect that evolution, working through natural selection, would favor smaller and more efficient genomes. Prokaryotes have very efficient genomes; almost all their DNA has a clear function. But eukaryotes do not.

Presumably, eukaryotes, including us, derive some evolutionary benefit to offset the cost of copying (and carrying around) all that apparently useless DNA. Scientists have speculated that the advantage may be that these repeats lead to more variation. That sounds plausible and, in the absence of other good explanations, it may be true. As Sherlock Holmes told Watson, "How often have I said to you that when you have eliminated the impossible, whatever remains, *however improbable*, must be the truth?"[12] At this point, though, it should probably be classified as a mystery, for which, in the words that conclude untold numbers of scientific articles, "further research is required."

2

REPRODUCTION: IN GENERAL AND IN HUMANS

To discuss the end of sex, we need to talk about what sex is. It is actually at least three different things—and each much more diverse in nature than we normally realize. First, sex is a method of reproduction in which the new organism gets a new mixture of genetic material from two parents. Second, sex is a condition of having either male or female reproductive organs—is an organism male or female? And, third, sex is a male and a female acting in ways that can, in the right circumstances, lead to reproduction. Each is first discussed below in the broad context of all life. The chapter will follow that discussion with a much more detailed discussion of how sex for reproduction works in humans.

Reproduction in General

At its most basic biological level, sexual reproduction means that a new individual organism is a mix of the genetic variations from two parent organisms. In humans that means that (about) half of one's genome comes from the mother and (about) half from the father. This seems deeply natural to us, but it is a minority approach in the biological world.

As noted in Chapter 1, the world of living things is now divided into the "domains" of bacteria, archaea, and eukarya. Bacteria and archaea are lumped together as prokaryotes. When we think of living things, we tend to think of our fellow mammals, with, perhaps, a few birds, reptiles, fish, and (maybe) amphibians. Those animals, the other main branches of the vertebrate subphylum, (almost) all reproduce sexually. So do most, but not all, other eukaryotes. But although most eukaryotes reproduce sexually, the vast majority of living organisms—and living species—on this planet do not use sex at all. All bacteria, all archaebacteria, and a significant number of eukaryotes reproduce by cloning. Each offspring has exactly the same set of genetic variations, the same DNA sequence, as its parent. When a microbe reproduces, it normally splits into two copies, each genetically identical to its parent. (Some bacteria do engage occasionally in something called "conjugation" that is vaguely like sex, but it is not an even split of the daughter organism's genetic variations and is not limited to genetic exchanges within the same species.)

Sex as a Method of Reproduction

So why does sex exist? As a biological question, that turns out to be surprisingly mysterious.[1] Sexual reproduction is clearly expensive. Organisms need to produce special cells that have only half the normal complement of DNA. These cells, called gametes (eggs and sperm) combine to make up a full genome, but the process of making gametes is complicated and not free. Even worse, a sexually reproducing species needs to have one of its gametes meet another gamete. A sperm needs to encounter an egg at the right time and place to give rise to another individual. That scenario is always going to be more complicated—and less likely—than having one cell double its complement of DNA and then keep dividing to create an identical daughter organism.

On the face of it, clonal reproduction looks better. No dating games or, in the nonhuman context, no need for eggs and sperm to meet. And the resulting organism will have the same genome as the parent organism, which, after all, managed to survive long enough to reproduce. A new genome created by mixing two parental genomes may not be as successful.

So why sex? Presumably, the answer is that, in the long run, the variation in genomes of the succeeding generations will help the species survive. Of course, variation can be helpful or harmful. Many variations thrown up by sexual reproduction will be harmful and hence will, presumably, not survive or survive as well. Many others will be neutral, neither positive nor negative in leading to survival of the next generation. A few will be positive. Sex only makes sense if the long-term benefits of those occasional positive changes outweigh the many costs of sexual reproduction.

There is a little evidence about this. Some species of rotifers, a tiny freshwater multicellular eukaryote, switch back and forth between reproducing clonally and reproducing sexually. Experimental evidence shows that when their environment is changing, these rotifers tend to reproduce through sex. When the environment is stable, they reproduce clonally.[2]

Of course, multicellular organisms cannot reproduce clonally as easily as single-celled organisms. They cannot just split in half and become two separate organisms. (Some plants do something similar to that, though, when they produce new copies of themselves by growing new aboveground portions from "runners" hidden underground, but is this reproduction or just growth?)

For rotifers, though, neither splitting in half nor growing new offshoots is a viable way of cloning. These organisms, and other clonal multicellular species, practice parthenogenesis. Parthenogenesis, from the Greek for "virgin birth" (the same Greek root that gave rise to "Parthenon," the name of the temple to the goddess Athena, who sprung, unconceived, from the forehead of her father Zeus), usually involves the growth of a new genetically identical animal from one cell of the old one. Often that cell will be an egg cell, but one that has not yet cut its number of chromosomes in half through meiosis. There are several kinds of parthenogenesis, but in this sort the egg begins to act as though it has been fertilized and starts dividing to form an embryo and eventually a new individual, genetically identical to its parent.

Natural parthenogenesis is unknown in mammals. It has been reported, though, in various fish, amphibians, reptiles, and even, rarely, in some birds. The Komodo dragon, the world's largest lizard, can reproduce either sexually or through parthenogenesis. Parthenogenesis is fairly common in insects and crustaceans. In the most common species of honeybees, for example, the queen can reproduce sexually, after mating with male drones, but both the queen and worker bees can reproduce by parthenogenesis to produce the drones and more workers.

But why do eukaryotes, and especially vertebrates, the animals we know the best, usually (and, in the case of mammals, always) reproduce sexually? Maybe these complex organisms confront a world so uncertain that sexual reproduction is always an advantage. Or maybe these species, for some reason, got "stuck" with sexual reproduction and could not easily change.

Sex as a Biological State

Organisms that never reproduce sexually will not have males and females. The terms are only meaningful in the context of sexual reproduction; bacteria do not come in "men" and "women". But, of course, to say that something reproduces sexually is still not to say that it reproduces from two distinctive sexual statuses, male and female. Some species are made up of hermaphrodites, organisms that produce both eggs and sperm. Many plants are hermaphrodites, as well as invertebrate animals (snails and earthworms, for example) and a few vertebrates.

Simultaneous hermaphrodites can produce either kind of germ cell (gamete) at any one time. Some simultaneous hermaphrodites are able to fertilize themselves. Unlike normal prokaryote reproduction, the products of this kind of hermaphroditic reproduction will not be clones. Each egg and each sperm has only a random half of the genetic variations of the parent organism. These self-fertilizing hermaphrodites thus get some of the advantages of the genetic variation produced by sex, but less than the variation produced by sex between different individuals.

But life is still more complicated. Some simultaneous hermaphrodites possess protections against self-fertilization. They only reproduce with another organism, but can provide either the sperm or the egg. Still other hermaphrodites, including many species of fish, are sequential, not simultaneous. These creatures change their sex over time. Some of them change just once; others can go back and forth between male and female many times. These sequential hermaphrodites are born with, or with the ability to create, both male and female reproductive organs and switch between them.

Hermaphroditic organisms are each sex, either simultaneously or at different times. But among organisms where individuals normally have one and only one sex, what is it that determines which sex will result? Generally, it is sex chromosomes. As noted in Chapter 1, humans have 22 pairs of chromosomes (the autosomes) plus two more mismatched sex chromosomes, found in humans and all other mammals, the X and Y chromosomes. Females have two copies of the X chromosome; males have one copy of the X and one copy of the Y. No humans have two copies of the Y chromosome without any X chromosome, for two separate but equally good reasons. The first is that everyone with a mother *must* have received an X chromosome because the mother has only X chromosomes to give. The second is that the X chromosome is large and contains many important genes; no viable human could be born without at least one X chromosome.

Of course, neither chromosomes nor genes themselves make reproductive organs. They work through the proteins that are produced when particular genes are turned on, or "expressed." In some species, environmental triggers can determine what sex a new individual will have. In alligators for example, the temperature when an egg is developing (in this case, the large, shell-enclosed spheroid that a female alligator has laid, not the egg cell) determines whether that alligator will be male or female.

Sometimes, in humans and in other organisms, something happens differently in the process that leads from sex chromosomes to sex organs. Occasionally people are born with at least some of the reproductive apparatus of both sexes. Traditionally referred to as hermaphrodites, though now generally called "intersex," such people are not true biological hermaphrodites and do not have two sets of functional reproductive organs. Their situation raises many difficult and complex issues, but they are beyond the purposes of this book.[3]

Furthermore, even with mammals that are solely biologically male or female, sometimes their physiological sex will not correspond to their sex chromosomes. The underlying sex of all mammals is female; genes on the Y chromosome must produce particular proteins at just the right times during prenatal development for the default female organism to become male. Those genes might end up accidentally moving to an X chromosome, making an X/X organism physiologically male, or, while staying on the Y chromosome, might become ineffective, making an X/Y organism physiologically female. These problems are rare, but not vanishingly so. They are among the reasons that, after a brief flirtation, the Olympic Games stopped using chromosome status to determine an athlete's "true" sex.[4]

Sex as an Activity

Our survey of sex in the biological world still has to deal with the *process* or, actually, processes of *having* sex. Again, in the biological world we usually think about, living creatures have sex through sexual intercourse. In cats and dogs, and in mice and men, a male uses a penis to introduce sperm into a female's vagina. The sperm then (sometimes) meets an egg and begins a process that leads to the birth of one or more new individuals.

But even among sexually reproducing organisms, sexual intercourse is far from universal. Consider sexually reproducing plants. They cannot, of course, have sexual intercourse; they aren't built for it. Instead, many of them scatter their sperm (in the form of pollen) to the winds, hoping that it will land in the right spot on the body of another member of their species, find an egg cell, and make seeds. Some plants cheat and use a mobile organism—insects, birds, bats, and other creatures (including, now, human farmers)—to transfer their pollen for them.

Even among mobile organisms that could have sex, many do not. Many species that live in water reproduce by spawning. They cast their eggs and their sperm into the water, to float around in the hope of meeting a complementary gamete. Some species, though, get more geographically specific. Female salmon, for example, dig little depressions in gravel beds, called "redds," where they deposit their thousands of unfertilized eggs. Male salmon next release their millions of sperm near those depressions. The female then covers the eggs and sperm with gravel, providing some protection for any fertilized eggs. Although not exactly "in vitro" fertilization, this is not "in vivo" fertilization. The fertilization occurs outside the female's body.

And, of course, if fertilization does not occur inside the female's body, embryonic and (in mammals) fetal development does not occur in that body. External fertilization does not lead to "live births." But even in species with internal fertilization, some go through some early stages of development inside the female but are then expelled to continue their development as eggs (classically but not always of the shelled type) on the outside. Most vertebrates—many fish and amphibians, all reptiles and birds—reproduce this way.

Only in mammals does the fertilized egg develop into a new organism while physically and biologically attached to the female. Pregnancy is uniquely mammalian and, across the whole spectrum of the living world, quite unusual (and arguably bizarre). Even with mammals, whether the offspring fully develops inside the mother can be doubted. For some placental mammal species, the newborn may be a viable individual immediately, or almost immediately. In others, and certainly in humans, the development of the new individual will continue outside the mother for a long time, sometimes decades, before the new one is able to survive on its own.

And, of course, there are intermediate versions. Some fish make eggs but keep them on the inside. In effect, they "lay" their eggs inside themselves. When the eggs hatch, the mother "gives birth" to the live young but the egg has been developing inside the female without a biological connection to it. On the other hand, marsupials, one type of mammal, give birth to extremely underdeveloped offspring that then continue to develop inside the marsupial's pouch, no longer biologically connected to the mother by a placenta, but still inside the mother.

Living organisms reproduce in a host of different ways, many of which do not involve sex in any of its meanings and certainly not sex as we humans know it. Taking a wide view, while sex may be "natural," what that means varies enormously across nature. And, as the next section shows, even the process of sexual reproduction with which we are most familiar—our own—is strange and complicated beyond our imaging.

Sex in Humans

You did not pick up this book to read about rotifers and Komodo dragons. Although the biological world contains many different ways to reproduce, unless humans use assisted reproductive methods on them, all mammals, including humans, reproduce sexually, as do all humans, through sexual intercourse between a male and a female that, on occasion, results in a pregnancy and a live birth. This is not the most common method of reproduction, but it is our method of reproduction. And it is this method of reproduction that, at least in some critical details, will be replaced in the coming decades. Before we talk about replacing it, though, we need to understand it. I will start by describing the basic process of going from cells with a full set of paired chromosomes to cells with only an unpaired set, a process called meiosis, and then discuss the origin of eggs and sperm, the effects of puberty, the fertilization that comes from the meeting of eggs and sperm, and what happens after fertilization.

Meiosis

Cells that have a full set of chromosomes—in humans, one pair each of chromosomes 1 through 22 (the autosomes) and two sex chromosomes—are called diploid. When a cell has only one chromosome from each pair and only one sex chromosome, it is called haploid. In mammals, only sperm and fully mature egg cells are haploid; all the other cells are normally diploid.

Cells almost always divide through a process called mitosis. The contents of the cell increase in number and amount and the chromosomes are copied, giving a human cell, temporarily, not 46 but 92 chromosomes. When the cell divides, each daughter cell receives 46 chromosomes, one copy of each of the original cell's chromosomes, with the same DNA sequence (except for occasional new mutations) as the parent cell's chromosomes.

Only eggs and sperm divide by something other than mitosis. The process for making gametes turns diploid gamete precursor cells into haploid gametes. It is called meiosis and, in humans, it is complex in men and amazingly complicated in women. In both men and women it has two phases, called meiosis I and meiosis II.

The diploid human reproductive cells that give rise to eggs and sperm will prepare for meiosis the same way other cells prepare for mitosis—they copy their 46 chromosomes, making them 92. Each of the original 46 chromosomes will now exist in two copies that are connected with each other at their centrosomes, the copies of the paternal version of chromosome 1, for example, connected to each other. In mitosis, these 46 sets of duplicated chromosomes line up in the cell and are pulled apart, one copy of each duplicated chromosome being pulled in the direction of each of the ends of the dividing cell. The result is that each daughter cell gets one copy of each of the preexisting 46 chromosomes.

In meiosis I, on the other hand, the now 92 doubled chromosomes pair off with their equivalents from the other parent, so the two connected copies of each paternal chromosome 1 pair off with the two connected copies of maternal chromosome 1, pulling all four copies of chromosome 1 together. At this point, a process called recombination happens. Chunks of DNA from a copy of one or both of the paternal copies of chromosome 1 swap with chunks of DNA from one or more of the paired maternal copies of chromosome 1. Typically, any one chromosome undergoing recombination will swap only a few chunks of DNA, usually about two to four pieces.

This recombination does not happen in mitosis, only in meiosis, and has important implications. Without recombination, each of us would inherit 23 chromosomes from each of our parents but each of those chromosomes would be identical to one inherited from the parents. Thus, a person's two copies of chromosome 1 might include one exact copy of his or her father's mother's chromosome 1 and one exact copy of the mother's father's chromosome 1. Because of recombination, a person's parents will each give a chromosome 1 that combines DNA from that parent's father's and mother's chromosomes 1, giving four sources for that chromosome's DNA, not just two. So there are now four different versions of chromosome 1 in the cell, still stuck together in two pairs.

These pairs now separate, but not the same way as in mitosis. The DNA in the cell is divided, producing two chromosome-containing cell nuclei. Each nucleus still has 46 chromosomes, but in pairs made up of two copies of what started as either a paternal or maternal chromosome, as modified by recombination. Each will have some mix of originally paternal and maternal chromosomes, so that, say, ten of the chromosome pairs in one daughter cell will be originally (and still mostly) maternal and thirteen will be originally (and still mostly) paternal, while the other will be thirteen to ten in the other direction. This is the end of meiosis I.

Meiosis II is then simple. The two daughter nuclei produced by meiosis I, with their 23 pairs of chromosomes, divide again, separating each pair. Each of these two nuclei is now truly haploid. They have only 23 chromosomes, one of each autosome and one sex chromosome. Each will have some originally paternal and maternal chromosomes, further mixed by the recombination process. If merged with a complementary haploid nucleus (that of an egg with that of a sperm), they will produce a new nucleus with 46 (and only 46) chromosomes, with one pair of each autosome and two sex chromosomes.

Making Eggs and Sperm

The story of just how the body makes, and matures, eggs and sperm is incredibly complex—in an earlier draft of this book it took up about twenty pages of text—and fascinating. But, for the purposes of this book, it is (alas) not important enough for that much detail.

In males, the cells that eventually become sperm lodge in what will eventually become testicles about halfway through fetal development. They stay there, quiescent, until puberty, when a cascade of hormones, starting with the brain and ending with testosterone from the testicles, leads some of them to start going through a roughly 60-day process involving their transformation through six differently named cell types before they have gone through both stages of meiosis (and thus become haploid, having only 23 chromosomes instead of 46) and become fully functional sperm. These sperm lodge in a structure on the outside of the testicle called the epididymis. Men start making sperm at puberty and continue, though in diminishing quantities, throughout their lives.

Women are different. They make all the cells that will become eggs (oocytes) that they will ever have by early in the third trimester of their fetal development; they are born with around one million.[5] Before birth those eggs begin to go through the first stage of meiosis, but then stall until puberty. The brain-released hormones that trigger puberty in men do the same in women, leading to the production not of testosterone but of several kinds of estrogens, notably estradiol. At puberty, women have, on average, about 400,000 eggs left.

With menarche, the start of the menstrual cycle, every month (roughly) one egg (usually) will complete a process that perhaps as many as a thousand would-be mature eggs started twelve months earlier. (Some sources say about a dozen, again illustrating the limits of our knowledge of this process.) That egg grows, finishes meiosis I, ejecting 23 chromosomes into something called the first polar body, and begins but does not complete meiosis II, all to become "ripe" or "mature." It then gets released from a large (about 0.8-inch) fluid-filled follicle (think of a blister) on the surface of one of her two ovaries. (Sometimes more than one egg ovulates in a month, which can lead to the birth of nonidentical twins if they both are fertilized and successfully implant.) This process continues until menopause, with the number of potentially viable eggs shrinking all the while.

How Sperm and Egg Meet

This is the stage where "sex," in the meaning of sexual intercourse, has traditionally been crucial. The sperm needs to get from the epididymis into the female reproductive tract and move upstream in that tract until it meets up with an egg, moving downstream since ovulation. Again, this simple-sounding process turns out to be much more complicated than we would imagine—and, in spite of years of research, its story still surprisingly lacks consensus on many details. Treat the following description (and particularly any numbers in it) as approximate and do not be surprised if you read somewhat different accounts in other sources.

The sperm wait in the epididymis until ejaculation. At that point, roughly 100 million to several hundred million of them (the amounts vary from man to man and time to time—and written source to written source) move down the length of the vas deferens (variously, and annoyingly, described in the literature as being about 12 inches, 17 inches, and

2 feet long) to the ejaculatory ducts, which open into the urethra, the same tube through which urine passes, just downstream from the bladder. Along the way, the sperm become just a small part (about 2 percent by weight) of the semen, a mixture of sperm plus contributions from other parts of the reproductive system.

The vagina is naturally very acidic; seminal fluid provides alkaline bases to the semen to counteract the acidity. The sperm will use a lot of energy moving through the vagina, cervix, uterus, and fallopian tube; the seminal vesicles contribute large amounts of fructose, a sugar, to help feed the sperm. The female reproductive system will mount an immune response against sperm cells; prostaglandins in the semen help repress that.

As the semen moves downstream it is propelled largely by muscles in the male reproductive system and not yet primarily by the sperm's tails. Eventually, the semen moves through the urethra, out of the tip of the penis, and, in heterosexual penile-vaginal intercourse, into the vagina. Most of these sperm actually have no chance of fertilizing an egg; they are misshapen or otherwise nonfunctional. But, functional or not, after their rapid journey of several feet, the sperm take a break for about a half hour, where they separate from the rest of the semen and begin various physical and chemical changes, including just warming up. (The temperature in the testicles can be as much as 7 degrees Fahrenheit cooler than normal body temperature.)

The sperm's first challenge is to get into, and through, the 3- or 4-centimeter-long cervix. Most of the month, the cervix is plugged by mucus. The hormones expressed around the time of ovulation thin that mucus, unplugging the cervix. However, there are no road signs, and, even if there were, sperm can't read—at this point sperm just move randomly, in part on their own and in part in response to contractions of the vagina. Some of them end up moving toward and then into the cervix; most do not. Only about 50,000 sperm in any ejaculation actually enter the cervix—about one in 5,000 to 10,000 of those in the initial ejaculate. And then only 10 percent of the sperm that enter the cervix manage to navigate successfully through it and into the uterus.

At some point, probably in the cervix, the sperm become activated and their tails hyperactive, allowing them to move much better on their own. The tail makes a propeller-like rotating motion to move a sperm forward at a rate of one to three millimeters per minute—a breathtaking four to six inches per hour.

Now, a few hours after ejaculation, a few thousand sperm will end up at one of two utero-tubal junctions—the places at the far end of the uterus where each of the two fallopian tubes enters the uterus. (Sperm can survive for several days in the woman's cervix or uterus, so fertilization does not necessarily depend on this first wave of sperm.)

The junctions are tiny, only two or three times the diameter of the sperm head, while the uterus is, by comparison, vast—from the top of the cervix to the oviducts is another 2 or 3 centimeters, about 500 times bigger than the sperm's head. On top of everything else, these utero-tubal junctions are on different sides of the uterus, even though, most months, only one of the fallopian tubes will ever contain an egg. The woman's immune system poses an additional problem, as white blood cells in the vagina try to destroy sperm. From any ejaculation, it is thought that only a few dozen sperm will actually get into the (appropriate) fallopian tube, which is about 10 centimeters (4 inches) long. Fallopian tubes are a very congenial environment for sperm, with nutrients, a comfortable pH level, and protection from the immune system. And there they may finally meet an egg, to whose progress we now return.

We last saw the egg during ovulation, being ejected from the ovary when the blister that the ovarian follicle has become, in effect, pops. There are difficulties here, as well. Oddly enough, the fallopian tubes are not actually attached to the ovaries. They float in the abdominal cavity with their openings, the oviducts, near to, but not quite touching, the ovaries. The oviducts are wreathed in fingerlike protuberances called fimbriae. Under the influence of hormones released by the ovary around ovulation, those fimbriae become larger. Each one is covered with cilia, little beating hairlike projections that help guide the released egg into and down the fallopian tube. Occasionally, eggs escape from the fimbriae and float, uselessly, into the abdominal cavity. Very rarely, these eggs will be fertilized by sperm and will establish a very dangerous—to the fetus and woman—abdominal pregnancy.

When the egg enters the fallopian tube, it releases a chemical that finally gives the sperm some direction. The sperm then head directly for the egg. Only a handful of sperm actually reach the egg—or, more accurately, the cumulus cell-oocyte complex.

Because what the sperm reaches is not just an egg, but an egg surrounded by several hundred small cumulus cells that form the corona radiata (the "radiating crown"). The egg alone has a diameter about twenty times larger than the diameter of the head of the sperm—if the sperm were the size of an adult human, the egg would be a sphere over 120 feet in diameter, taller than a ten-story building. The sperm binds to the corona radiata, which both helps prepare it for fertilization by physically "roughening" it and puts it into contact with the zona pellucida (the "clear zone"), a gooey layer surrounding the actual egg. A molecule on the sperm head binds to a receptor molecule on the zona pellucida, which in turn leads to the "acrosome reaction," the bursting of a region in the head of the sperm that contains enzymes that help make a hole in the zona pellucida. Once the sperm penetrates the zona pellucida, it then easily enters the egg's membrane. As soon as that happen, the egg produces a signal that causes the zona pellucida to harden, usually preventing fertilization by more than one sperm.

The sperm that fertilizes the egg may have been ejaculated anytime from a few hours to several days earlier. The egg, on the other hand, had to have ovulated within the past day. Once ovulated, eggs deteriorate after about twenty-four hours. All of us conceived the traditional way exist only because a sperm ejaculated a few hours or days earlier managed to find an egg within twenty-four hours (or less) of ovulation. It is no wonder that healthy fertile couples trying to get pregnant take about five months on average to succeed. The wonder is that this process ever works.

After Fertilization

The egg and sperm have now met and merged—they have become a zygote. But there are no guarantees that any babies will be born from this zygote (fewer than half of zygotes become babies), or that only one baby will be born from it (identical, or monozygotic, twins form from one zygote); in fact, that zygote does not immediately act like it is a single cell. The former egg and former sperm still have some separate work to do.

For one thing, at this point the genome of the new possible person is not yet fixed. At fertilization, when the sperm is absorbed into the egg, that egg has not yet completed meiosis. It still has 46 chromosomes and is still stuck in the middle of meiosis II; the chromosome pairs are lined up but still attached. Only at fertilization does the egg complete meiosis II. Those chromosome pairs now separate, randomly keeping one of each pair of recombination-modified chromosomes in the egg's nucleus (called now, and for just a little longer, the pronucleus). The fertilized egg segregates the other copies of its chromosomes into the so-called second polar body, which it expels.

The sperm has been absorbed into the much, much larger egg. (The egg's volume is about 10,000 times greater than the sperm's.) The sperm's tail and midpiece degenerate, destroying the mitochondria in its midpiece. This is why, in humans at least, all of our mitochondria, and hence our mitochondrial DNA, come from our mothers. Then the membrane around the sperm's head dissolves, leaving its chromosomes as the male pronucleus.

The completion of meiosis II by the former egg and the formation of the male pronucleus from the former sperm take up about eighteen hours after fertilization. The two pronuclei migrate toward the center of the zygote. During this time the chromosomes inside each pronucleus are duplicating, preparing for the zygote's mitotic cell division. Thus, before they meet, the sperm's pronucleus and the egg's pronucleus each contain 46 chromosomes, two identical copies each of one of each of the autosomes and one sex chromosome.

As the two pronuclei approach, the membranes holding their chromosomes together dissolve and the chromosomes from the egg and the sperm finally meet, completing the process known as syngamy. They come together as part of the act of the zygote's first cell division, moving directly into mitosis and preparing for each of the zygote's daughter cells identical copies of each chromosome inherited from the egg and from the sperm. Nuclear membranes now form around these two new cell nuclei. The nuclei migrate to different ends of the zygote, which then splits to form a two-cell embryo.

So now where there was one zygote, there is a two-celled embryo. The two cells of the embryo begin their own process of mitosis, splitting into four cells, which split again into eight cells. The timing of these divisions is not tightly controlled enough to make the divisions of different embryonic cells occur simultaneously, so not all early embryos have exactly two, four, eight, or sixteen cells, but that is generally the path they take. It usually takes about three days for a zygote to become an eight-cell-stage embryo by completing three divisions.

To this point, though, the embryo is not really growing; it is "cleaving." The same amount of material—the contents of the original egg along with the tiny contribution from the sperm, almost entirely in its chromosomes—is being divided into smaller and smaller compartments. The size of the embryo at this point is the same as the size of the egg and its eight cells are not really bound to each other. They are just held together within the zona pellucida. And, for the most part, no genes are being transcribed and no new protein is being made. The embryo is living off the resources from the egg.

That begins to change after the eight-cell stage. The embryo begins the process of compaction; by the thirty-two-cell stage (two more divisions) the cells are bound tightly to each other in a sphere. (At this stage, the embryo is called a morula.) At about the same time, widespread gene activation starts and the embryonic cells begin to use their own genomes to make their own proteins.

Shortly afterward, the embryo begins to differentiate. The cells on the outside of the sphere begin to pump fluid inward, forming a fluid-filled cavity in the middle of the sphere. This marks the beginning of the blastocyst stage of the embryo, a hollow, fluid-filled sphere. The outer wall of the blastocyst, looking like a lumpy soccer ball, is made up of cells called trophoblasts and is itself called the trophectoderm, which eventually becomes the placenta and other supporting tissues for the pregnancy. The middle is filled with fluid and a few other embryonic cells, poetically named the "inner cell mass." It is the inner cell mass that later becomes the fetus and ultimately the baby. In humans the blastocyst forms about five days after fertilization.

Meanwhile, hormones produced in the ovary stimulate the lining of the uterus, called the endometrium, to thicken, providing a welcoming site for the blastocyst. This thickened endometrium will be the site of implantation, if the egg is successfully fertilized and then arrives at the uterus at the right time. In the course of a menstrual cycle, the thickness of the endometrium goes from microscopic to about a quarter of an inch. It is at its largest for about four days, starting roughly seven days after ovulation.

Fertilization usually occurs high in the fallopian tubes, close to the ovary. After fertilization, while the former egg, now zygote, is cleaving and then growing to a morula and then a blastocyst, it is also moving down the fallopian tube, pushed toward the uterus by the cilia lining the tube. By the sixth day or so, it will have reached the uterus, where the endometrium will have grown in preparation to receive it.

The next step is the implantation of the blastocyst into the endometrial lining of the uterus. Unless this occurs successfully, the blastocyst will die and be flushed out with the next menstrual cycle, which, unless implantation succeeds, will begin about twelve or thirteen days after fertilization. The blastocyst spends about one day, on average, in the uterus before it implants. During this time the blastocyst must "hatch" out of the zona pellucida. Enzymes in the uterus begin to break down the zona pellucida and the blastocyst oozes out of it.

About a week after fertilization, the blastocyst, now wholly free from the zona pellucida, needs to make contact with the enlarged endometrium (also called the decidua) to begin to form the placenta. The blastocyst is now secreting several important proteins. Some of them stimulate cells from the blastocyst to invade the endometrium. Others make that invasion easier by breaking down the bonds between endometrial cells. The blastocyst also secretes several different proteins that inhibit the woman's immune system. After all, the blastocyst only shares half its genes with its host. That host's immune system will see the blastocyst (with some justice) as a parasitic invader and, unless prevented, will attack it.

Finally the implanted blastocyst secretes human chorionic gonadotrophin (hCG), which tells the body that a pregnancy has started. Modern pregnancy tests look for hCG to see if a pregnancy has begun. It can be detected almost immediately after implantation (and hence before any missed menstrual periods), but its concentration increases more than a thousandfold during the first six weeks of pregnancy. Therefore, home pregnancy tests typically advise waiting for a week after a missed period.

And only now, with a successful implantation, can a pregnancy truly be said to have started.

If implantation did not occur, and take, no hCG is produced and, about two weeks after ovulation, the endometrium begins to slough off. Menstrual fluid is a mixture of the endometrium with blood; the first day of menstrual bleeding marks the first day of a new menstrual cycle. And, once again, a new group of immature eggs will start the roughly year long process of competing to become the dominant follicle, to ovulate, and to have their chance to lead to a baby, just as another group is nearing the end of its race, which will result in the ovulation of one egg (usually) in about two weeks.

And so it will continue, roughly every month, unless or until interrupted by pregnancies, oral contraceptives, disease, or menopause. An average healthy woman who never becomes pregnant will go through about thirty to forty years of menstrual cycles, ovulating around 420 times. Of the millions of egg precursors that the fetus started with, and the hundreds of thousands of primary oocytes each baby girl had at the beginning of her life, only those roughly 400 eggs have a chance to continue the cycle of life. Healthy men continue to produce sperm until death, making hundreds of billions—perhaps a trillion—sperm over their lifetimes, but for only a few men do more than a handful of those sperm produce children. And, of course, those few "lucky" eggs and sperm may produce children who are healthy or sick, happy or miserable.

The process of human reproduction is wasteful, expensive, and bizarrely complicated. Such a process surely must be the product of evolution, because no one would have designed it this way. And, for many people who want to be parents, it does not work. Ultimately, this book is about the ways we are likely to redesign that system, to make it less wasteful, expensive, and complicated. But we will start, in the next chapter, with how we currently try to fix it for prospective parents for whom it does not work.

3

INFERTILITY AND ASSISTED REPRODUCTION

This book is about the broad future use of techniques of assisted reproduction by fertile people who don't "need" to use them, but that future will strongly rely on techniques established to help infertile people have "children of their own"—children made from their own DNA. This chapter starts with some background on infertility and the history of treatments for it until 1978. It then describes in some detail the history and practice of in vitro fertilization (IVF), the crucial method that will lead to the end of sex.

Infertility and Its Treatments Developed before 1978

Doctors define infertility as a condition occurring when at least one year of unprotected sexual intercourse between a man and a woman between menarche and menopause does not produce a pregnancy. On that definition, an estimated six million American women between the ages of fifteen and forty-four—about 10 percent of the women in that age range—are infertile, as are about 14 percent of couples.[1] Of course, some of those people who missed during the first year will become pregnant in the thirteenth month, or the fourteenth. But there will also be women who become pregnant but cannot sustain the pregnancy to a live birth, as well as people who become pregnant once, or more, but then never become pregnant again.

Having just read so much about the things that must go right to establish a pregnancy, it should come as no surprise that many different problems can cause infertility. The most common cause of infertility in women is a lack of ovulation, often caused by polycystic ovarian syndrome or by primary ovarian insufficiency—or by aging. About one-third of couples where the woman is over thirty-five are infertile. Blocked fallopian tubes are also a substantial contributor to female infertility, along with physical problems with the uterus.

In men, the most common problem is absent, insufficient, or badly formed sperm. Another very common reason for male infertility is varicocele. In this condition the veins in the testicles are too large, causing the testicles to become too warm and killing or damaging the sperm. (Lowering the temperature for sperm is presumably the reason men's testicles hang outside the body in a dangerously, and sometimes very painfully, exposed way—though assigning reasons to evolution is always speculative.) Various diseases, from mumps to cystic fibrosis, can also cause male infertility.

For couples, infertility stems from problems on the man's side about one-quarter of the time and on the woman's side about one-half of the time, with about half of women's infertility caused by ovulation problems and the remainder coming from fallopian tube blockages or other causes. In the remaining quarter of cases, the cause is a mystery.

Infertility is certainly not new; it features prominently in the book of Genesis and has regularly caused dynastic changes in monarchies, as well as divorces, beheadings, and other dramatic and unpleasant events. (See, e.g., Henry VIII.) Less dramatically, it has also caused great grief to people who have desperately wanted to become genetic parents.

Treatments for Male-Based Infertility

Effective medical treatments for some kinds of infertility are also not new.[2] The first artificial insemination in nonhuman animals (dogs) occurred in 1784. The first successful human artificial insemination is thought to have occurred sometime in the late eighteenth century; the details, including the year, vary from source to source. Perhaps more clearly, John Pancoast, an American doctor, performed artificial insemination in 1884, leading to the birth of a child, though apparently without the knowledge or consent of either his patient or her (known to be infertile) husband.[3]

Soviet experiments with livestock in the 1930s led to the common use of artificial insemination for livestock in the United States in the 1940s and 1950s.[4] This increased the interest in using the technique in humans, but that proved intensely controversial because it was generally attempted using donated semen when the husband could not produce his own. A British commission recommended that it be made a crime, and the pope declared that it was sinful. In 1953 a court in Chicago held that a married woman who received artificial insemination (presumably with donor sperm), even with the husband's consent, was guilty of adultery and that the resulting child was illegitimate.

The tide soon turned. In 1964, Georgia became the first U.S. state to legitimize the children of donor artificial insemination in marriage, as long as both husband and wife had consented in writing. And in 1973 the National Conference of Commissioners on Uniform State Laws approved the Uniform Parentage Act, which expressly approved donor artificial insemination (at least in married couples), designating the husband as the (only) father. In spite of their names, "uniform acts" are only recommendations to state legislatures for action; they have no legal force in themselves. The act, updated in 2000 and again in 2002, has never been universally adopted—or adopted exactly the same way in the states that have accepted it. But for our purposes, the 1973 act is important as establishing broad social acceptance of donor artificial insemination. This climate made possible the opening of the first sperm banks, storing and selling donated sperm for artificial insemination.

Today, artificial insemination in humans involves taking sperm, either from the man who wants to be the child's father or from a donor who does not intend to play that role, and inserting it into the woman's reproductive tract either into the cervix or the vagina, or by putting "washed" sperm directly into the uterus or into the fallopian tubes (or both). (The "washing" involves several processes to make this sperm more similar to the condition of sperm that reaches the vagina in the normal course of events; it also removes components of the semen that are left in the vagina in normal intercourse and that can cause problems in the uterus.)

Artificial insemination clearly can help when the prospective father cannot make his own functional sperm, but it does not always have to involve donor sperm. Sometimes, artificial insemination will be used when the man can produce sperm but cannot ejaculate it. Some men, for example, make sperm but it all gets trapped in their testicles. That sperm can be removed using a long hypodermic needle; the photos are disconcerting (to a man, at least), but the procedure is safe. Men with only a few sperm or with sperm that, for whatever reason, cannot normally reach and fertilize an egg usually need another procedure, ICSI, but as that requires IVF, I will discuss it later.

Treatments for Female-Based Infertility

Artificial insemination can seldom help infertile couples where the problem is on the woman's side (or on both sides), but starting in the 1960s, medicine began to offer real relief to some infertile women through fertility drugs, tubal surgery, and endometrial treatments.

The first approach increases ovulation, particularly in women with highly irregular or absent menstrual periods. Women receive drugs that, in effect, stimulate the hormonal system that naturally leads to ovulation. The FDA approved clomiphene, the first drug for use in increasing ovulation, in 1967. Others have followed. Infertile women who are thought to have ovulation problems can be prescribed pills containing these drugs. If that does not work, doctors may recommend daily hormone shots.

For some women, blocked fallopian tubes can prevent the sperm and the egg from meeting (at least at the right time) or the embryo from getting to the uterus. The blockages often arise from pelvic inflammatory disease, often caused by chlamydia, a sexually transmitted infection. Specialists estimate that somewhere between 20 and 40 percent of female infertility comes from blocked fallopian tubes. If the blockage is minor, fallopian tubes can sometimes be opened by surgery. The first fallopian tube surgeries were tried at the end of the nineteenth century with no success, but the development of microsurgical techniques in the 1960s and 1970s made this procedure useful.

The endometrium is the lining of the uterus, which expands and contracts dramatically during the menstrual cycle. Sometimes those endometrial cells end up colonizing the wrong parts of the female reproductive system, such as the ovaries or the fallopian tubes. This condition, called endometriosis, can cause significant pelvic pain, as well as fertility problems. Doctors estimate that 5 to 10 percent of women of reproductive age have endometriosis. Fertility problems from endometriosis can sometimes be treated by laparoscopic surgery to remove the misplaced endometrium.

These treatments are still widely used in couples for whom they can work. But for some couples, none of them work. About a quarter of the time, infertility has no identified cause. Even when it does have a known cause, those older treatments may not cure it. These kinds of cases led to the truly revolutionary development of IVF and the birth of Louise Brown, the world's first "test tube baby" on July 25, 1978.

IVF—Its History and Current Status

Human in vitro fertilization may be the most important medical advance of the twentieth century to have taken place without substantial support, financial or otherwise, from either corporations or governments. Although the preparation took place in the shadows, the first successful case generated a glare of publicity and controversy, but today, over thirty-five years later, it is practiced worldwide. It has brought to life millions of people who would not otherwise have been born—an estimated five million by 2012 with another 400,000 or so added each year.

Getting to Human IVF

An Italian priest named Lazzaro Spallanzani claimed to have performed what we would now call in vitro fertilization in the 1780s, but he did it the easy way, using organisms where fertilization naturally occurs outside the body. Mixing frog eggs and frog sperm in a Petri dish is not much different from the way those eggs and sperm mix in nature. IVF in animals where fertilization normally takes place internally proved much harder. Various researchers experimented with IVF or its close relatives in mammals starting in the late nineteenth and early twentieth centuries, but with very limited success.

Substantial fictional progress in IVF was made, though, in 1932 when Aldous Huxley published *Brave New World,* his famous dystopian novel. In the world of his novel, ovaries are surgically removed and the eggs are artificially matured, using "Podsnap's Technique." The ripened eggs are then examined under a microscope for imperfections. Those that pass are fertilized with sperm from selected men. Fertilized eggs destined to become members of the two highest castes, the Alphas and Betas, are moved to artificial wombs called "bottles." Fertilized eggs intended to become Gammas, Deltas, and Epsilons are instead subjected to the "Bokanovsky Process," where they "bud" into clones, as many as ninety-six, though seventy-two is the average. These are then also moved to bottles, where they develop, under close and modifying control, to birth.

The initial reviewers were not kind to Huxley's book,[5] but it thrived and put the idea of assisted, or artificial, human reproduction firmly into the public imagination.

Actual IVF proved much harder than the fictional version. Early researchers did not understand the importance of the maturity of the oocytes or the changes seminal fluid and sperm go through between ejaculation and fertilization. In 1934, however, two researchers from Harvard, Gregory Pincus and E. V. Enzmann, got attention by claiming, perhaps accurately, in the *Proceedings of the National Academy of Sciences* to have successfully performed IVF on rabbits.[6]

Three years later, in 1937, this work reemerged in an editorial in the *New England Journal of Medicine* called "Conception in a Watch Glass," which hailed the research, opening with a direct reference to *Brave New World*.[7]

The immediate excuse for the editorial was the publication, in that issue, of a report, building on earlier work in (of course) rabbits, that electrical signals could be detected in humans at the moment of ovulation. (It later turned out not to work in humans.) The last paragraph of this two-paragraph editorial is worth quoting in full for its prescience:

> Contemplating this new discovery, one's mind travels much farther. Lewis and Hartman have isolated a fertilized monkey ovum and photographed its early cleavage in vitro. Pincus and Enzmann have started one step earlier with the rabbit, isolating an ovum, fertilizing it in a watch glass, and reimplanting it in a doe other than the one which furnished the egg, and have thus successfully inaugurated pregnancy in an unmated animal. If such an accomplishment with rabbits were to be duplicated in human beings, we should, in the words of "flaming youth," be "going places." The difficulty with human ova has been that those recovered from tubes have regressed beyond the possibility of fertilization in vitro. But by utilizing the electrical sign we may be able to obtain them from the follicle at the peak of their maturity. If the new peritoneoscope can be developed along the lines of the operating cystoscope, laparotomy may even be dispensed with. What a boon for the barren woman with closed tubes! Walton is quoted as saying that it is theoretically possible to separate male-determining from female-determining spermatozoa. Will it be possible to obtain son or daughter, according to specifications, and even deliver them of women who are not their mothers? Truly it seems as if the forge were being warmed, and another link may be welded in the chain by which mankind strives to hold nature under control.[8]

It was not until 1951 that real progress was made in mammalian IVF, as two researchers independently discovered capacitation, essential changes in sperm to enable fertilization. It then took eight years before the first clear case of a live birth from mammalian IVF took place. In 1959 Min Chueh Chang, one of the discoverers of capacitation, combined rabbit eggs and sperm in a flask for three or four hours, eventually producing some four-cell embryos. He transferred thirty-six of them into female rabbits and they yielded fifteen healthy live bunnies. IVF had finally been achieved but, unfortunately, only in a species not generally thought to have *in*fertility problems.

The issues in moving to humans were many. Human eggs had to be successfully retrieved and in a time frame that respected their short fertile lives. This meant not only the development of surgical procedures for the egg retrieval (sometimes called "harvest"), but ways to know exactly when ovulation was going to occur. Human sperm had to be successfully capacitated in vitro. The human eggs and sperm had to be kept alive and viable in vitro while waiting to be combined; then the fertilized eggs had to be kept alive in vitro long enough to determine which ones were progressing normally. And the time, place, and manner of transferring the resulting embryos into a woman's reproductive tract had to be worked out. All of this had to be done with more difficult experimental subjects than rabbits, while facing serious ethical, religious, political, and funding challenges.

Laparoscopic surgery, abdominal surgery performed through a small incision with the aid of an inserted camera to allow the surgeon to view the internal area, solved the problem of getting human eggs, starting in 1961. Only later did it become possible to time the laparoscopy to catch the eggs just as they were about to ovulate, through using either hormone levels or ultrasound.

The 1960s and 1970s were largely spent in achieving IVF in other mammals, which required learning the surprisingly varied quirks of reproduction in different species. Hamsters, mice, rats, sheep, pigs, guinea pigs, cats, and dogs were all successfully produced by IVF, eventually. As a nice example of the difficulties, though, hamster eggs were successfully fertilized and brought to the two-cell stage by 1963, but the first hamster pups were not born from IVF until 1992. (Ironically, IVF also proved unusually difficult in nonhuman primates.)

In spite of the obvious applications of such a technology to treating infertility, especially infertility caused by blocked fallopian tubes, human

IVF was not a well-funded or closely followed area of research. Only a handful of researchers around the world pursued human IVF, led by Patrick Steptoe and Robert Edwards in England. The first convincing report of in vitro fertilization of human eggs did not come until 1969 because finding the right medium in which to sustain (to "culture") the eggs, before and after fertilization, had proven difficult. Edwards finally cracked that problem and reported, with understatement, "There may be certain clinical and scientific uses for human eggs fertilized by this procedure."[9] In 1970 and 1971 Steptoe and Edwards reported getting normal cleavage of fertilized eggs in culture, but then they got stuck again. An Australian group in Melbourne made the first efforts to transfer human embryos into women to try to initiate pregnancy in 1973, possibly with a day or two of success. Three years later, Steptoe and Edwards achieved a nonviable pregnancy, an "ectopic" pregnancy where the embryo implanted outside the uterus. Then, shortly before midnight, July 25, 1978, their work produced Louise Joy Brown, the first baby born as a result of IVF.

From Louise Brown to Today

The full story of the process leading up to Louise Brown has been told in impressive and moving detail by Robin Marantz Henig in her book *Pandora's Baby*.[10] The Browns were infertile because both of Lesley Brown's fallopian tubes were blocked. Happily for Louise Brown, her parents, the field, and the millions of children born through IVF thereafter, Louise was a healthy, normal infant. The method that produced Brown was not entirely the modern version of IVF, which evolved over a dozen years of rapid creative change, as new techniques were tried and, sometimes, perfected. In 1978, the Australian researchers began using clomiphene to stimulate ovulation, instead of relying on natural ovulation cycles. In 1979 several teams began using ultrasound of the ovaries rather than hormone levels to determine when ovulation was about to occur. In 1980 the culture medium used for embryos was further modified. And other groups began to produce IVF births—the Melbourne group in 1980, Howard and Georgeanna Jones in the United States in 1981, and in 1982 other groups in France, Sweden, and Austria. (Howard Jones died at the age of 104 just two weeks before I finished the submitted manuscript of this book.)

1983 saw the first pregnancies and births from egg donation, where the gestational mother was not the genetic mother. The early cases involved women who could not produce their own viable eggs and carried the pregnancies using donated eggs, but the method opened the way to the first case of gestational surrogacy in 1984, where the prospective mother provided the egg that, after fertilization, was transferred into the uterus of another woman for gestation. Also in 1983 the first baby was born from a frozen embryo, setting up a major component of modern IVF, the freezing of "extra" embryos.

The decade ended with the first use, in 1989, of preimplantation genetic diagnosis (PGD), a technique we will discuss further in Chapter 5. The next major development took place in 1991 and 1992 with the development of ICSI—intracytoplasmic sperm injection. The technique developed from late 1980s work in Australia on injecting sperm into the zona pellucida. In ICSI a microscope-guided needle "grabs" a single sperm, breaks its tail, and injects it directly into the egg, producing much higher rates of fertilization than the earlier method. ICSI greatly increased the chances of establishing pregnancies from men with very low sperm counts or with dysfunctional sperm.

By 1992, IVF had fully emerged in its modern form. Although there have been refinements since then, today's process is basically the same.

IVF Today

IVF can overcome most forms of infertility. It requires a viable egg, sperm that can fertilize the egg, and a woman whose uterus can maintain the pregnancy. IVF, however, is not an entirely benign procedure; apart from the pregnancy itself, IVF is expensive to all parties and, to the woman providing the eggs, at best uncomfortable and at worst dangerous.

The most common IVF protocol today starts with ten to fourteen days of injections to stop the prospective mother's menstrual cycle. This assures the physician, usually a reproductive endocrinologist, that, however regular or irregular the prospective mother's menstrual cycles, the maturation of the woman's eggs can be synchronized by starting this particular menstrual cycle at a precise time.

That's what comes next, the starting of a new menstrual cycle by injections of another hormone to stimulate and, indeed, hyperstimulate the follicular process of ripening eggs. In normal reproduction, several follicles reach the tertiary stage before each new menstrual cycle, but only one receives enough hormonal support to become the dominant follicle and eventually ovulate. In IVF, the woman is provided with enough of the hormone to allow many follicles to mature and to ovulate. These injections must be closely monitored. Too much ovarian stimulation can be dangerous; in fact, about 1.5 percent of women who undergo egg retrieval end up with ovarian hyperstimulation syndrome, although only about 0.4 percent require hospitalization.[11]

This series of shots lasts for about ten days; near the end of this period of stimulation, women also receive shots of yet a third human hormone to prevent natural, and therefore less controlled, ovulation. Once the physician decides the follicles are sufficiently developed, the doctor orders the injection of a fourth hormone as "the trigger shot" to induce ovulation.

Egg retrieval originally required an invasive surgery, but is now done through the vagina. The doctor inserts a needle, guided by ultrasound, through the vaginal wall to reach the ovaries. The needle punctures each follicle and draws its fluid, which includes the mature egg, into the syringe. Egg retrieval takes fluid, and eggs, from as many follicles as are mature, usually between ten and thirty. The egg retrieval process takes about ten to twenty minutes; the patient may be under conscious sedation or general anesthesia. The fluid is taken to the laboratory where the eggs are identified and removed from the aspirated fluid. If ICSI is to be used, the eggs are also stripped of their surrounding cumulus cells.

Meanwhile, moving back to the man, he will have been asked to provide a sperm sample shortly before the egg retrieval process takes place. This is almost always done through masturbation, with or without visual aids. The semen is treated by removing the seminal fluid and nonfunctioning sperm cells and producing "washed" sperm.

In some IVF cases, eggs and sperm are brought together in a heated culture medium for about eighteen hours, with about 75,000 sperm for each egg. In other cases, single sperm will be picked out for ICSI. Although ICSI was developed for use when the fertility problem was, at least partially, the man's, it is now used in nearly 70 percent of IVF cycles in the United States.

Each fertilized egg, now technically a zygote, is picked out and moved to a special growth medium, where it develops (or not) for the next forty-eight hours, until it reaches the eight-cell stage. How long the embryos are left to develop depends on the lab, often influenced by different national practices. In Europe, embryos usually are allowed to develop in vitro for only two days; in the United States, Canada, and Australia, development goes on for at least three days and often for five or six days.

However long the in vitro culture process continues, it usually ends with embryo selection. Often there are more embryos than are needed immediately. The number of embryos to be transferred in IVF varies from country to country, clinic to clinic, and patient to patient. In terms of safety, transferring one embryo would clearly be best; carrying multiple fetuses is dangerous to the mother as well as the fetuses (and eventual children). Still, in 2012, the last year for which data are available, only 19.5 percent of U.S. cases transferred only one embryo. Two embryos were transferred 55 percent of the time, three embryos 17.7 percent, and four or more just under 8 percent of the time.[12] The number of "good looking" embryos available has some influence, but so does the age of the mother (and her eggs), her past history of pregnancy attempts, and, perhaps, the prospective parents' desperation (emotional and financial). And, unstated but well understood, couples paying out of pocket for IVF may feel financial pressures to avoid multiple expensive IVF cycles by transferring more embryos to increase their chances of success.

If there are more embryos than will be immediately transferred, someone has to decide just which of the developing embryos will be transferred to a woman's uterus for potential implantation, fetal development, and birth. This selection is usually done by the IVF lab, which tries to pick the embryos that are most likely to lead to successful pregnancies, based mainly on examining the embryos under a microscope. Clinicians look at the embryos and "score" them based mainly on their shapes and sizes for their stage of development (their time since fertilization). No standard scoring system is widely used; IVF clinics tend to use their own idiosyncratic systems, often with quite a bit of seat-of-the-pants empirical "art."

One clear reality is that some of the best-looking embryos do not lead to pregnancies, while, on occasion, very low-grade embryos become healthy children. Clinics and researchers continue to try to find better ways to predict which embryos will be successful. At least one company, Auxogyn (newly renamed Progyny after a merger), has recently started marketing a method of embryo selection that relies on time-lapse photography of the developing embryos to see which ones are most likely to be viable.[13]

Disputes continue over whether it is best to transfer at day three or wait until day five. In 2012 in the United States, the latest year for which data are available, for IVF cycles using fresh (not frozen) nondonor (made without donor gametes) embryos, about 46 percent of the cycles transferred cleavage-stage embryos on the third day and about

45 percent transferred blastocysts on the fifth or sixth day. Three years earlier, day three transfers were 55 percent of the total.

Any extra embryos can, especially if they "look good," be frozen (at the temperature of liquid nitrogen, -340 degrees Fahrenheit or -196 Celsius) to be used in a later pregnancy attempt, thus avoiding another egg retrieval. Historically, rates of pregnancy with frozen embryos have not been as good as with fresh embryos but have not been much worse and usually the frozen embryos had not looked "as good". (In recent trials when embryos of equivalent quality are transferred or frozen, the frozen embryos seem to be more successful than the fresh ones, perhaps because freezing gives the prospective mother's reproductive system time to recover from the hormones used in IVF.) It is unknown how long frozen embryos can remain viable; babies have been born from embryos frozen for twelve years and the success rates for using frozen embryos do not seem to decline with the length of time spent frozen.

Once the embryos have been selected, the actual transfer is straightforward. A thin plastic catheter is run through the vagina and cervix and into the uterus of the woman who is to carry the pregnancy. One or more embryos are then put into the catheter and flushed with fluid through the catheter into the woman, where it is hoped that implantation will follow. Sometimes the clinic will make a weak spot or opening in the embryos' outer shells to making it easier for them to "hatch" out of the shells.

About nine days after a blastocyst transfer (about eleven days if the transfer was of a three-day embryo) and then again two days later, the woman is carefully checked for the spike in her levels of the hormone that signals implantation. The clinicians will want to see hCG levels in the normal pregnancy range on the first test date and, equally important, will want to see that they have increased by at least 60 percent in the next two days. Those levels and growth rates would indicate that implantation has started. At four or five weeks after fertilization pregnancy can be confirmed by an early ultrasound and by listening for a fetal heartbeat. Only at that point can the prospective parents be sure that a pregnancy has been started, as well as how many fetuses are involved.

IVF Risks, Costs, and Successes

The hormones given as part of IVF—to repress ovulation, to stimulate follicle growth, to trigger ovulation, and then to prepare for and maintain pregnancy—almost always have side effects that, although usually not serious, are unpleasant. They include bloating, cramping, and mood swings. And, of course, few people will enjoy receiving (or giving themselves) scores of injections in the course of five or six weeks.

The biggest medical risk from IVF for the woman is that the egg retrieval process will trigger ovarian hyperstimulation syndrome, in which the stimulated ovaries grow larger. About 30 percent of women going through egg retrieval will have some symptoms of this condition, mainly mild symptoms such as bloating, nausea, diarrhea, and abdominal pain or tenderness. Moderate hyperstimulation syndrome can include vomiting, rapid weight gain, increased size of the abdomen (from fluid collecting inside it), and decreased or stopped urination. This requires bed rest and close monitoring of electrolytes, blood counts, and fluids going in and coming out of the woman's body. In severe cases, the fluid imbalances can trigger heart, lung, and kidney problems, as well as, in a few cases, the very dangerous condition called adult respiratory distress syndrome. Severe cases can be complicated by either the rupture or the twisting of ovaries.

In addition to hyperstimulation risks, the egg retrieval process can lead to infection or to the needle puncturing the wrong things. IVF also has a higher risk of ectopic pregnancies than those that occur naturally. About 1 percent of traditional pregnancies take place, very dangerously, outside the uterus, in the fallopian tubes or on the cervix, fallopian tubes, or abdominal wall. In IVF pregnancies, the ectopic pregnancy rate is between 2 and 5 percent. (This may not be a result of the IVF process itself, as damage to the fallopian tubes that can cause ectopic pregnancies is also associated with fertility problems—women who need IVF to get pregnant may just be, on average, at higher risk for an ectopic pregnancy.)

The bottom line is that almost all women who undergo the procedure have pain and discomfort along the way, mainly from the hormones they are given, but also from the egg retrieval process itself. Up to 1 percent of women who undergo egg retrieval end up hospitalized as a result, generally from complications of ovarian hyperstimulation syndrome. On occasion, women die directly from ovarian hyperstimulation syndrome; the British literature reports at least four deaths.[14]

How much does IVF cost? In some countries it is covered by national health plans—for some kinds of people, for some procedures, and for a particular number of cycles. In the United States, by contrast, it is not covered by public health plans (and it is not mentioned in the recent health reform legislation). Most private coverage also excludes IVF though it may cover some limited fertility services. In around fifteen states, state law requires private insurers to cover fertility treatments; they tend to offer fairly minimal coverage, and employers can avoid even these state mandates by offering self-insured health plans to their employees and families.

The lack of health coverage for IVF not only makes the prospective parents bear the entire cost of the treatment, but it also increases the price of those treatments. In most medical procedures, health coverage can serve to drive down the price as big insurers negotiate with doctors and hospitals for favorable rates. Without the purchasing power of insurers to limit them, IVF clinics in most places are free to charge whatever the market will bear. Given the desperation of many infertile people, the market will bear a lot.

The minimum price of one cycle of IVF in the United States in 2016 is probably between $12,000 and $15,000. Various bells and whistles—ICSI, assisted hatching, embryo freezing, and PGD, among others—can easily add another $5,000 to $15,000. On the other hand, using frozen embryos is much cheaper, around $3,000 per cycle, because the costs of egg retrieval and early embryo culturing have already been paid. Using "donated" gametes adds a few thousand dollars for sperm and $15,000 or more for eggs. Someone seeking to use a paid gestational surrogate—a woman who will carry the fetus—will probably pay another $40,000 to $60,000 for her services (and her health care), at least if a pregnancy is established and goes to term.

The price for one IVF cycle will vary from case to case and region to region but $20,000 is probably a fair estimate for the median price of a basic IVF cycle. And, of course, that is the price for one IVF cycle, not for having a baby. That will usually cost more—and not just for the prenatal and delivery costs. Unfortunately, most of the time one cycle of IVF does not yield a baby.

The success rates for IVF vary dramatically depending on the reason for infertility, the age of the woman who provided the eggs, the patient's past history with IVF, and which procedures are being used. The U.S. Centers for Disease Control and Prevention (CDC) collect data on IVF success and try to enforce standardized definitions. The latest data, published in November 2014, contains fascinating information, in great detail, both at the national level and down to the individual IVF clinic on 456 IVF clinics (more than 90 percent of the total).

Overall in 2012, the reporting clinics started 158,000 cycles in the hope of immediately transferring one or more embryos for possible birth. Those 158,000 cycles led to over 51,000 live births and over 65,000 babies. These babies made up about 1.6 percent of all the children born in the United States in 2012—not a large percentage, but not trivial. (There are more babies than live births because a live birth is a pregnancy that leads to the delivery of at least one living infant; over 45 percent of IVF births produced multiple births, mainly twins, but about 3 percent of the live births are triplets or higher-order births.) The overall success rate—the percentage of cycles that led to a live birth—was just over 32 percent.

On average, then, statistically average prospective parents should expect to undergo—and pay for—two cycles of IVF in order to have a baby. At roughly $20,000 per cycle, that is over $40,000 in IVF costs alone. And, of course, the "about two cycles" is just a statistic; any individual couple might be successful in their first cycle—or might go through five, six, seven, or more cycles, with their attendant financial costs, medical risks, and emotional pain without success. In 2012, about 58 percent of women who attempted fresh, nondonor cycles were going through an IVF cycle for the first time, but some were doing their fourth or more.

Not all IVF cycles have the same chances of success. The CDC report breaks down the live birth success rates by age and by procedure type. Maternal age is crucial, at least when nondonor eggs are used. In 2012, for women under thirty-five, the rate of live births for each time embryos were transferred into the uterus was about 47 percent. For women thirty-five through thirty-seven, the live birth rate was about 38 percent. For women thirty-eight to forty, the rate fell to 28 percent, then to 16 percent for women forty-one or forty-two, about 6 percent for women forty-three or forty-four, and around 3 percent for women over forty-four.

In general, cycles using donor eggs were more successful than those using nondonor eggs and cycles using fresh embryos were more successful than those using frozen embryos. In the second case, the difference could be a result of harm caused by the freezing and thawing processes, though, as noted above, it may well be a result of the "best-looking" fresh embryos being transferred and only the "leftovers" being frozen.

In the thirty-seven years since Louise Brown was born, the procedure that allowed her to exist has produced millions of children around the world—and has become a multibillion dollar industry (an estimated $3.5 billion per year in the United States alone).[15] But it remains expensive, uncomfortable, risky, and not very efficient. As a result, it is very much a minority taste, generally accounting for less than 2 percent of births even in rich countries. Developments in genetic testing and in human stem cells will change that, so to those areas we now turn.

4

GENETICS

After starting the book with a chapter on the basics of DNA, chromosomes, genomes, and genes, I have spent the last two chapters on reproduction—it is now time to return to genetics. The power of genetic testing to tell us about our future children will drive the revolution in human reproduction. Understanding that revolution will require some background on genetics.

DNA makes RNA makes proteins. That is the central dogma of molecular biology. Although exceptions and qualifications have been found, it remains largely true. It is also true that variations in DNA can lead to variations in RNA, which can lead to variations in protein, which in turn can lead to variations in organisms. Gregor Mendel knew nothing about DNA, which was not identified until four years after his (now) famous publication, but ultimately DNA variations drove the differences he saw and recorded for his pea plants that led to his discovery of the principles of genetics. The variations in DNA that affect traits come in many different types. For this book, I will focus on four categories—Mendelian traits, non-Mendelian traits, chromosomal abnormalities, plus a smaller "other."

Mendelian Traits

It is never entirely clear what someone means by Mendelian traits, but it means something like "traits that are caused or very strongly influenced by a variation in a single gene." And the term "Mendelian" is also bound up with ideas of dominant, recessive, autosomal, and X-linked inheritance, among others.

Before Mendel, people knew that organisms inherited some characteristics from their parents. Thousands of years of human plant and animal breeding had relied on this fact and the idea that human children normally resemble their parents was not new. No one, however, saw how this was done. In fact, Darwin, the first edition of whose *On the Origin of Species* was published just a few years before Mendel's unheralded publication, continued through all six editions of his masterpiece to worry that the lack of any mechanism for inheritance of parental traits was a weakness for his theory. The general idea was that children inherited a blend of parents' traits, but this made it hard to understand why offspring did not eventually all inherit an average of all the original values of each trait.

In careful work with various traits in peas, Mendel saw that crosses of parents whose seed colors were yellow and green did not end up with chartreuse or olive seeds, but yellow or green seeds.[1] He derived two principles from his work, sometimes called Mendel's Laws—the law of segregation and the law of independent assortment. The first law said that inheritance is determined by particular factors passed on by each parent to its offspring and that each offspring got only one of the two factors the parent possessed. Thus, when pollen from one pea plant was mated with eggs from another pea plant, each parent gave the resulting plant one discrete factor determining, say, pea color.

The second law said that, for any two traits, the inheritance of the factors for one was independent of the inheritance of the factors of the other. Thus, whether the offspring pea plants inherited a yellow or a green seed factor from one parent was independent of whether that parent gave it a factor for a smooth or a wrinkled seed. In addition to these principles, he worked out the idea of dominant and recessive inheritance based on his statistical analysis of the results of his pea breeding.

Although these principles have been modified over the years, for some associations between genes and traits (including diseases) they hold true; these traits are now often called "Mendelian" or "simple Mendelian" traits. A trait called Mendelian is entirely (or very largely) determined by the variations an individual has inherited in its two copies (one from each parent) of a single gene. (I will mainly use the common term "alleles" for different variants of genes hereafter.) For most Mendelian traits, one allele is "dominant" over the other, which is "recessive."

Mendel looked at seven traits in his pea plants, including whether the ripe seeds were smooth or wrinkled. When he started his crosses with one parent from a long line of smooth-seeded plants and a second from a long line of wrinkled-seeded plants, his first generation of crosses all had smooth seeds. But when he crossed those first generation plants with each other, he found that three-quarters of the offspring had smooth seeds and one-quarter had wrinkled seeds. Although Mendel did not know it, in terms of DNA, the smooth-seeded parent had two copies of the smooth-seeded allele; the wrinkled-seeded parent had two copies of the wrinkled-seeded allele. All of their offspring, which each got one copy of the gene from each parent, had to have one copy of the smooth allele and one copy of the wrinkled allele. But when they were crossed with each other, the offspring had a 50/50 chance of getting a smooth or a wrinkled seed DNA allele from each parent. That meant that, on average, one-quarter of the offspring would get one smooth-seeded allele from both parents, one-quarter would get a smooth-seeded allele from their "father" and wrinkled-seeded allele from their "mother," one-quarter would get a wrinkled allele from their father and a smooth allele from their mother, and one-quarter would get a wrinkled allele from both parents. The result would be three smooth-seeded plants for each one wrinkled-seeded plant—Mendel had discovered the classic 3:1 inheritance pattern for "Mendelian" traits.[2]

Many human genetic traits or diseases are Mendelian. Take, for example, the Rh blood type. We all are either Rh positive or Rh negative, depending on the DNA sequences of the two copies of the Rh gene we inherited from our parents. This gene instructs cells to make a protein that sits on the outer surface of the membrane of red blood cells. The gene has two main alleles: Rh positive and Rh negative. A person can therefore have two copies of the Rh positive allele, two copies of the Rh negative allele, or one copy of each. Someone with two copies of either allele will have that trait—either Rh positive or negative. But what about someone with one copy of each? A person who inherits one copy of the Rh positive allele and one copy of the Rh negative allele will be Rh positive. The Rh positive allele is dominant over the recessive Rh negative allele. If each of two parents had one copy of each allele, on average three-quarters of their children would be Rh positive—the quarter who got two copies of the Rh positive allele and the two-quarters who got one copy of each—and one-quarter would be Rh negative, the quarter who got two copies of the Rh negative allele.

In humans, not just Rh negative blood type, but many other traits, including thousands of genetic diseases, are inherited in a Mendelian recessive way. Cystic fibrosis, sickle cell anemia, beta thalassemia, and phenylketonuria are a few of these recessive diseases. People only have one of these diseases if they have two copies of the recessive (disease-causing) allele. A parent with one "normal" and one "disease" allele will not have the disease, but will have a 50 percent chance of passing that disease allele on to any children. He or she would be called a "carrier" of that disease. Normally that will not matter because the child will get a normal version from the other parent. But if the other parent is also a carrier, their children have a 25 percent chance of having the disease (inheriting two copies of the disease allele), a 50 percent chance of being carriers themselves (inheriting one copy of each allele), and a 25 percent chance of having two normal copies (inheriting two copies of the normal allele).

Other traits are dominant Mendelian traits. A person with one normal allele of the *Huntingtin* gene and one copy of the allele that causes Huntington disease will get the disease. Each of his or her children will have a 50/50 chance of inheriting his normal allele or his disease allele. Assuming the child's other parent does not have the disease allele (which, given its happy rarity, is usually safe), the child's chances of inheriting the disease are 50/50.

Recessive diseases or traits are often caused by DNA alleles that make the gene ineffective. It is the absence of any working copy of the gene that codes for the CFTR protein, for example, that causes cystic fibrosis. A person with one normal and one disease-associated allele will have, in each cell, one copy of the gene that makes the proper protein and that almost always provides enough protein to allow for a normal life. The recessive allele does not do anything affirmatively harmful; it just does not provide a gene able to play an important, sometimes vitally important, role. These are "loss of function" variants.

Dominant diseases are usually caused by alleles that affirmatively do bad things. The *Huntingtin* allele that causes Huntington disease makes a variant of the protein that causes brain neurons to die. Having one good copy does no good; the one "bad" copy causes the disease. These are "gain of (bad) function" variants.

Thus far we have been talking about Mendelian traits that are inherited on the autosomes, chromosomes 1 through 22, where everyone inherits one copy from each parent. Some diseases, however, are caused by genes found on the sex chromosomes, X and Y. Almost no diseases, except for a very few related to male fertility, are found on the Y chromosome. Those very few can only be, and necessarily will be passed down from father to son; mothers have no Y chromosome to give and daughters cannot have received a Y chromosome from anyone.

On the other hand, many important genes, and so many important disease genes, are found on the large X chromosome. These include, among others, genes that cause hemophilia, fragile X syndrome, several forms of muscular dystrophy, and color blindness. Like autosomal recessive traits, these diseases are caused by loss of function alleles, but X-linked traits are inherited in a different fashion, one that puts boys and men at a serious disadvantage.

Take hemophilia, the common forms of which are caused by genes on the X chromosome. A woman has two copies of the X chromosome, one from her mother and one from her father. To have hemophilia, she would need to inherit one disease copy from each parent. But a man has only one copy of the X chromosome, which he *had* to inherit from his mother, who had only X's to give. If, let us say, one in every ten thousand X chromosomes has the hemophilia allele, one man in ten thousand would inherit the disease but only one woman in one hundred million—ten thousand times ten thousand—would have inherited two X chromosomes with the disease-causing allele. These classic X-linked traits, then, appear (almost) entirely in men but will not be inherited by their sons, who cannot get their father's X chromosome and be male. The daughters of affected men have to get their father's X chromosome and must be carriers. Each of the daughter's children will have a 50 percent chance of getting her X chromosome with the hemophilia allele. If a daughter, she will be a carrier; if a son, he will have the disease. Queen Victoria carried the hemophilia gene on her X chromosome and several of her male descendants suffered from hemophilia, including Alexei, the son of Tsar Nicholas and Tsarina Alexandra (Victoria's daughter's daughter).[3]

In addition to the autosomes, the X chromosome, and the (very small) Y chromosome, human DNA comes in one other place—the mitochondria. Remember, from Chapter 1, that the mitochondria have their own small circular genome. Although it does not have many genes, some of those genes are strongly associated with disease. But, remember also from Chapter 1, our mitochondria come solely from our mothers. The few mitochondria in the sperm when it finally fertilizes an egg are generally destroyed and are always grossly outnumbered by the mitochondria in the massive egg. That means that mitochondrial diseases cannot be inherited from the father (who contributes no mitochondrial DNA) but if the mother has the variant, it will *always* be passed on to all of her children.

So to sum up these Mendelian traits or diseases, they can be inherited as autosomal dominant or recessive (if on chromosomes 1 through 22), as Y-linked (passed only and always from father to son, but these are few and rare), as X-linked (recessive and rare in women but more common and effectively dominant in men), and as mitochondrial (passed only by women but to all their children). Each of these inheritance methods produces a distinctive pattern on a family pedigree chart—at least, if the family is big enough and its disease history well enough known. On average (but only on average), an autosomal recessive disease, if it appears at all in a set of children, will appear in one-quarter of them, an autosomal dominant disease will appear in half of them, a Y-linked disease can only come from the father and will always pass to a son but never to a daughter, an X-linked disease for which the mother is a carrier will appear in half of the boys and none of the girls, and a mitochondrial disease (in a mother) will appear in all her children.

Non-Mendelian Traits

Logically, if one set of traits is Mendelian, "non-Mendelian" should describe everything else. But, like the law, the life of a language is not logic but experience.[4] Thus, the term "non-Mendelian" is not used to describe all traits which are not Mendelian. Some traits are caused by "unusual" DNA with abnormalities in the number or arrangement of the chromosomes. We will deal with those later as "chromosomal abnormalities." Non-Mendelian traits, on the other hand, are made up of traits where a single gene has some influence—in combination with many other genes, the environment, and chance.

What we call Mendelian traits (and most of the chromosomal abnormalities) are always or very often present when the relevant DNA variations are present. In the language of geneticists, their penetrance is very high: the percentage of people with a particular configuration of DNA (a genotype) who have (or "express") the relevant trait (a "phenotype") is close to 100. After Mendel's laws were rediscovered in 1900, early and enthusiastic human geneticists applied them indiscriminately to any traits that seemed to run in families, from working on the ocean (thalassophilia) to changing homes frequently (nomadism).[5] (Interestingly, as genome sequencing is giving us more data about Mendelian "disease genes" in healthy people, we are learning that many of them are less penetrant than we had believed.)

Although DNA variations are associated with many traits or diseases, highly penetrant genetic associations are uncommon. Many things about people are affected by their genes, but they are often also affected by the interplay of many different genes as well as by the environment (understood very broadly) and chance. For example, an average American woman's chance of being diagnosed sometime in her life with breast cancer is about 12 percent and with ovarian cancer, about 1 percent. A woman who inherits a disease-causing (pathogenic) allele of the *BRCA1* or *BRCA2* gene has a chance of being diagnosed sometime in her life with breast cancer of roughly 55 to 85 percent and with ovarian cancer, roughly 30 percent. Her risks are higher, but not 100 percent.[6] For other traits, we know of alleles that raise or lower the risk by much less, taking, say, the roughly 0.7-percent average risk of Crohn's disease down to 0.5 percent or up to nearly 3 percent.[7]

The same disease may be, for some people, caused by a completely penetrant allele, but others with the disease may have a high-penetrance or low-penetrance allele or no alleles linked to the disease at all. Alzheimer disease fits that description. About one person in one thousand has a completely or very highly penetrant allele of a gene called *Presenilin 1*—or *PS1*—found on chromosome 14, that will cause early onset (ages forty to sixty) Alzheimer disease in everyone who carries one "bad" copy (and lives long enough).[8] This is a Mendelian autosomal dominant condition. For these people, Alzheimer disease is inevitable and entirely genetic.

On the other hand, we all have a gene on chromosome 19 called *ApoE,* which makes a protein called apolipoprotein E. In humans, this gene comes in three common alleles: *ApoE2, ApoE3,* and *ApoE4.* From 3 to 4 percent of people inherit two copies of *ApoE4,* one from each parent. These people have a very high chance of developing Alzheimer disease, somewhere between 50 and 80 percent. About 20 percent of people inherit just one copy of *ApoE4.* They have two to three times the normal risk of developing Alzheimer disease, but that risk is still somewhere between roughly 20 and 45 percent.[9]

And about half of all the people with Alzheimer disease have two normal copies of *PS1,* no copies of *ApoE4,* and no other known genetic cause of their disease. Why do they get the disease? We do not know. It may be the current unknown effect of other DNA variations that we have not yet identified, some aspect of their environment, pure (bad) luck, or any combination of the above. Alzheimer disease thus can be wholly genetic, largely genetic, significantly genetic, or apparently not genetic at all.

Most human traits and diseases seem to be associated, in one way or another, with genetic variations. We see this through what are called "concordance studies." These studies, ideally, compare identical (monozygotic) twins, who have exactly the same genomes (and hence the same alleles) with nonidentical (dizygotic) twins or ordinary siblings, who share 50 percent of each other's alleles. If identical twins share a particular phenotype (trait or disease) more commonly than nonidentical twins do, it is hard not to conclude that the difference is caused by their much closer (very nearly identical) genotypes. That's still not entirely definitive—maybe they share the same trait because their parents (and others) treated the identical twins more alike than they treated nonidentical twins—but when the differences are large, it is hard to draw other conclusions.

And sometimes those differences are large. For a wide variety of conditions, such as schizophrenia, autism, or height, identical twins will be much more similar than nonidentical twins. If one identical twin is diagnosed with schizophrenia, the other twin has nearly a 50 percent chance of being diagnosed with the disease; if one nonidentical twin has the diagnosis, the other twin has about a 15 percent chance.[10] These concordance studies lead to statistical estimates of "heritability." Heritability roughly means the proportion of an observed difference between individuals that is due to their DNA. This turns out to be much trickier to calculate, or even to understand, than it seems, but for our purposes, heritability is mainly interesting as a mystery.

Many phenotypes appear to be highly heritable, but thus far careful genetic research has found only a small part of the DNA variations that explain the differences. Thus, height clearly is strongly heritable. About 90 percent of the difference in the height of adults in any given population seems to be the result of their parents' heights (and presumably their genes); the rest is due to environment and chance. But so far more than fifty different regions of DNA have been associated with variations in human height. Added all together, they account for only a small fraction of the overall inherited variation in height. If height were a simple Mendelian trait—like, say, Rh positive or Rh negative blood type—we would know by now exactly which genes are responsible for the differences between people. But for height, as for numerous other traits, many different genes seem to contribute—probably more than fifty—and their combined contribution is small—perhaps as low as 20 percent of the inherited portion.[11]

This so-called "missing heritability" is a major mystery in current genetics.[12] Are these traits, in fact, highly heritable so that, with time and effort, we will find alleles that fully explain them? Are these traits highly heritable but based on such complicated interactions between different alleles that we will never fully understand them? Or are our measures of heritability somehow wrong? We may (or may not) know the answer to those questions sometime in the next few decades. For now, the important thing to remember is that even for traits that we have reason to think are strongly associated with genetic inheritance, our ability to predict the trait from the genes—the phenotype from the genotype—is often bad. This is true of many fairly common diseases, from asthma and coronary artery disease to schizophrenia and autism, as well as many nondisease traits, from height and weight to IQ test results and musical ability.

Chromosomal Abnormalities

DNA can lead to traits and diseases in another way, not involving Mendelian inheritance, but through chromosomal abnormalities. The chromosomes are, after all, just very long DNA molecules. When the chromosomes are abnormal—too many, too few, with missing stretches, extra stretches, or rearranged stretches—bad things often happen. The most dramatic and fairly common problem is to have the wrong number of chromosomes. A person with 46 chromosomes (22 pairs of autosomes, plus two sex chromosomes) is called "euploid," with the *eu* from the Greek for "good." A person with a different set of chromosomes is "aneuploid," or "not" euploid.

The most common aneuploidy involves having an extra copy of chromosome 21. People with three copies, instead of the normal two copies of chromosome 21 have what is called trisomy 21—also known as Down syndrome. The extra copy of chromosome 21 causes problems, notably a mild to moderate intellectual disability but also an increased chance of a variety of physical difficulties such as heart problems, intestinal blockages, and visual deficits. About one baby in 700 in the United States is born with Down syndrome; that's about 6,000 each year.[13]

In most cases aneuploid embryos and fetuses do not develop successfully enough to lead to live births or even well-established pregnancies. There are very few, if any, live births, for example, of children missing one of the two copies of any of the autosomes and there are not many more of people with an extra copy. In addition to trisomy 21, trisomy 13 (Edwards syndrome) and trisomy 18 (Patau syndrome), where infants have three copies of chromosomes 13 or 18, respectively, each occurs in around one of 5,000 live births in the United States.[14] Median survival is one or two weeks although 10 percent or more survive more than one year. Those who do survive have very serious mental and physical problems. It is not clear that any human is ever born alive whose cells all have extra copies of any of the autosomes other than chromosomes 13, 18, and 21.

Apart from those three chromosomes, only two other chromosomes are found in unusual numbers of copies in living humans: the sex chromosomes, X and Y.[15] There are a variety of nonlethal sex chromosome aneuploidies, not all of which have terrible outcomes. These include Turner syndrome, where women are born with only one sex chromosome, an X; Klinefelter syndrome, in men with two X chromosomes and one Y; and XYY syndrome, in men with one X and two Y chromosomes. Each of these three conditions occurs in around one birth in a thousand. These conditions lead to some unusual phenotypes but are relatively benign. Other, rarer, sex chromosomes aneuploidies, such as babies born with three or more X chromosomes, are more disabling.

In addition to people born with missing or extra entire chromosomes, other people are born with substantial chunks of a chromosome missing, duplicated, moved around, or even just inverted—running in the opposite direction from normal. Cri du Chat syndrome, for example, is caused by a deletion of the end of the short arm of chromosome 5; the name ("cat's cry") comes from the distinctive sound made by the malformed larynx of an affected person. Trisomy 9p is a rare syndrome where children have an extra copy of some or all of the short arm of chromosome 9; serious intellectual disability is one of the consequences. Sometimes these abnormalities involving parts of chromosomes have serious or deadly consequences; sometimes they have no discernible consequences at all.

At a smaller scale, scientists have recently discovered substantial chromosomal variation among people in the form of "copy number variation" (CNV). Unlike aneuploidies, which involve whole chromosomes and other large chromosomal changes smaller than whole chromosomes, chromosomal abnormalities also occur that involve the duplication or deletion of much shorter pieces of chromosomes. When these changes are very small—and so were not visible until recently—we call them CNVs. Once the technology made it easy to see CNVs, geneticists were surprised to see how much CNV variation there was from person to person. At least some CNVs have been linked to diseases; for example, a CNV is one of the few specific DNA variations strongly linked to some cases of autism (one of those conditions that has seemed to be highly heritable until you look for specific DNA locations that cause it).[16] Just how important CNVs turn out to be for diseases or other traits remains to be seen.

What causes chromosomal abnormalities? At the level of aneuploidies, it is usually a mistake in "disjunction," the separation of chromosome pairs during cell division. Sometimes two copies of a chromosome will go one way instead of the appropriate one. Small chromosomal abnormalities are probably smaller errors as the chromosomes line up next to each other, either during meiosis or in the early rounds of mitosis that egg and sperm predecessors go through. One piece from one copy of chromosome 8, let's say, accidentally gets incorporated into the other copy, leaving one daughter cell with a duplication and the other with a deletion. Or the piece that switches chromosomes during recombination might end up stuck in backwards, leading to an inversion.

One More Thing

Almost every trait or disease that has been linked to inherited variations in the genome can be, fairly easily, classified as Mendelian, chromosomal (including copy number variations), or non-Mendelian. But geneticists have been surprised by genetics (and genomics) in the recent past and may well be again in the near future. At least one "other" thing may be, at least sometimes, important—an oddball form of inheritance sometimes called "epigenetics," though "inherited epigenetics" might be more accurate.

The basic idea is not controversial. Not all genes are expressed in all cells, in the same way or at all. Something "epi," Greek for on top of, the genes, determines what genes will be "turned on" or "turned off" in certain kinds of cells or at certain times. We are beginning to understand a variety of mechanisms through which epigenetics works, including processes like methylation—adding methyl groups (one carbon with three hydrogen atoms) to DNA.

We have long known that epigenetic differences can be inherited from cell to cell. When a liver cell divides, it stays a liver cell because the epigenetic factors regulating the expression of the genes in both of the daughter cells are the same as those in the parent cell. What has some people excited is the possibility that these epigenetic changes—changes to the DNA and its chromosomes but not of the DNA sequence—might be inherited by children from their parents (or even grandparents). This would open up the possibility, for example, that changes made to gene expression during someone's lifetime could be passed on to one's children, which begins to sound like an old idea called, after the French scientist who promoted it, Lamarckian evolution—that organisms inherit characteristics their parents have acquired.

The often-used example for Lamarckian thinking is that ancestors to giraffes kept stretching their necks to reach higher leaves during their lifetimes, leaving their necks taller than their parents' and that they passed their taller necks on to their descendants, who stretched their necks as well, passing on still taller necks to *their* descendants and so on. Lamarckianism was not banished as a heresy until well into the twentieth century; Darwin himself thought it was a possibility. But it was ultimately rejected, making the current discussion of the possibility of transgenerational inherited epigenetic changes quite unsettling to many geneticists.

One form of epigenetics is called imprinting. With imprinted genes, contrary to the conventional wisdom about all genes, it matters whether the disease allele is inherited from the mother or the father. This has been demonstrated with several traits in mice and in a few rare genetic diseases in humans. Thus, a baby whose zygote lost the 15q11–13 region (bands 11 through 13 of the long arm of chromosome 15) will have a genetic disease, but which disease depends on whether the lost band is on the chromosome inherited from the baby's father (in which case the disease is Prader-Willi syndrome) or the baby's mother (in which case the disease is Angelman syndrome). These two syndromes are quite different; the first involves poor muscle tone, learning disabilities, obesity, underdeveloped gonads, and an odd physical appearance; the second brings severe developmental delays, serious speech impediments, a very small head, epilepsy, tremors, and a constant smile. (An older term for Angelman syndrome, now, happily, discarded, is "happy puppet syndrome.")[17]

Whether, in humans, imprinting and other forms of inherited epigenetics are vanishingly rare, quite common, or somewhere between is a very hot topic right now in genetics. And, like CNVs, further research will be required to answer that question.

5

GENETIC TESTING

All this information about the relationship between DNA and various diseases and traits is only useful if we can know not just the relationships but exactly which DNA variations particular people, fetuses, embryos, or gametes have. Genetic testing is the process of finding that out. Genetic testing of humans has taken place for over sixty years. During that time, the methods used have varied enormously but have consistently become more precise and less expensive. This chapter will first explain five methods relevant to this book: karyotyping, fluorescence in situ hybridization (FISH), array comparative genomic hybridization (aCGH), SNP chips (often called genotyping), and sequencing. Applying those tests to fetuses or embryos is complicated; the chapter will end by discussing four ways to do that, including the current version of the star of this book, PGD.

Methods of Testing

Karyotyping, the classic method to test for chromosomal abnormalities, literally looks at chromosomes. Cells are taken and grown in culture until some of them have reached the stage in the cell cycle when they were about to divide. At this point, the chromosomes condense and form the rodlike structures that, after staining, show bands of alternating colors. These structures can be seen with a moderately powerful microscope. The chromosomes can then be photographed, cut out (literally in the early days, figuratively now with computerized images), and each moved next to its paired chromosome. The result is a karyotype—a lineup of 22 pairs of autosomes and two sex chromosomes.

At least, that is how the karyotype looks in a euploid person. An aneuploid person does not have 46 chromosomes. If three copies of chromosome 21 are visible, the person has 47 chromosomes, called trisomy 21, and hence will have Down syndrome. If there is only one sex chromosome, an X, the person will have Turner syndrome. Karyotypes can also sometimes make visible smaller problems, such as the deletion, copying, or rearrangement (called a translocation) of various fairly large chunks of chromosomes. Karyotyping has several problems. First, it takes time. The cells have to be grown in culture until they reach the phase of the cell cycle when their chromosomes are condensed. And then some skilled work needs to be done to pick out, line up, and assess the paired chromosomes. But, perhaps more importantly, it only shows quite large changes in chromosomes; it will miss many changes that could have serious ramifications, including the sequence changes responsible for Mendelian diseases.

Beginning in the 1990s, karyotyping to test for aneuploidy was largely replaced by another process, fluorescence in situ hybridization (FISH). In this process, probes are created that are made up of sequences of DNA known to bind to other DNA sequences found only on a specific chromosome. These DNA probes are combined with fluorescent dyes and then mixed with the sample and allowed to combine with the chromosomes. When the mixture of sample and probe is then viewed under a microscope and illuminated with a fluorescent light, the selected chromosomes will glow with the appropriate color. So, for example, a probe can be made of a particular stretch of DNA found only on chromosome 21 and labeled with a green dye. When the probe is mixed with the sample from a cell, it will attach only to that section of chromosome 21. If viewed under a fluorescent light, a normal cell will show two green dots, one for each copy of chromosome 21. A cell from someone with trisomy 21 will show three dots, meaning that it has three copies of the chromosome. Although some of the methods of FISH continue to be widely used in clinical tests and research, as a clinical genetic test FISH has almost entirely been replaced by a different method of looking for differences in the number of chromosomes, or increasingly small parts of chromosomes, known as array comparative genomic hybridization (aCGH).

This method was developed around 2005 and uses a chip (or array) approach. A glass slide is divided into rows and columns. At each of the spots defined by the intersection of a row and a column, a specific DNA "oligonucleotide"—a stretch of DNA about ten to twenty base pairs long—would be placed. The sample to be analyzed would be broken down into small pieces of DNA that would be tagged with a fluorescent dye. These would be put into a fluid and floated onto the chip. Pieces of sample DNA that were perfectly complementary to the oligonucleotide in the first spot would latch on at that spot. A fluorescent light would then reveal whether any DNA sequences had combined at that spot and, if so, roughly how many.

Like karyotyping or FISH, aCGH is useful for detecting aneuploidies and other chromosomal abnormalities, including copy number variations. Today, aCGH is commonly used for medical purposes, including prenatal diagnosis, but also for understanding mutations in the chromosomes of cancer cells, which often are wildly abnormal.

I discussed aCGH above because it replaced the earlier FISH technology for testing, but historically aCGH itself grew out of a testing approach using so-called SNP chips that was invented in the 1990s and caught fire in the early 2000s. Karyotyping, FISH, and aCGH look at the number of and, to some extent, the structure of chromosomes. They do not help with the variations in genetic sequence. SNP chips do not generally look for the amount of a particular variation found in a DNA sample and so do not provide information on chromosome structure. Neither do they provide base-by-base sequence data along a chromosome, but they do offer cheap and fast information about a certain kind of sequence variation, the single nucleotide polymorphism.

This method is not generally used to test for specific genes, but to look for "markers" in an individual's genome. It provides information about the sequence of the DNA, but only at about one base pair in 3,000. That's a disadvantage, but SNP chips had the advantage of being cheap and easy, thus allowing researchers and clinicians for the first time to test hundreds of thousands—and ultimately more than a million—different genetic variations in a sample in one cheap assay.

A SNP chip might contain, say, one million locations, ordered in 1,000 rows and 1,000 columns. Many small DNA molecules, usually about 25 base pairs long, with an identical sequence would be put on each specific spot, say, row 263, column 840. The next spot over would have its own set of DNA molecules with their own specific sequence, perhaps the same except for one change, say a G for a T.

The two hypothetical sequences are almost entirely identical. They vary in only one spot, a single nucleotide (the middle one in this case), a T in the first example and a G in the second. These stretches of DNA that vary in only a single nucleotide are called SNVs, single nucleotide variations. Some of these SNVs are common among humans. It might be the case that the first of these sequences, the T variant, is found in 60 percent of people and the second, with the G variant, is found in 40 percent. When each of two variants is found in a substantial portion of a population, the SNV gets promoted to the status of a single nucleotide polymorphism (SNP).

No universally agreed-upon standard exists for how common an SNV must be to rate being called a SNP, but whatever the cutoff, the human genome has millions of SNPs—millions of places where most people have a T but a significant percentage of others have a G (or other combination). That makes SNPs good markers; the arrays or chips make it cheap to analyze those markers in volume.

With improvements, before long, hundreds of thousands, and then more than a million SNPs could be detected from any single sample for less than $100. Scanning SNPs stretched across the entire human genome became a standard research approach, using SNP chips to do a genome wide association study (GWAS). These studies were called "association" studies because they were not necessarily looking at the actual DNA variations that caused the trait. Most just identified SNPs that were close to the locations of the causative variations. And because of the way recombination works in meiosis, if a SNP is close to a disease-causing location in the parent, it will likely be close to the same location in a child who inherits that chromosome. So, on average, SNPs close to mutated versions of *BRCA1* should run in a family along with *BRCA1*. If one did not know that *BRCA1* was a cancer-causing gene, finding SNPs associated with the disease would give researchers a clue to look at nearby genes, and thus might help find some of the missing heritability.

That, at least, was the hope. And it has worked, but somewhat disappointingly. SNP chips have pointed to many regions that are statistically associated with particular traits—but too many regions and each associated with the trait only to a very small extent.

As SNP chips became more common in research, they also made their appearance as consumer products. Starting shortly after 2000, several firms offered ancestry information based on Y chromosome and mitochondrial DNA variations. In late 2007, they were joined by a firm called 23andMe that began to offer ancestry information based on SNP chips, a move widely followed in the industry.[1] This gives people some rough guesses about the percentage of their ancestors from various continents, but also a chance to identify otherwise unknown close genetic relatives, including, sometimes, the biological parents of adoptees or the children of sperm or egg donors.

The genetic genealogy business has been successful—over a million people (including me) have paid companies for this information—but 23andMe had other aspirations from its beginning. That firm, and many competitors no longer in the market, such as Navigenics, deCODEme, and Knome, quickly began to provide health-related information directly to consumers, largely bypassing physicians, geneticists, or genetic counselors. Most of the other firms had pulled out of this market by 2012, but 23andMe continued to expand its testing.

By late 2013, 23andMe was offering its customers genetic risk assessments of 254 health conditions based on the almost always very weak associations from SNP chip studies. The original 23andMe product cost $999; it can now be purchased for far less—at one point it cost as little as $99—and currently costs $199, but as a much less informative package than originally offered. In November 2013 the U.S. Food and Drug Administration (FDA) ordered 23andMe to stop providing health information.[2] 23andMe continued to provide ancestry and non-health-related trait information while worked its way back into the FDA's good graces. In February 2015 the FDA allowed 23andMe to sell a test to consumers for their carrier status for Bloom syndrome, a rare genetic condition. In October the FDA expanded its approval to include 35 such carrier status tests. None of these tests tell consumers their own health risks; just their chances of having a child with these conditions, which range from uncommon to vanishingly rare. The firm still cannot provide any genetic information that is directly about the consumers' health.

The big problem with SNP chips is that the information they provide is usually just not very powerful. A powerful effect for SNP chips might have a relative risk of 2.0, meaning that a person with a particular set of SNPs associated with a given disease would have twice the normal risk. But if that normal risk is, say, 1 percent, knowing that your risk is 2 percent, or one-half of 1 percent, is not very interesting—or actionable. SNP chips continue to be widely used in research because they provide broad coverage of the genome for very little money, but as sequencing becomes less and less expensive, their future is uncertain.

Karyotyping, FISH, and aCGH look at the number of and, to some extent, the structure of chromosomes. They do not help with the variations in genetic sequence that cause Mendelian traits or diseases or that influence non-Mendelian traits. The other method, SNP chips, provides information about the DNA sequence in some widely spaced locations over the entire genome, but not the whole sequence in any one location (such as a gene). For analysis of a particular stretch of DNA, as opposed to wholesale chromosome analysis or checking a SNP every few thousand base pairs, other methods have been necessary. For the last forty years, most of those methods have involved sequencing the DNA—determining the order of all its A, C, G, and T combinations.

The first DNA sequencing was done in the early 1970s, but very slowly and laboriously in academic laboratories.[3] A friend of mine recently told me about his PhD thesis in the 1970s, for which he spent several years sequencing about 100 base pairs in one gene. By the late 1970s, two groups had developed (comparatively) rapid sequencing methods—Frederick Sanger and his group at Cambridge University and Allan Maxam and Wally Gilbert at Harvard. Their sequencing methods went back and forth in popularity, with Sanger sequencing, in regularly improving versions, ultimately winning out. But Sanger sequencing, though better, was still expensive and slow.

In the 1990s the Human Genome Project drove innovation in sequencing technologies, leading to a wide variety of "next generation" sequencing techniques that have continued to proliferate. These methods have resulted in a remarkable decline in the price of sequencing, which, for an individual gene, even one tens or hundreds of thousands of base pairs long, can now be done for several hundred dollars. But the real excitement has been about doing "wholesale" DNA sequencing, either of the entire human genome or of the exons, the roughly 1.5 percent of it that "codes" for proteins, collectively called the exome.

The first fairly complete human genome sequence, finished in the early 2000s, cost several hundred million dollars. ($500 million is the canonical number, but, accounting being the darkest of dark arts, it should be called anywhere from $100 million to $3 billion.) By 2011, the price was under $5,000 to sequence a whole human genome for research. A whole exome sequence for research (the sequence of the protein-coding exons) cost about $1,000. By 2015, those figures were about $1,500 to $2,000 for a whole human genome and about $1,000 for a whole human exome. (Clinical sequences are more expensive because of the increased need for accuracy and interpretation; a clinical whole genome sequence today still costs about $5,000.) The $1,000 research whole genome sequence has been announced as imminent since 2012; it hasn't quite arrived (depending on your accounting principles) yet, but is genuinely expected soon.

Testing for Genetic Variations before Birth

These five different methods of genetic testing (and others not mentioned here) can be applied to adults, children, fetuses, and embryos. There is nothing special about how to test a prenatal sample; the special problem has been how to get that prenatal DNA sample. It is not easy to test DNA from fetuses, embryos, or zygotes because it is not easy to get DNA from them; it is (usually) hidden away inside a woman. The story of prenatal genetic testing has largely been the story of different methods for getting access to the embryo or fetus's DNA. The rest of this chapter will explain each of these approaches: invasive testing (amniocentesis and chorionic villus sampling), screening, noninvasive prenatal testing, and PGD.

Invasive Testing

Amniocentesis was the first method developed to obtain tissue samples, including DNA samples, from a fetus. It involves using a needle to draw fluid from the amniotic sac that surrounds the fetus inside the pregnant woman's uterus. The amniotic fluid can then be examined for signs of infection or other problems; it can also be used as a source for cells the fetus has sloughed off, whose chromosomes and DNA can then be tested. Genetic testing using amniotic fluid drawn by amniocentesis was first reported in 1956; cultured fetal cells from the amniotic fluid were first used for karyotyping in 1966. By the early 1970s, amniocentesis for genetic testing was well established.

Today, the procedure usually is not performed until the fifteenth week of pregnancy and most often is done between the sixteenth and eighteenth weeks; before then, the needle used in the procedure is more likely to damage the fetus. The doctor usually starts by administering a local anesthetic in the abdomen, then inserts a long needle, guided by ultrasound, through the abdominal wall and the uterine wall, into the amniotic sac, entering it as far away as possible from the fetus. The doctor will then withdraw about roughly two-thirds of an ounce of amniotic fluid. The fluid is sent to a clinical laboratory that extracts fetal cells from the fluid, which are then grown in cell culture either for karyotyping or for direct DNA analysis.

The actual procedure takes only about a minute or two; the most frequent complications are soreness from the site of the needle's insertion, as well as the possibility of cramping. Amniocentesis does, however, carry a small additional risk of causing a miscarriage. The exact size of the risk has always been controversial, but originally it was thought to be as high as 0.5 percent, added to the roughly 1 percent miscarriage risk at that stage of pregnancy. The extra risks of miscarriage from amniocentesis are now viewed as being substantially smaller, in the range of 0.1 percent, at least when done by experienced practitioners. The procedure is invasive and must be performed by a physician; between the cost of the procedure and the cost of the analysis of the material removed, amniocentesis currently costs about $1,500 to $2,000 in the United States.

Because of the risks of doing the procedure earlier, amniocentesis is not normally done before the middle of the second trimester, well after the pregnancy will have started "showing" and around the time the pregnant woman will have begun to feel the fetus move. Aborting a fetus at that stage is harder—medically, psychologically, and socially—than doing so when the fetus is smaller and the pregnancy less obvious. The desire to be able to test earlier in pregnancy led to the development of another method, chorionic villus sampling (CVS).

The chorion is a membrane that develops in pregnancy between the amniotic sac, on one side, and the placenta, lodged in the uterus, on the other. A chorionic villus is a fingerlike or hairlike projection from the chorion that burrows into the endometrium layer of the uterus, but the villi are made up of fetal cells, not maternal cells. In CVS, the fetal sample comes from these chorionic villi and not from the amniotic fluid. The sample can be taken in one of two ways, through a catheter inserted through the vagina and cervix into the placenta or through a long needle inserted through the abdomen directly into the placenta (more like amniocentesis). Ultrasound is used to guide either form of the procedure; the actual invasive procedure takes only a minute or two.

The main side effects to the woman are cramping, the possibility of some vaginal bleeding, and a small risk of infection. Like amniocentesis, however, CVS does carry a somewhat uncertain risk of causing a miscarriage, which was originally thought to be somewhere between 0.5 percent and 2 percent. That risk is now much lower, though still somewhat higher than the miscarriage risk of amniocentesis. On the other hand, CVS can be done much earlier than amniocentesis. It is usually performed in the tenth to thirteenth week of pregnancy, about six weeks earlier than amniocentesis. Like amniocentesis, it costs about $1,500 to $2,000.

Screening

Amniocentesis and CVS are diagnostic tests for aneuploidy (and other conditions) because they examine fetal cells. Due to their risks and costs, however, amniocentesis and CVS are not recommended for every pregnancy. Instead, pregnant women are screened to see if, in their cases, the procedures are likely to be worthwhile. Several different methods, from simple to high tech, are used to predict whether a particular fetus's risks are high enough to justify these invasive procedures.

One screening test is simplicity itself: maternal age. The age of a pregnant woman correlates very strongly with increasing risk for three of the most common serious genetic conditions, notably Down syndrome (trisomy 21) but also Edwards syndrome (trisomy 18) and Patau syndrome (trisomy 13). For younger women, the risk of having a fetus with trisomy 21 is about one in a thousand, but that rate begins to climb in a woman's thirties. By the time a woman is thirty-five, the chance of a trisomy 21 pregnancy is about 0.4 percent, by forty it is close to 1 percent, and by forty-five it is over 3.5 percent.[4]

In the 1970s researchers noticed a biochemical method that might be used to screen for Down syndrome. Alpha-fetoprotein (AFP) is a human blood protein that the fetus produces most actively during the first trimester of its development. Some AFP crosses the placental barrier and gets into the pregnant woman's bloodstream. Researchers noticed that if the level of AFP in the pregnant woman's blood (the maternal serum alpha-fetoprotein, or MSAFP) were unusually low, the chance that the pregnancy would lead to a baby with Down syndrome was increased. If, on the other hand, the level of AFP were unusually high, children would be at increased risk for one of a set of usually nasty but nongenetic conditions called neural tube defects.[5]

By the early 1980s pregnant women were encouraged to undergo MSAFP screening tests. These tests require only a blood draw from the pregnant woman—no fetal cells or DNA is being tested, just the level of AFP in the pregnant woman's bloodstream. An unusually low level, suggestive of Down syndrome, could then be followed by amniocentesis or CVS as a diagnostic test for trisomy 21—or not, as the pregnant woman chose.

Over the years maternal serum screening for Down syndrome and neural tube defects has evolved to become both more complicated and more accurate.[6] Today, women of all ages are offered an integrated screening for Down syndrome, which also provides risk information about two other chromosomal abnormalities (trisomy 13 and trisomy 18) as well as neural tube defects. This screening involves two blood serum tests, one in the first trimester and one in the second, plus a screening ultrasound in the first trimester.

The integrated Down screen sets its positive level at a risk of 1 in 270 that the fetus has trisomy 21—about the same level as the average risk of a thirty-five-year-old pregnant woman and roughly the average risk of a miscarriage from amniocentesis. A woman whose pregnancy screens positive is still highly unlikely to have a Down syndrome pregnancy—that 1 in 270 risk is about 0.4 percent, meaning 99.6 percent of the time the fetus will not have Down syndrome. About 5 percent of screened women will screen positive but only about one in a thousand will actually be carrying a Down syndrome fetus. Women who screen positive will be offered amniocentesis or CVS to determine whether the fetus actually has the extra chromosome. A woman whose pregnancy screens negative has less than the 1 in 270 chance of having an affected fetus and typically will not be offered one of the invasive diagnostic tests. Of course, some fetuses that screen negative will, in fact, have Down syndrome; the screening is thought to be able to detect about 90 percent of all Down syndrome fetuses—which means that it does fail to detect about 10 percent of the cases.

One result of all these screening tests is that many women over thirty-five, who previously would have been urged to receive amniocentesis or CVS because of their age, screen negative and do not undergo the further testing. Although good data are hard to find, it appears that by 2012 only about half as many pregnant women were undergoing those invasive procedures as was the case twenty-five years ago before prenatal screening was more widely used.

These screening tests are good only for Down syndrome, trisomy 13, trisomy 18, and neural tube defects. None of those are Mendelian diseases; the first three are chromosomal abnormalities, and the latter are only weakly associated with DNA, if at all. There is no equivalent to either maternal age or the biochemical screen for Mendelian traits and diseases, though there is one fairly simple screen: the genetic characteristics of the genetic parents, as shown through themselves or through their earlier children. This is, in effect, "preconception" genetic testing.

Women who have already had one or more children with a particular inherited genetic disease may be at higher risk for having another child with the same disease. If, for example, a couple has already had a child with Tay-Sachs disease, there is one chance in four that any subsequent fetus will also carry the genetic variations that cause this disease. Starting in the 1970s, though, some groups started doing population-wide preconception testing to see whether individuals in a particularly high risk group were carriers for some autosomal recessive diseases, before they established any pregnancies.

The most successful of these carrier screening drives involved Tay-Sachs disease and Ashkenazi (Central or Eastern European) Jews, in whom the disease-linked allele is found at a relatively high rate.[7] Thousands of Jewish young adults participated in voluntary screening to find out their carrier status. When they started having children, they could compare carrier status with their mates and see if their children would be at risk, information they might use to decide whether to have amniocentesis or CVS and possibly abort any affected fetuses. Some orthodox Jewish communities took the screening even further and used the carrier status to match couples so that they *could* not have an affected child.[8] The number of children born today in the United States with Tay-Sachs disease is about one-third of what would otherwise be expected but for genetic screening and testing, and most of those children are born to non-Jewish couples.

Thanks to the decreased cost of genetic testing, some couples now get screened before pregnancy for their carrier status for scores of genetic diseases, without necessarily having a population-based higher risk for the disease. A company called Counsyl, for example, will test prospective parents for their carrier status for about 100 diseases, mainly autosomal recessive or X-linked, for about $350.[9] If the screening tests are accurate (and currently they are substantially but not perfectly accurate), the parents will only have to worry about a child having one of those diseases if both of them are carriers for that disease. Hence, they will only have to do amniocentesis or CVS if they are at risk and then only for the disease or diseases for which they are both carriers.

Noninvasive Prenatal Testing

Since late 2012, several firms have developed noninvasive prenatal testing (NIPT), a new method for doing noninvasive prenatal genetic diagnosis. Different companies offer different versions of this testing, but they all share the same important characteristics—they test the fetus's DNA by finding small pieces in a blood sample drawn from the pregnant woman. NIPT thus avoids the risks and discomfort of invasive diagnostic tests.

These NIPT tests rely on the fact that some fetal cells get through the placenta and enter the pregnant woman's bloodstream. Most of them quickly die and break apart.[10] Cells die and fall apart all the time in and around our bloodstream. When the dead cell's DNA is dumped into the blood, enzymes in the blood break it apart into small pieces, about 100 base pairs long. This cell-free DNA then gets recycled or excreted. We all constantly have billions of these small pieces of cell-free DNA in each ounce of our blood. Pregnant women have not only cell-free pieces of their own DNA, but pieces of the fetus's DNA. By the eighth week of pregnancy, about 5 percent of the cell-free DNA in a pregnant woman's blood serum will be from the fetus. This fraction can then be tested.

The simplest test is just to look for bits of DNA sequence that the pregnant woman does not have, but that her mate does—and hence that her fetus might. If, for example, there is Y chromosome DNA in the pregnant woman's bloodstream, it cannot come from the woman and must mean that the fetus has a Y chromosome—and will therefore be a boy. Similarly, if the father carried a genetic variation linked to an autosomal dominant disease and the mother did not, finding bits of that genetic variation (or not) in the pregnant woman's blood would reveal whether or not the fetus had inherited its father's allele and therefore the disease.

Aneuploidies and recessive diseases can also be tested for by the simple expedient of counting. A fetus with trisomy 21 has an extra copy of that chromosome. The pregnant woman's serum will thus contain a few percentage points more bits of cell-free DNA from chromosome 21 than they otherwise would.

Similarly, consider a case where both the mother and father are unaffected carriers of an autosomal recessive disorder, say, cystic fibrosis. If 90 percent of the cell-free DNA in the pregnant woman's blood comes from her own cells, both the "normal" and the disease-linked allele of the *CFTR1* gene will be present in half of that cell-free DNA, making 45 percent of each. The remaining 10 percent, which comes from the fetus, will either be all normal (the fetus got a normal copy from each parent), half normal and half abnormal (the fetus got one of each and will be an unaffected carrier), or all abnormal (the fetus got two disease alleles and will have the disease). Those three possibilities would correspond to the total percentage of normal alleles in the cell-free DNA in the pregnant woman's blood being 55 percent (normal fetus), 50 percent (carrier fetus), or 45 percent (affected fetus).

As of this writing, four companies have already each introduced a fetal cell-free DNA test for Down syndrome in the American market; those companies and others are using NIPT in other countries. Their tests take 10 to 20 milliliters of the pregnant woman's blood (around 0.5 ounces) and cost between $800 and $2,400 in the United States. Right now, the tests are being sold as screening tests, which means any positive results would still have to be followed by invasive diagnostic tests. They are better than existing screening tests for Down syndrome because they have a much lower false positive rate and hence a much higher "positive predictive value," the percentage of times a positive test will actually turn out to be positive. Far fewer women will have the anxiety of a positive test or need the follow-up invasive confirmation test. The use of NIPT has boomed in the United States; by 2015 it was used in several hundred thousand American pregnancies. And as NIPT is used more often, invasive tests are becoming even less common.

In the longer run, NIPT may be able to do more than look for chromosomal abnormalities. It should be able to test for single gene mutations or even provide a whole genome sequence. By making use of information about the parents' DNA, NIPT has already been used in a proof of principle experiment to infer a fetus's entire genome.[11] This is currently too expensive and uncertain for clinical use, but that will change.

Preimplantation Genetic Diagnosis

NIPT can let most pregnant women check for aneuploidies without the discomfort and risks of amniocentesis or CVS and it opens the prospect of doing the same for many genetic diseases. But, like those invasive tests, it only tells you about a fetus's DNA after the pregnancy has started. At that point, the prospective parents' options in case of a bad test result are to abort the fetus or to continue the pregnancy and prepare for the birth of a child with a genetic disease or trait. Preconception screening does not have that problem, but, as a result, it also does not tell you a lot about a particular pregnancy. It just gives you statistical information about your odds with future pregnancies.

Enter preimplantation genetic diagnosis (PGD), which can give you precise genetic information about any given pregnancy before it has started. PGD is genetic testing done on a particular embryo after it has begun to develop but before it has implanted.

PGD originally used three-day-old eight-cell embryos; now, five-day-old embryos (blastocysts) are increasingly being used. At five days the embryo is a few hundred cells in size and differentiated between the inner cell mass (which becomes the fetus) and the trophectoderm (which becomes the placenta and other supporting tissues). The increased size allows clinicians to take more cells from the blastocyst, making the analysis easier and more accurate. I will only talk about blastocyst biopsy here.

Lab technicians put each blastocyst on a slide and examine it under a microscope. Using micromanipulators, they make a small hole in the zona pellucida and, using a suction pipette, carefully draw about five cells from the trophectoderm. At that point, the cells are sent off for genetic analysis.

Time is a big problem with using blastocysts. Clinics greatly prefer to transfer on the fifth day compared with the sixth day. When three-day embryos were used for PGD, the genetics lab had about forty-eight hours to receive the cell, do the analysis, and report the analysis back to the IVF clinic, which then needed to interpret it and discuss it with the prospective parents. The genetic analysis had to be performed very quickly in order to allow transfer even on the embryo's sixth day, which is the limit for current IVF practice. With blastocysts, there is (almost) no time for genetic analysis. As a result, blastocysts that receive PGD are usually frozen after the biopsy, extending the analysis time indefinitely, though incurring the possible risks (or benefits) of the freezing process.

PGD is not perfect. Genetic analysis of one cell is difficult and even five or six cells can give bad results. The possibility of contamination leading to false results is significant. Even more troubling, sometimes not all cells in the embryo have the same genome. This is called mosaicism, because the embryo, or the organism, is like a mosaic, made up of different types of pieces. In our adult bodies, mosaicism is inevitable—our cells have accumulated miscellaneous mutations in the thousands of divisions most of them have gone through since we were zygotes. As long as the mutations do not lead to a disease, like cancer, these DNA variations make no difference.

In an early embryo, though, a few neighboring cells might be aneuploid or have a mutation that other cells in the embryo do not share. If the sampled trophectoderm cells have different DNA from the cells in the inner cell mass, blastocyst PGD could lead to false positives (the sampled cell is abnormal but the rest are normal) or false negatives (the sampled cell is normal but the rest are abnormal). As a result, doctors advise women who have become pregnant with embryos after PGD to follow up the PGD with more standard prenatal genetic tests.

Although it sounds like science fiction, PGD was first performed on human embryos around 1989, taking one cell from an eight-cell embryo, with the first birth in 1990. The babies who had had one-eighth of their three-day-old embryos ripped away were not missing arms, legs, or other body parts and were, in fact, quite normal. Thousands of babies are born every year, around the world, after their early embryos have gone through PGD. Studies show no statistically higher rate of problems than for other babies born after IVF.[12] (Like all IVF babies, though, this cohort suffers from a higher rate of problems caused by pregnancies that involve two, three, or more fetuses.)

About 5 percent of the IVF cycles in the United States in 2012 used PGD, yielding about 3,000 "post-PGD children."[13] It had been thought that PGD might become much more frequent because of the hope that it could improve IVF's success rate, through screening out embryos that, because of aneuploidy or other genetic reasons, had little to no chance of implanting successfully. Thus far, studies have not shown that PGD does improve the success rate, though the reasons for this failure remain unclear—it really should work. PGD costs about $3,000 to $5,000 in the United States. It will sometimes be covered by insurance if it is being used to avoid a known high risk of a genetic disease in the child.

But, of course, that cost is in addition to the costs of IVF, because the biggest problem of PGD is that it can only be used effectively in conjunction with IVF. The embryo has to be readily available; an embryo at some unknown spot inside one of a woman's two fallopian tubes does not qualify. And, as we saw in Chapter 3, IVF is expensive, unpleasant, and, for the woman who provides the egg, physically risky. Widespread use of PGD will have to wait for a time when there is an easier way to obtain eggs for fertilization—which leads us to our last background chapter, on stem cells.

STEM CELLS

Stem cells provide the last piece of the puzzle that assembles into Easy PGD. This chapter provides background to stem cells and their possible uses. It will first outline what stem cells are, then describe the development of human embryonic stem cells (hESCs), and, after a brief digression into the immune system and human cloning, explain human induced pluripotent stem cells (human iPSCs).

Stem Cells and the Hayflick Limit

Our bodies are all about cells. Except for water-based fluids, our bodily components are either made of cells or, in the unusual instances where they are not made up of cells, like cartilage, bone, or hair, they are made by cells. Each of our cells is descended from the zygote produced from the fusion of our father's sperm into our mother's egg and each of those cells carries DNA that, except for the occasional mutation after that fertilization, is identical. That's as true of a newborn baby as it is of a wizened centenarian.

But few cells in that newborn will still be alive when she turns 100. Cells can, of course, die or be killed prematurely—think about an amputation—but they also have their own natural lifespans. These vary dramatically from cell type to cell type. None of the cells of the very early embryo will survive intact until late in life, because each of them will have been transformed, in their daughter cells, into other, more specialized cell types; nevertheless, some cells will last from before birth until death. Some neurons, for example, in the brain and elsewhere, will survive until the end, as will some cells in other tissues, like the heart. Other cells have fleeting lifespans. Skin cells, blood cells, and most of the cells lining the lungs live for only a few weeks or months before dying. Some intestinal cells have only a forty-eight-hour lifespan, less if you eat spicy food. (Perhaps not surprisingly, cells exposed to the outside world—remember, the gut and lungs are, in a sense, "internal parts" of the outside world—tend to have the shortest lifespans.) Most other human cells survive inside us for more than years but less than centuries.

Of course, our cells cannot only survive inside us but, it became clear in the twentieth century, could survive outside us given the right conditions. The ability to keep cells alive in vitro and then to watch them grow and change, called "cell culture," was a revolutionary development in the early twentieth century. Several researchers, including particularly Ross Harrison, were important to this development but it is most associated with Alexis Carrel, a French surgeon who moved to the Rockefeller Institute in 1906.[1]

Carrel won the Nobel Prize in Physiology or Medicine in 1912 for discovering a few years earlier how to stitch together blood vessels safely, thus paving the way (eventually) for organ transplantation. Carrel became interested in growing organs outside the body, which in turn led him to an interest in keeping cells alive outside the body.

Cells in culture were, and to a large extent still are, finicky. They were, and are, subject to sudden or lingering death and to contamination from other cells and from infection. Particularly before antibiotics, keeping cell cultures alive was hard. Carrel's lab was the center of successful cell culture. It invented devices and techniques to improve cell culture, as well as protocols to try to avoid or minimize contamination, infection, and other problems. And it was the Carrel lab that recognized that, to keep them vigorous, cell lines had to be divided and subcultivated—broken back down into individual cells that were in turn put into new dishes to begin the growing process again.

The star of the Carrel laboratory was a cell culture made from fragments of an embryonic chicken heart. It was begun in January 1912 and, supposedly, it survived until 1946 (two years longer than Carrel), being regularly subdivided and "repotted" in another Petri dish hundreds of times.[2] This seemingly immortal chicken heart was famous in its day—its birthday was celebrated every year in a New York newspaper and immortalized (in another sense) in a 1966 Bill Cosby comedy skit.[3]

The "supposedly" in the last paragraph is there for a reason. It actually seems unlikely that the chicken heart tissue culture from 1912 actually survived until 1946 without getting fresh chicken cells from time to time, because it turns out that normal cells of birds and mammals (including humans) will not divide indefinitely. Leonard Hayflick, a scientist at the Wistar Institute in Philadelphia, realized this in 1961 in work done with Paul Moorhead.[4] Hayflick and Moorhead took human fibroblasts, a type of cell from the skin, and "explanted" them into two clean culture flasks. Eventually the cells grew to cover the entire interior surface of those two flasks, at which point they stopped dividing. The researchers then took those cells and subcultivated them into two new flasks. The cells would grow and divide and ultimately cover the surface of the two new flasks, often in a few days, only to be subcultivated again. But not indefinitely. Hayflick and Moorhead found that human skin cells eventually divided more and more slowly—took more time before they covered the flask surface and required subcultivation—and eventually stopped. The fibroblasts in their 1961 experiments went through, on average, about fifty doublings before they stopped growing.

Such a limit was not expected. Hayflick and Moorhead pursued it with other human cell types and with various cell types from other species. The limit to the number of doublings varied from cell type to cell type and from species to species, but it was (almost) always there. Alas for poor Moorhead—this finding has become known as the Hayflick limit. Most human cell types will divide somewhere between forty and eighty times before they stop. The reasons for the Hayflick limit remain unknown, as do its consequences for aging. Do we age, in whole or in part, because some or all of our cells hit their Hayflick limits and no longer are renewed by dividing into two new cells? No one knows, but research is continuing.

The Hayflick limit, though, raised a more immediate question. As noted above, some of our cell types are short-lived; they die quickly and are replaced. Skin cells live for about two to four weeks. They start out in the lower layers of the skin and gradually move toward the outermost layer, pushed up by new cells being born below and old cells being sloughed above. By the time they reach the surface and form what we *think* of as our skin (the outermost layer of the epidermis), they are dead.

Let's say a skin cell has a Hayflick limit of 100 cell divisions and that the average cell lives for four weeks—generous estimates for each. That means that any skin cell should stop dividing after 400 weeks, or less than eight years. But most of us are more than eight years old—how can we still have skin?

The answer is stem cells. Stem cells are able to divide indefinitely. Typically, they divide in an unequal way. One of the two daughter cells remains a stem cell, able to keep dividing indefinitely, while the other one becomes a more differentiated cell, in this case a skin cell, that is launched on a more limited career. For skin, these stem cells are found at the base of the epidermis and are called (reasonably enough) basal cells. They seem to live forever. To be more precise, we know they can live at least 122 years, 164 days. We know that because Jeanne Calment, the person with the longest well-documented lifespan, lived that long.[5] Although we have no direct evidence that stem cells can survive longer than that, they are sometimes thought of as being, and called, "immortal." And stem cells can be kept alive in vitro well past the Hayflick limit, forming continuous cell lines.

It is only fair to point out another type of human cell that can be, effectively, immortal, because it, too, does not obey any known Hayflick limit: a cancer cell. Cancer cells will divide indefinitely. It is their division, growth, and spread that makes cancer an often fatal disease. It also means that cancer cells can be kept alive indefinitely in vitro. The first commonly used long-living human cell line is the HeLa cell line, derived in the early 1950s from a particularly virulent cervical cancer in a woman named Henrietta Lacks, recently the subject of a compelling book.[6] Cancer cells are the evil twins of stem cells.

Some of our tissues have stem cells throughout our lives, typically in the places where cells have short lifespans and need to be replaced constantly. The skin, the bone marrow (where blood cells are formed), the linings of our digestive tracts and our lungs, and other places have stem cells. In other organs, there appear to be no or, at least, very few stem cells. Our heart muscle cells scarcely grow and divide after our youth. If our hearts are damaged, they do not get significantly repaired with new, living cells; they just stay damaged. (Until recently it was thought they never divided; now it appears there may be some very small continuing cell division.)[7]

Human Embryonic Stem Cells and Jamie Thomson

The various stem cells in our adult bodies have the disadvantage that they are either somewhat, or completely, committed to making one kind of cell. Basal skin stem cells just make skin cells. The stem cells in the lining of the colon just make cells for the lining of the colon. Blood-forming stem cells, found mainly in the bone marrow, make lots of different cell types, but they are all different types of blood cells. Skin stem cells do not make blood; blood-forming stem cells do not make heart muscle cells.

But there are cells that make all of those cell types, and all other human cell types, directly or indirectly. We know the cells in the early embryo give rise to every kind of human cell type because they did, in every one of us. And, more specifically, we know that in the blastocyst, the cells inside the blastocyst, the inner cell mass, form every kind of cell type in the fetus and, ultimately, in the adult, because they did.

Some researchers thought to take inner cell mass cells out of an embryo and turn them into cell lines, lines that would both last indefinitely and that could be kept undifferentiated, preserving their potential for forming all of the different cell types. These embryonic stem cells quickly became a reality—in mice. In 1981, two groups, one at Cambridge and one at UC San Francisco, independently established the first successful cell lines of mouse embryonic stem cells.[8] Such cell lines quickly became useful research tools—but only in mice.

It proved difficult to do the same thing with humans. Part of the difficulty was scientific; different species react differently in many complicated and unpredictable ways. The conditions necessary to keep extracted inner cell mass cells both alive and undifferentiated in other species, including humans, were not the same as in mice. But, real as the scientific problems were, they were not the biggest problems.

In 1981, when mouse embryonic stem cell lines were first established, IVF was first entering clinical practice, and research with early human embryos was first becoming possible. But research aimed at making embryonic stem cells necessarily required destroying embryos in order to extract their inner cell masses. While no one cared about destroying mouse embryos, to many people, in the United States and elsewhere, destroying embryos was killing babies—or, at the very least, destroying something whose potential to become a baby made its destruction wrong.

As result, National Institute of Health (NIH) funding, the life's blood of American academic biomedical research, was not available for research trying to develop hESC lines. Dr. James Thomson of the University of Wisconsin nevertheless took on the task of trying to develop such cell lines. Thomson was originally trained as a veterinarian, then added a PhD in molecular biology. One part of his postdoctoral training took him for two years to the Primate In Vitro Fertilization and Experimental Embryology laboratory in Oregon. He moved to the University of Wisconsin in 1991, where he continued research on nonhuman primate embryos, including efforts to make embryonic stem cell lines from those species.[9]

In August 1995 Thomson reported in the *Proceedings of the National Academy of Sciences* that he had successfully created embryonic stem cell lines from macaques (also known as rhesus monkeys).[10] He had taken the inner cell mass from macaque blastocysts, put them into laboratory equipment, and, after much tinkering, found the conditions that would both keep those cells alive and keep them from spontaneously differentiating. This last point was the most important; embryonic stem cells do not want to stay undifferentiated—they want to turn into other kinds of cells.

Thomson's 1995 publication points out that macaque embryonic stem cells could be a better model for how human cells develop than mouse embryonic stem cells but that, of course, the best model would be hESCs. He turned his research toward humans, using "excess" embryos from IVF clinics to try to create hESCs. In November 1998 he announced success in the pages of *Science*.[11] Thomson transferred the inner cell masses of fourteen embryos and, using the same methods he had successfully used with nonhuman primates, managed to create five hESC lines. Thomson's human research had been partly funded, starting in 1995, by a private biotechnology firm named Geron, which provided money in return for preferential access to the research results and intellectual property resulting from it.[12]

Thomson's discovery immediately touched off great scientific excitement. Basic science researchers were fascinated by the possibility of taking hESCs and using them to make all human cell types, including those many types for which we do not have adult stem cells. They could then study how those cell types developed and how they reacted to different conditions. Clinical researchers, doctors, and the public got excited about another possibility—using hESCs to make new cells that could be transplanted into patients to replace cells that had been damaged or destroyed or even to regrow whole human organs. Widely discussed possibilities included making new beta cells for the pancreas to treat type 1 diabetes, making new cardiomyocytes to repair heart damage from heart attacks, making new spinal cord cells to repair spinal cord injuries, or making new brain cells to treat Parkinson's disease or Alzheimer disease.

The discovery also touched off, a little more slowly, a political firestorm. Federal financial support for hESC research became a key issue in political races in the early 2000s, with many Republican candidates adopting the pro-life movement's opposition to any research in which babies (embryos) were killed or harmed and many Democratic candidates arguing that this research was necessary to cure many diseases of living children and adults. On August 9, 2001, President Bush announced what he termed a compromise—he would allow federal funding for hESC research on cell lines that had been developed before that date—that is, on cell lines where the embryos had already been destroyed.[13]

Although some in the right-to-life movement were disappointed by his decision to allow funding for any hESC research, it was largely seen as a victory for the opponents of the research. Several (Democratic) states passed their own statutes to provide state funding for research; in November 2004, an initiative in California authorized the sale of $3 billion in state bonds to support that work.[14] Several (Republican) states passed statutes banning state funding of the research, discouraging it in other ways, or, in South Dakota, making the research criminal.[15]

The ebbs and flows of the stem cell controversy could fill (and indeed have filled) whole books. For our purposes it is only important to know that research continues with these cell lines. So far no wonder cures have emerged. On the other hand, some promising results have been obtained and a great deal has been uncovered about the factors that drive the differentiation and development of stem cells. Scientists are getting better at learning how to turn hESCs into particular cell types, how to purify the resulting cells to make sure that they are all of the right cell type, and how to transplant them successfully so that they do what we want them to do, where we want them to do it. The U.S. FDA has authorized several clinical trials of hESCs (or cells derived from them) for medical conditions, including type 1 diabetes, spinal cord injuries, and two related forms of blindness, age-related macular degeneration and Stargardt's disease (a rare juvenile form of this condition).[16]

The Immune System, Cloning, and Dolly

Even if we become able to cause hESCs to differentiate safely and reliably in the particular cell types we need for transplantation, a problem remains—the immune system. Each patient needing a cellular transplant has an immune system that distinguishes between "self" and "non-self"—and attacks nonself. In humans the proteins involved in this system are referred to as the HLA (human leukocyte antigen) system.[17] (Across other species, it is called the major histocompatibility complex, or MHC—HLA is just the human version of the MHC.)

There are more than twenty different proteins in the HLA system, each encoded in a different gene, but six are considered the major antigens. (ABO blood type is also an independently important consideration in organ transplantation.) These six are taken into consideration when deciding whether someone is a "match" for purpose of organ transplantation. Unfortunately, these proteins exist in many different versions within human populations—some of the HLA markers have thousands of known variations; all have at least a dozen common families of variants. So making a match is not easy.

The best way to have a good match for organ transplantation is to have an identical twin, who will be a perfect match (and who, as a result, were the subjects of the first successful cases of organ transplantation). Most of us are not so lucky. Close family members are the next best chances for a match, but even then the chance that a sibling will be a perfect match is low. Just how low depends on how good a match is required; in clinical practice, different levels of matches are sought for different organs and even imperfect matches can be successful with the use of drugs that suppress the recipient's immune system—although those drugs bring their own hazards. For kidney transplantation, for instance, doctors would like to see matches on all six antigens. Siblings who share the same genetic mother and father have only about a 25 percent chance of being an acceptable six-antigen match.[18]

So here's the problem. If a person needs, say, heart muscle cells to repair a heart damaged by a heart attack, the hESCs used to make those cells will have their own set of HLA genetic variations. If those variations are not identical to the patient's, the patient's immune system is likely to attack the new cells, making the transplant useless or, because of side effects of the immune system attack, perhaps even affirmatively harmful. The HLA variants in any hESC line will be determined by the versions of the HLA genes that embryo received from the egg and sperm that combined to form it. There is no particular reason to believe the embryo's HLA types will match any particular patient who needs cells derived from it.

One possibility is to make lots and lots of hESC lines, each carrying a different set of HLA markers so that compatible cells could always be generated. Given the degree of variation in HLA types, that could be billions of different embryonic stem cell lines, although, in fact, a few thousand lines with carefully chosen different HLA markers would probably provide "good enough" cells for most patients.

Another possibility is to make cells that are perfect matches, because they have exactly the same HLA variants as the patient who needs them. Even if you could get sperm and eggs from the parents of the patient, the chance of getting an embryo with even a six-marker match (let alone a perfect match) would be about 25 percent—and how often will the patient have living parents who are both willing and still able to provide both eggs and sperm in order to make embryos from which such a cell line might be derived?

Dolly the sheep provided a possible answer. Dolly, whose existence was announced about twenty months before Thomson's publication on hESCs lines, was a clone made by somatic cell nuclear transfer (SCNT).[19] In SCNT, an egg whose nucleus has been surgically removed will be fused with a somatic cell of another individual of the same species. In Dolly's case an enucleated egg from one breed of sheep, a Scottish Blackface, was fused with a cell taken from previously frozen mammary tissue of a ewe of another breed, a Finn Dorset.[20] The researchers, Ian Wilmut, Keith Campbell, and colleagues in Scotland, gave the resulting cell an electric shock and it began to divide, ultimately becoming an embryo that was transferred into the uterus of a third sheep. The embryo implanted successfully and about seven months later Dolly was born. Dolly was the one successful birth from nearly 300 attempts in this SCNT experiment, but she was one more birth than most people, even scientists in the field, had expected.

Dolly was, of course, not the world's first clone—as Chapter 2 pointed out, most life on earth reproduces clonally. Nor was she the first mammalian clone—remember identical twins—or the first mammal born as a result of artificial human intervention, or even the first mammal born from SCNT.[21] Dolly was, however, the first mammal ever born as a result of SCNT performed with cells from an adult. As such, she made scientifically plausible the scenarios for making cloned embryos of living human beings that could then be transferred for potential implantation and eventual birth—human reproductive cloning (though, of course, she only *proved* the possibility of sheep reproductive cloning).[22] These were the implications that brought front page headlines, intense controversy, and legislation in several nations and states banning or limiting human reproductive cloning.[23]

The research kicked off by Dolly has helped illustrate, yet again, how different reproduction is in different species. Over twenty mammalian species have been successfully cloned to live births, some easily and some only after years of hard work.[24] Cats and mice, for example, were easy; fairly closely related dogs and rats turned out to be difficult. SCNT in primates did not lead even to any successfully cloned embryos until 2007; it still has not produced a born SCNT-cloned primate of any kind.[25] The first successfully cloned human embryos were not made until 2013 (and, in spite of some early claims to the contrary, no human babies are known to have been created through SCNT).[26]

But Dolly also made scientifically plausible scenarios for making cloned embryos of living human beings and destroying them at the blastocyst stage to create hESC lines, lines that would be perfect genetic matches for (clones of) the living human whose somatic cell was used in the SCNT. In that case, a patient who needed new heart muscle cells might be able to donate a skin cell that could be fused into an enucleated human egg and turned into an embryo, which would in turn give rise to an embryonic stem cell line. That line could then be used to make heart muscle cells that, as far as the patient's immune system was concerned, came from an identical twin.

The prospect of this "nonreproductive human cloning" was very exciting. It generated a great deal of research, an example of huge scientific fraud (in 2004 by Dr. Woo-Suk Hwang in South Korea), and much controversy and concern.[27] It was not until 2013 when a group at Oregon Health Sciences University announced that it had succeeded in creating human cloned embryos, using fetal cells or the cells from an eight-month-old baby. It then made embryonic stem cells from some of those embryos.[28] It was another eleven months before two groups successfully made cloned human embryos and stem cell lines using cells from human adults.[29] Human SCNT cell lines had finally become a reality, about seventeen years after Dolly's birth was announced.

Induced Pluripotent Stem Cells and Shinya Yamanaka

In the meantime, in Kyoto, Japan, Shinya Yamanaka decided to try a different path to making personalized embryonic stem cells. Rather than making personalized embryos through SCNT, he wanted to try to take somatic cells and cause them to de-differentiate—to become more primitive and, eventually, pluripotent. In effect, he tried to transform adult, differentiated cells into cells that were just like embryonic stem cells and could then be redifferentiated into other cell types.

In 2006 he succeeded, using mouse cells.[30] He followed that announcement with a November 2007 paper in *Cell,* revealing that he had used the same method to do the same thing with human cells.[31]

Yamanaka's method started with human somatic cells. In his first work he used human fibroblasts, a common skin cell. (Rather than taking the fibroblasts from a particular human donor, he used some readily available fibroblast cell lines.) He then employed viruses to introduce additional copies of four human genes, genes mainly expressed during embryonic development, into the fibroblasts, genes that were transferred complete with genetic instructions that turned them on. He cultured the resulting cells and, after a few weeks, saw that some of them began to look very similar to embryonic stem cells.

Laboratories all over the world quickly replicated Yamanaka's results. Since Yamanaka's initial announcement, cells produced by his method, induced pluripotent stem cells (iPSCs), have been one of the hottest topics in biology and for good reason. These cells hold out the hope of providing personalized embryonic stem cells without needing to make an embryo. This not only sidesteps the difficulties in using SCNT to make human stem cells, but also the practical (and ethical) problems of obtaining human eggs and the ethical and moral issues around destroying human embryos to make stem cells.

Yamanaka and other researchers have developed refinements to the stem cell derivation process, so that now they can make iPSCs by adding proteins to cells rather than adding less controllable genes. Making iPSCs through one or more of these routes has now become routine. Some researchers have even reported success using variations on Yamanaka's method to change one human cell type directly into other human cell types without first taking them back to the undifferentiated, pluripotent stage.[32]

More generally, researchers are eagerly trying to see whether iPSCs really do work the same as, or as well as, hESCs. Right now, the answer seems to be that the two cell types are not quite the same. Among other things, they express some different genes. What this means for the future clinical use of iPSCs remains unclear. If they are not exactly the same as hESCs, can they be made the same? And, even though they are not exactly the same as hESCs, can they be used safely and effectively to treat disease? We do not, as yet, have any examples of using hESCs safely and effectively to treat disease; only a few human clinical trials have started. The FDA has approved a few human clinical trials for iPSCs, as have Japanese regulators,[33] but the results are not yet known. Meanwhile, very exciting research, both basic and preclinical work putting human iPSCs into laboratory animals, continues.

FIRST INTERLUDE

EASY PGD: THE POSSIBILITIES

The first six chapters of this book have given you some background on various parts of biology and human medicine—molecular biology, reproduction, infertility treatments, genetics, genetic testing, and stem cell research. Those threads will soon be woven together to make a new pattern, one that will change fundamentally how our species reproduces.

I call the process that will emerge from these developments Easy PGD. It is basically just an extension of preimplantation genetic diagnosis, but an extension that will turn PGD from an uncommon curiosity to the way many if not most babies will be conceived, at least in the rich world. Two technical advances will lead to this change: the accelerating improvement of whole genome sequencing and the ability to turn iPSCs into gametes, particularly into eggs. The first makes PGD more useful; the second makes it much easier.

PGD is currently used only to look at a few genetic or chromosomal issues. It can tell prospective parents about whether their embryos are aneuploid or euploid, of concern in part for Down syndrome and in part for the likely success from transferring any particular embryo. It can also be used to look for a genetic disease that runs in the family—Huntington disease, sickle cell anemia, Tay-Sachs, or some other dread condition.

But our ability to do genetic analysis—cheaply, accurately, and quickly—is expanding dramatically. Sequencing one whole human genome cost $500 million a decade ago, but was $50,000 in 2009. It cost about $1,500 in 2015. The cost will soon be in measured in hundreds, not thousands of dollars. And in twenty to forty years, that cost may well be expressed in scores—or fewer. Doing whole genome sequencing from one embryonic cell will be not only feasible but cheap. With sufficient precautions, it should also be accurate. When PGD can look at a whole genome and not just karyotypes or one particular genetic disease, it will become much more interesting to prospective parents.

At the same time, our knowledge of the connections between genetic variants and traits or diseases is already large. It will become vast. It is unlikely ever to be a perfect predictor for most of the things parents will care about, but it already is perfect, or near perfect, for some things parents prize, like "boy or girl?" Combine whole genome sequencing with a stronger ability to predict phenotypes from genotypes and PGD will only become increasingly attractive. And not just to parents, but also to health insurers or government health programs that can project saving money on the care of some sick children by paying for prenatal genetic diagnosis to avoid their births.

But no matter how attractive PGD becomes, its current form has a serious problem. It requires IVF and IVF is not attractive. It is expensive, unpleasant, emotionally trying, and physically risky. People use IVF because it is the only way they can hope to become pregnant or because they know their children are at very high risk for a terrible genetic disease. The problems of IVF, though, are almost all the problems of egg retrieval. Getting human sperm is usually easy; getting human eggs is always hard. Egg retrieval accounts for about 80 percent or more of the cost of IVF, almost all of the discomfort, and all of the health risk.

"Easy" PGD will be easy because it will avoid egg retrieval. Instead, prospective parents will provide some of their cells—probably skin cells—that a clinical laboratory will transform into undifferentiated iPSCs. These iPSCs will in turn be redifferentiated into gametes: eggs, primarily, but also sperm when necessary. These gametes will be made from the prospective parents' own cells and own genomes. They will hold out the prospect of having "a child of our own" without the difficulties of egg retrieval—and with the advantages of PGD. Eggs from iPSCs will make PGD easy and Easy PGD will change our species.

With Easy PGD, it will not just be the infertile or those haunted by a family genetic disease who will use PGD. It will be almost anyone. And it will open some brand new possibilities. Infertile people who do not have their own eggs or sperm will have, for the first time, a chance to have a "child of their own." And, in the not unlikely event that iPSCs can make not only eggs from women and sperm from men, but sperm from women and eggs from men, gay and lesbian couples will, for the first time, have a chance to have "a child of their own."

The science for safe and effective Easy PGD is likely to exist sometime in the next twenty to forty years. The scientifically possible does not always happen or, if it does happen, persist—remember flying cars or supersonic commercial aircraft. Easy PGD, though, will have favorable medical, economic, social, legal, and political factors, at least in the United States. Its reception will prove more negative in some countries, but more positive in others. Overall, I expect that within forty years, around the world among people with good health care, half or more will use Easy PGD to conceive their children, selecting them at least in part on their DNA and the traits and risks it predicts.

The next six chapters of this book expand on this rough outline, specifying the scientific progress—and, equally importantly, the social factors—that will transform Easy PGD from science fiction into an important reality.

PART II

THE PATHWAY

You have just read my vision of Easy PGD as a large part of the future of human reproduction. That is not my view of what would be good or bad, but my best guess about what will actually happen. Getting there, though, will require some important steps: it will have to become not just scientifically plausible but *real,* and in ways that are both safe and effective enough to reach the clinical world, to become a technology. This will involve advances in both genetic and stem cell technologies. And Easy PGD will have to benefit from medical, economic, social, legal, and political factors that allow the technology to be widely adopted. This section's six chapters discuss all those issues, as well as some other plausible technological alternatives to Easy PGD. You will not, I trust, be surprised to learn that I think Easy PGD will pass those tests. One reason that runs as a thread through this part is that Easy PGD will benefit from other uses for its component parts. It will not emerge so much as the end of specific efforts to create it, but as a "secondary" use, or effect, of many other developments.

The first two chapters in this part, Chapters 7 and 8, talk of the needed improvements in the two fields that will lead to Easy PGD: genomics and stem cells. Chapter 9 looks mainly at money—for research, for commercial firms, and for paying to use Easy PGD. Chapters 10 and 11 look at legal and political issues, respectively. And Chapter 12 takes a walk on the even wilder side of reproductive alternatives.

7

GENETIC ANALYSIS

For Easy PGD to become a clinical reality, PGD will not only need to get easier, but also to get better. For that to happen, two genetic technologies will have to be improved—DNA sequencing and DNA interpretation. This chapter will discuss each.

DNA Sequencing

Each embryo has a set of genetic variations that will provide information—sometimes powerful, sometimes weak—about the traits of the person that embryo might become. Those variations need to be determined through analyzing the embryo's DNA. Genome sequencing is the best tool we have for this kind of analysis.

How we have done genome sequencing has changed dramatically over the past forty years and is continuing to change. Today, many different methods (perhaps as many as ten) are being actively explored as paths forward to so-called "next generation sequencing."[1] I will not discuss the methods used, either by current sequencing or by the various next generation alternatives; however whole genome sequencing is done in twenty to forty years will no doubt be different from what is done today or even tomorrow. Instead, I will discuss what those methods will need to do to enable Easy PGD.

The technical challenges to using genome sequencing for Easy PGD are substantial. The method will have to sequence 6.4 billion base pairs with high accuracy. The method will have to be very inexpensive, particularly if many embryos are created (and need to be tested). And the process will have to take into account both the best time to take cells from the growing embryo and the embryo's timetable—it must be transferred for implantation, or frozen, by its sixth day. That's a tall order, but one that is clearly on the route to being filled. This section will discuss first what "whole genome sequencing" means and then its accuracy, cost, and speed.

The "Whole Human Genome"

The first "complete" human genome was sequenced around 2003. The exact date depends on definitions. President Bill Clinton and Prime Minister Tony Blair announced completion of "the human genome sequence" in a joint press conference on June 26, 2000, but that was a "rough draft."[2] *The* human genome was declared finished in 2003.[3]

None of those whole human genomes, or any completed since, have actually sequenced every bit of "the human genome." This is true for two reasons. First, no one person's whole genome has ever been sequenced. Some areas of our genome are very hard to sequence. These include the ends of chromosomes, their telomeres, and their middles—the centromeres—as well as regions that are "locked up" by what is called "heterochromatin." Also hard to read are regions with gene families, stretches with many slightly different versions of one ancestral gene. The good news is that these regions are thought to provide (almost?) no useful information.

But, second, no one has sequenced "the human genome" because it doesn't exist. All 7.3 billion humans (apart from identical twins) have somewhat different genomes. We all have (almost) all the same genes, as well as the vastly more common DNA between genes, but in slightly different versions. Is *the* human genome yours or mine, George Bush's or Barack Obama's?

Even identical twins, formed from the same egg and sperm, will have subtly different genomes. Every time a cell divides, the two daughter cells end up with slightly different genomes. Errors in DNA duplication mean that most normal cell divisions cause some new mutations. The cells of each twin will have acquired somewhat different mutations after their separation (before the twelfth day after fertilization) and during their subsequent development, during pregnancy, and throughout life.

More subtly, none of us has our own single genome. Skin cells on the palm of my right hand will have subtly different genomes from skin cells on the palm of my left hand or on the backs of either. Each cell took a different path, with different ancestral cells, from the first few weeks of embryonic development to its current place and time. In following those paths, each cell accumulated different mutations. This is called mosaicism—an organism or a tissue has not one genome but is a mosaic of somewhat different genomes. These mutations almost always make no difference. They are in areas of the genome that do nothing, or areas that do nothing in that particular kind of cell. If they do cause a serious problem, that cell dies, unmourned and rapidly replaced. Only when a mutation, whether caused by mistakes in division, radiation, chemicals, or just bad luck, sets off a cancer is its result important. Happily, mutations rarely give rise to malignant cancers, but it is no coincidence that most common cancer types (lung, colon, breast, skin, and blood) occur in tissues where cells divide frequently and hence pick up more "natural" mutations.[4]

But when we talk about sequencing "the whole human genome," we really mean sequencing the easiest-to-sequence 90 percent or so of an individual's "average" genome. That we can already do; going further is neither necessary nor particularly useful.

Accuracy

Getting the cost to a reasonable level is only one challenge for the genome sequencers. The accuracy will need to be very high. Currently, if you give a whole genome sequencing process a stretch of DNA that contains a sequence of a hundred known base pairs, it will, on average, get about ninety-nine of them right. With children at stake, 99 percent is not good enough. Making a mistake one time in a hundred means tens of millions of mistakes in the whole genome, which would make it highly likely that some prospectively healthy embryos will have been labeled diseased, and not used, and, more painfully, some prospectively unhealthy embryos will have been labeled healthy and possibly become children with genetic diseases. This basic accuracy needs to improve, but there is no reason to think that, in the next twenty years, that will not happen.

But accuracy is an issue in other ways. Current methods of whole genome sequencing have various systemic weaknesses. Three problems stand out currently: indels, "phasing," and repeats.

Indels, short insertions or deletions, can wreak havoc, especially if they are found in exons and are of a size not divisible by three, so they change how the codons are read. Researchers have run experiments giving the same DNA samples to different sequencing systems. The systems agree on well over 90 percent of the base pairs, but they agree on indels at much lower rates, about 25 percent of the time in one study and just over 50 percent in another.[5] If they do not agree, one process or the other (or both) must be wrong.

"Phasing" means determining which of the two copies of a chromosome (the one from the mother or from the father) a variation is found on. Current whole genome sequencing techniques cannot reveal the phase, which can be a serious problem, especially with autosomal recessive diseases, which occur when a person has no "good" copies of the genes and so the normal version of the protein is not being produced at all. The problem happens when a person's sequence shows two harmful mutations in different sites in one gene, along with benign versions at those sites. Maybe the embryo has one harmful mutation on each chromosome, giving it no functional copies and thus the disease. But maybe one of its chromosomes has both mutations and the other one is perfectly normal, making it an unaffected carrier. You cannot know what the result will be without knowing whether the two disease-causing variations are on the same or different chromosomes—and that current whole genome sequencing technologies cannot do.

Finally, repeats are a problem for most current methods of whole genome sequencing. These rely on a "shotgun" method that breaks a DNA sample into small pieces, which it sequences over and over again. A shotgun sequence might be made up of hundreds of millions of "reads" of, say, 100 base pair sequences.[6] Some diseases involve very long stretches of repeats and shotgun sequencing is not good at finding them. For example, Huntington disease is caused by a person having too many copies of a CAG repeat in the *Huntingtin* gene, located on chromosome 4. In a particular stretch of *Huntingtin,* for unknown reasons, the sequence repeats CAG several times. Usually there are four to ten copies of this CAG repeat. Some people have more alleles with more copies. People with up to thirty-seven copies of the CAG repeat on both of their copies of *Huntingtin* are fine; people with more than about thirty-seven copies on either copy seem to be doomed.

But if each small chunk of DNA that is read is only about 100 base pairs long, how can you tell how long the repeats are? Consider a person whose two alleles of the *Huntingtin* gene have eight and forty CAG repeats, respectively. Sequencing small pieces of gene containing the eight-repeat allele should not be a problem. Each will have twenty-four bases in the repeat sequence and most pieces will be fully contained within a 100 base pair read. But forty CAG repeats takes up 120 base pairs. If you get a 100 base pair read that is all CAG repeats, that's not quite thirty-four repeats. There is no way to know how long the repeated sequence really is.

These issues are real problems for whole genome sequencing accuracy—today. But there is no reason to think they cannot, and will not, be solved in the next twenty to forty years, especially given how important the accuracy of whole genome sequencing will be for many applications of genomics to medicine, from newborn screening to "personalized" or "precision" medicine.[7]

Cost

But now let's talk money. As mentioned in Chapter 5, the first "whole human genome" is said to have cost somewhere around $500 million. The exact cost is unknowable. Not only was it the product of two different efforts, one funded by several governments and charities and the other by a private firm, but figuring out which costs to attribute to *the* sequence involves inherently arbitrary accounting decisions. It might have been a few hundred million dollars lower, it might have been a few billion dollars higher, depending on how the costs are allocated. In any event, we can safely say that it cost "a lot."

By 2009, genome sequences were much cheaper. That year, Professor Steve Quake, a bioengineering professor at Stanford who, among other things, designs sequencing machines, sequenced himself.[8] To be more precise, he and his students, postdocs, and staff, using one of his machines, sequenced his genome. Using some unusual accounting, he reported the cost at $48,000—the cost of the materials used in the sequencing. He added nothing for the costs of the sequencing machine or the time of his students, postdocs, and staff; students and postdocs are cheap, but not free!

By the fall of 2010, the price had fallen again. In a meeting at Cold Spring Harbor Laboratories, a Bay Area firm, Complete Genomics, let it be known that it would provide a complete human genome sequence for under $5,000. This was not the "rack rate" but a package deal—ten genomes for $50,000. By 2015, Complete Genomics (purchased that year by the Chinese genomics company, BGI), Illumina, and other firms are providing whole human genome sequences for, in some cases, around $1,500 to $2,000.

For several years firms have been announcing their upcoming $1,000 genomes, using some quite different approaches. Whole genome sequencing may or may not reach the $1,000 level in 2016, but the field firmly expects that the $1,000 genome will be available soon. Prospective parents may not be willing to sequence fifty embryos at $1,000 each—at least, not many prospective parents—but there seems no reason to believe that the price decline will stop at $1,000. If it goes as low as $200, twenty embryonic genomes could be sequenced for $4,000, around a quarter of the current cost of IVF. It seems entirely plausible to expect a $200 genome within five or ten years. It is impossible to know how low the cost might be in twenty to forty years; it surely will not go to zero, but at, say, $50 a genome, 100 embryos could be wholly sequenced for $5,000.

Speed

The speed of sequencing may, or may not, be a problem. Currently it takes less than three days of round-the-clock machine use to sequence a human genome. (Three years ago it took ten days.) PGD does not allow three days; at best, as currently performed on blastocysts, it barely allows one day. And even after the sequencing is done, the sequence must be interpreted and explained to the parents for their decision on which embryos to transfer. For Easy PGD to be feasible, one of three things must happen: sequencing must be moved back from five-day blastocysts to three-day embryos, the blastocysts must be frozen during the sequencing, or sequencing must get much faster. Each has possibilities—and problems.

Using three-day embryos, as has been done for most of the history of PGD and is still done frequently, buys time. The reason the field has moved toward blastocysts, though, is that several cells can be taken from the 100- to 200-cell blastocyst; only one can be taken from a three-day-old eight-cell embryo. The more cells that are sequenced, the better the accuracy—and the more likely the results apply to the whole embryo and not just the sequenced cell(s). It is hard to see how improved technology could eliminate this disadvantage of the three-day embryo.

Freezing embryos, on the other hand, makes the time problem disappear. The analysis, interpretation, and explanation to the parents can take as many days—or years—as desired. The problem is that frozen embryos have not been as successful at yielding babies as fresh ones. Some embryos cannot be successfully thawed and for those that are thawed, success rates have been lower.[9] On the other hand, some recent work has shown that frozen embryos of equal quality to fresh embryos may actually be more effective at producing babies.[10] Just how big a drop in success rates would come from using frozen embryos twenty to forty years in the future cannot be known. It does seem clear that if the couple has more embryos available to them, which Easy PGD will allow, a lower success rate may not be as important.

The third possibility is that, over the next twenty to forty years, the time to produce whole genome sequences will fall to a few hours. That could make it feasible (barely) to use five-day-old blastocysts and implant them on the sixth day.

One step to such greater speed would be having the parents' whole genome sequences. If we know every genetic variation in the parents, we know (except for the rare new mutation) every genetic variation possible in the embryo. Sequencing the prospective parents, which may well be a routine part of medical care twenty years from now, should allow faster and more accurate sequencing of their embryos. Indeed, when using parental genomes, it may prove faster, cheaper, and more accurate not to do the entire genome sequence for the embryos but to look in just enough detail to be able to infer the embryo's sequence from the parental sequences, as well as doing one rough pass to find any major problems that were not present in the parents' genomes.

A similar kind of advance preparation might cut the time needed to discuss the results with the parents. Talking with the parents before the sequencing about the most likely genetic results could make their choices after sequencing easier and faster.

For Easy PGD to work, our sequencing technologies will have to become more accurate, cheaper, and faster. Within twenty to forty years (and maybe much sooner), that should happen. And when it does, PGD will become much more attractive. Today, PGD can look at chromosomes, seeking a few aneuploidies (and the embryo's sex) or it can look for one or two specific genetic variations linked to disease or other traits (like the HLA system). In the future, prospective parents will be able to use PGD to look for everything genes can tell us about an embryo's fate. But what, exactly, does that amount to?

Genetic Interpretation: What Can We Learn, and How?

After you have sequenced the embryo's DNA, you still need to interpret it. It is no good to have a $200, two-hour whole genome sequence if the interpretation takes ten days and $100,000. Genomic interpretation is proving much more complicated and difficult than improving the sequencing processes. Even in twenty to forty years we may not be able to say much with any confidence about, say, an eventual child's intelligence after assessing a few cells from a blastocyst.

Yet we already can interpret enough to make Easy PGD very attractive to many prospective parents, especially if the costs and risks of egg harvest can be avoided. We can already tell them at least something about five categories of conditions or traits linked to the inherited genes: serious early onset genetic diseases, other diseases influenced by DNA variations, cosmetic traits, behavioral traits, and, last, easiest, but certainly not least, "boy or girl."

Early Onset, Highly Penetrant, Serious Diseases

We know of about 4,000 powerful and early onset genetic conditions, mainly Mendelian or a result of chromosomal abnormalities. Individually these diseases range from uncommon to vanishingly rare but collectively they affect somewhere around 1 to 2 percent of births today.[11] Currently every American state screens newborns for thirty to fifty of these diseases, ones for which some useful, or possibly useful, medical interventions exist. Using Easy PGD to extend that screening to in vitro embryos and expanding it to all highly penetrant early onset genetic diseases would be very valuable. Parents will never have a guarantee that their children will be healthy, but they can be guaranteed that their children will not have Tay-Sachs disease, Lesch-Nyhan syndrome, trisomy 13, or a host of other terrible conditions.

Other Diseases

With our present interpretative abilities parents would also be able to choose embryos based on knowing something about "other" diseases, a large category of diseases that are not early onset, highly penetrant, and serious. This group includes diseases for which our ability to predict may be quite strong, such as Huntington disease or early onset Alzheimer disease, but for which the expected onset comes late in life. It also includes diseases that may happen at any age but for which our predictions are not very strong. We can say something about a person's risk of being diagnosed with juvenile or adult onset diabetes, breast or colon cancer, or regular onset Alzheimer disease, but not with great confidence. And finally there are genetic diseases we may be able to say a lot about but that, frankly, are not very important, like having slightly different color vision. Like about 6 percent of American males, as a result of an allele of a gene on my X chromosome, I am deuteranomalous, which means (I am told) that I see shades of green as less intense than most of the rest of you. But it is no big deal (except occasionally in choosing socks).

Our ability to interpret embryonic DNA sequences to predict all of these kinds of disease risks for children (or the adults they will become) will only get better. We will learn more and more about the disease risks associated with various genetic variations, not for the purpose of perfecting Easy PGD, but to learn more about the diseases that afflict already living children and adults—what causes them, how they progress, and how they might be prevented or treated. Every new discovery about, say, genetic variations that predispose adults for type II diabetes will have implications for the diabetes risks of the people who would ultimately be born from sequenced embryos. How well we ultimately will be able to predict from their DNA which people—and hence which embryos—are at how much risk for these diseases remains unknown, but we can say something today and will certainly be able to do a better job in twenty to forty years.

It is also important to note that improved genetic interpretation may be quite useful not just for highly penetrant genetic traits but also for traits or diseases that have a lower penetrance because of a substantial environmental component. For example, phenylketonuria (commonly called PKU) is an autosomal recessive genetic disease. Children born with PKU will inevitably become severely intellectually disabled—unless their environment is changed by putting them, quite quickly, on a diet low in the amino acid phenylalanine. We have known this for nearly forty years and so almost every child born in the United States (and in many other places) in the last thirty-five years has been tested for PKU shortly after birth.

Improvements in genetic interpretation over the next several decades should provide other examples of similar traits. A known genetic predisposition may not now appear to be highly penetrant, because only a fraction of those with the predisposition have the environmental exposure that converts the predisposition into a disease. But knowing that an embryo has such a predisposition could allow the parents either to select a different embryo or to protect the eventual child from the triggering exposure.

Cosmetic Traits

It is not clear just how good our ability to predict disease risks will become, but the ability to predict at least one particular kind of trait should become much better—cosmetic traits. We are confident that skin, hair, and eye color, as well as hair type, nose shape, male pattern baldness, early gray or white hair, and many other traits are very strongly influenced, if not almost entirely determined, by genes. Thanks to sun exposure, hair dyes, colored contact lenses, and other interventions, we know that they often are not completely determined by genes, but the genetic variations associated with these cosmetic traits should be largely discoverable.

They have not been found yet, at least not very often, because they are not diseases. Most research in human genetics is aimed at diseases. It is funded by the National Institutes of *Health* or by pharmaceutical or biotech companies or by disease organizations. These groups are not interested in hair color—and neither, frankly, would that be a prudent investment of limited research funding. Some work has been done on these traits, often justified by diseases or disorders in coloration, such as albinism, but, by and large, working on those issues is not a great way to get funding.

The first powerful hair color gene we learned about with confidence actually helps prove this point. We can strongly predict whether an embryo will have red hair and a freckled complexion—a look common in Scotland and Ireland—because we know those traits are caused by a mutation in the gene that controls the body's production of melanin, the main protein for skin and hair coloring.[12] Rather than make the normal dark melanin, people with two copies of this gene make a much redder form of the protein. They also are at much higher risk for the dangerous skin cancer melanoma. It was the search for genetic risk factors for melanoma that uncovered this particular redhead allele.

The value of investing in a search for cosmetic genes is not likely to become much higher, but the cost of that investment will become much lower. As sequencing becomes cheaper, more and more complete genomic sequences can be analyzed and compared. The cosmetic traits will be, by and large, easy to detect, define, and enter into computer databases—easier than the diagnoses (or even the actual identities) of many diseases. I expect our ability to predict many of these cosmetic traits to improve dramatically over the next few years, without any intense effort.

Behavioral Traits

Another kind of genetic prediction will also be of great interest to parents—behavioral traits. Will Embryo #4 grow up to be good at math, at music, at sports, at filling in the right circles on the SAT test? Will Embryo #12 turn into a shy child (and adult) or an extroverted one, a diligent and duty-bound person or a carefree one? Researchers believe, based largely on concordance studies, that genetic variations must be associated with these and other behavioral traits.[13] Thus far, except at the pathological extremes, we have had great difficulty in identifying those variations.

Maybe the most powerful, and most disconcerting, example of the power of behavioral genomics is Lesch-Nyhan syndrome. About a dozen children with this disorder are born each year in the United States. The gene is on the X chromosome, so the mother is an unaffected carrier and half of her daughters are unaffected carriers, but half of her sons will have the condition. Children are born apparently normal, though close examination would detect excessive levels of uric acid. By three to six months, they are showing signs of moderate intellectual disability. In the first few years, movement problems develop that keep most patients confined to wheelchairs. But the worst symptom appears at about age three—self-mutilating behavior. About 85 percent of affected children will start biting their lips and tongue, then move on to biting fingers and toes, as well as banging their heads. So far, there is no useful treatment for the neurological and behavioral symptoms.

Or take another, more common behavior—intelligence. We do know quite a few "intelligence" variations in the human genomes. Unfortunately, they are variations that lead to pathologically low intelligence. Trisomies 13, 18, and 21; four or more copies of the X chromosome; fragile X syndrome (the consequence of a vast expansion of a gene on the X chromosome); PKU; and a host of other (happily rare) genetic variations cause intellectual disability. We know almost nothing about genetic links to differences within, or above, the normal range of intelligence.

So we know that some genetic variations are linked to some behaviors, but mainly pathological behaviors. Our knowledge of genetic variations linked to normal, and above-normal, behaviors is, so far, almost nonexistent. We believe from a variety of studies that a large part of the variation in human intelligence, as measured by IQ, comes from genetic variations but discovering which variations has proven nearly impossible. Intelligence, whatever it means, appears to be a classic example (at least currently) of missing heritability.

Will that change in the next twenty to forty years? Maybe—perhaps even probably. But how much? If we could (honestly and accurately) tell prospective parents, "Embryo #7 will score 1550 on the two-part SAT" (the most common U.S. college admission test), at least some parents would be very interested. If we were able to say, "Embryo #19 has a 58 percent chance of being in the top 50 percent of the IQ distribution and a 12 percent chance of being in the top 10 percent," how useful would they find that information? How much power we will have to predict IQ—or any other behavioral trait—remains unknown. I suspect we will never be able to pinpoint an embryo's future SAT score with decent accuracy, but I would also bet we will find some genetic associations with higher intelligence. My own guess is that, even in twenty to forty years, the predictions will be more like those for Embryo #19 than for Embryo #7, but that is just a guess. Obviously, the more accurately Easy PGD could predict behavioral traits, from intelligence to personality to particular talents, the more interest at least some prospective parents would have in it.

Sex

One other genetically associated trait (or, perhaps, bundle of traits) requires special mention. It is both extremely easy to determine and powerfully important for many people—boy or girl. A few people already use "hard PGD" for sex selection. In some places many "people regularly" use prenatal testing followed by abortion of fetuses of the "wrong" sex. And, of course, historically, infanticide or neglect of children of the wrong sex, almost always girls, was not uncommon. Some parents might care enough about it—for their first child or for their second, third, or fourth child after a string of children of one sex—to use the more benign Easy PGD just for this purpose.

In the long run, it remains unclear just how many more characteristics or conditions, and with what power, genetic interpretation will let Easy PGD tell us about for embryos in twenty to forty years than PGD can do today. We can safely say that however much that is, it will be more than we can accurately say today—and with the use of whole genome sequencing, it will include at least every genetic trait we know about today and not just the handful PGD can currently probe.

How Will We Interpret Whole Genomes?

In 2009 and 2010 I was a very small part of a group of thirty-two Stanford researchers who interpreted the medical significance of our colleague Steve Quake's genome. The interpretation took a long, long time, both to catalog what was known about disease-causing DNA variations and then to apply that catalog to Quake's genome. But, once it was done, one path forward was clear: take that catalog and turn it into software. Then let the software scan the genome for disease risks.

This is already being done today, although with substantial human oversight of the findings. The software will only get better, the human oversight less necessary. The catalog, however, cannot be a static project. Every day brings, and will likely continue to bring for many years, new publications about the association, or nonassociation, of particular DNA variants or networks of variants with diseases and other traits. It is unclear who will pay for, or undertake, the necessary constant curation and editing of this kind of catalog. It may be one centralized service or a host of private firms; it might be transparent or it might be opaque, held as proprietary trade secrets. (Myriad Genetics, for a long time the exclusive source for *BRCA1* and *BRCA2* testing in the United States, is currently using its proprietary database of variations as a competitive weapon.)[14] But whoever creates such an interpretative catalog and the software to use it, it will happen. Again, it will not happen in order to allow Easy PGD but in order to interpret the genetic risks of living people. Once it is available for that purpose, however, its application to Easy PGD is simple.

A Word of Caution

I must end with an important qualification about the accuracy of the Easy PGD process. Even if the DNA sequencing were perfect and even if our ability to interpret the whole genome sequences of multiple embryos were perfect, PGD, at least as currently practiced, is not entirely accurate. Some of the mistakes are, no doubt, human errors. In any process involving humans, people will make mistakes; human errors can be minimized but probably never eliminated. Other mistakes, though, may come from the nature of the early embryo.

The cells of a blastocyst, though all derived from the same one-celled zygote, will already be somewhat different. These differences might be a result of mutations in the DNA that occurred in the first few cell divisions or they might be bigger problems of chromosomal translocations or even aneuploidies. If most of the blastocyst's cells have 46 chromosomes, but the ones you sampled have 47, with an extra copy of chromosome 18, the embryo would, if it were transferred and became a fetus, probably be a normal 46-chromosome fetus; the aneuploid cell would be "outcompeted" by its normal siblings. (This is particularly true when, as with PGD at the blastocyst stage, the cells come from the trophectoderm, the shell of the blastocyst, and not from the inner cell mass, which becomes the eventual child.) If the cell selected for PGD is the aneuploid cell, a likely normal embryo may be discarded as likely to have an extremely severe and drastically life-shortening condition. On the other hand, if the inner cell mass cells of the blastocyst have trisomy 13 but the trophectoderm cells do not, selection of normal cells for sequencing would lead to a false negative—a seriously abnormal embryo being diagnosed as normal. Adding cells reduces the chance of mosaicism but taking them from an already somewhat specialized tissue increases that chance.

Today, parents who have used PGD to start a pregnancy, whether with three-day embryos or blastocysts, are encouraged to use prenatal fetal diagnosis, through amniocentesis or chorionic villus sampling (CVS) later in the pregnancy to check the later fetus. NIPT (noninvasive prenatal genetic testing, discussed in Chapter 5) is likely to take the place of both of these methods by the time Easy PGD is widely used. It would likely be a backstop to Easy PGD, though it could only "fix" one kind of mosaicism-caused PGD mistake, false negatives. It could not fix the mistake of the false positive—the healthy embryo that is never transferred and thus never has a chance to become a fetus because the cell whose DNA was analyzed by PGD was a sport. And it could only "fix" the problem of the false negative—the fetus with a dread disease, or other undesired trait, that was not picked up in PGD because the tested cell was normal—by leading to termination of the pregnancy, and thus negating one of the benefits to prospective parents of Easy PGD.

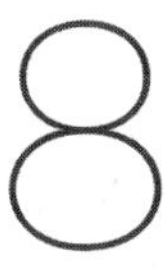

MAKING GAMETES

If the science stops with the genetics advances detailed in the last chapter, we will have substantially improved the cost, speed, and effectiveness of PGD, but as long as PGD requires egg retrieval, it will remain an uncommon procedure. The cost, discomfort, and risks of egg retrieval make it unlikely ever to be popular. How to get eggs without retrieving them? There are several possibilities, some easier and some harder. The best solution would be to make eggs from iPSCs, but other possibilities exist. And some of them offer other benefits—an effectively unlimited number of eggs (and sperm) that could be created at any stage during (or after) an individual's life.

This chapter looks at four of those options, from the nearest at hand—in vitro ripening of oocytes from ovarian samples or frozen slices—to progressively more challenging options—eggs from hESCs, eggs from iPSCs, and eggs from cloned embryos.[1] Each of these technologies faces more daunting challenges than the problems faced by whole genome sequencing, but most of those challenges should be surmountable. We will then look at one intriguing possibility that could come from the last three methods, cross-sex gametes and their potential offspring, the "uniparent."

In Vitro Ripening of Oocytes from Ovarian Slices

As discussed in Chapter 2, most researchers believe women are born with all the eggs they will ever have. These oocytes exist in primordial ovarian follicles, hollow balls of cells each surrounding one egg, until, with the start of menstrual cycles, a few of these follicles move toward maturation each month. At puberty, the average woman is thought to have about 200,000 ovarian follicles in each of her two ovaries. The ovaries, one on the right and one on the left, are the size and shape of almonds. (Interestingly, almost all the references just call them the size of an almond, without specifying just how big that is; substantial digging was required to find out that they are about 4 centimeters, or roughly 1.5 inches long, which seems like an awfully big almond.)[2] At puberty, and for thirty years or so thereafter, each month one of the two ovaries will release, on average, one "ripe" egg that has thrived out of those that had, long before, started to ripen.

Human ovaries release only around 420 mature eggs during a woman's reproductive period. Estimates of how many ovarian follicles are "used up" every month vary extremely widely. Lower estimates are around twenty, in which case the ovaries will use only around 9,000 oocytes, about 5 percent of those with which they start puberty. Other estimates range as high as 1,000 per month, which would imply that all the eggs available at puberty, and more, are used up. But at least until late in a women's reproductive life, she has an extra supply of immature ovarian follicles, which provide an opportunity. Right now, egg retrieval relies on powerful hormones to "hyperstimulate" ovaries, forcing them to ripen extra eggs, but also causing the woman uncomfortable, and occasionally dangerous, side effects. And when the eggs are just about to pop out of the ovary, a surgical procedure is necessary to retrieve them.

Why not ripen the eggs outside the woman's ovaries, eliminating the side effects of hyperstimulation? Instead, do one laparoscopic surgical procedure and take a slice out of one ovary. If a doctor removed 5 percent of one ovary—a slice about 1 centimeter high, 1 centimeter long, and 0.5 centimeter wide—it should contain about 3,500 ovarian follicles, enough to make lots of embryos.

This ovarian slice could be frozen and parts thawed throughout (and beyond) the woman's lifetime, whenever her eggs were wanted. If fifty eggs were thawed for each attempted IVF cycle, in theory this small slice could provide enough eggs for seventy attempts at IVF. Once thawed, though, the eggs would have to be ripened before they could be fertilized, subjected to PGD, and possibly used to start a pregnancy. There are, therefore, two challenges—successfully freezing (and thawing) ovarian slices and then ripening them in vitro. Work proceeds on both.

Audrey Smith, sometimes called "the mother of cryopreservation," made the first efforts to freeze ovarian tissue slices in 1951, using rat ovaries. She was trying to understand how to freeze tissues effectively.

She showed that frozen slices that were then thawed and transplanted into rats whose own oocytes had been destroyed by radiation could give rise to pregnancies and healthy rat pups.[3] Since then, ovarian slice freezing has been successfully demonstrated in many species, including humans. By 2010, two successful human births had been reported from vitrified (fast frozen) ovarian tissues, although these tissues were transplanted into the mother's ovary and not stimulated in vitro, and, by 2014, more than thirty births from slowly frozen ovarian tissue had taken place.[4] (Note that this is not egg freezing, which just freezes eggs obtained from conventional egg retrieval.) Just how successfully frozen ovarian tissue slices can restore fertility is not known. The FDA does not regulate this procedure and has not required any proof of safety or efficacy before it is tried.

But transplantation of an ovarian slice back into the donor's ovary is of no interest to us. That would, once again, leave the ripened eggs inside the women and difficult to hyperstimulate and to retrieve. Instead, for this process to make IVF (and hence PGD) easier, eggs from the slice would need to be ripened in the laboratory, where they could then be combined with sperm to make zygotes, then embryos, and eventually babies.

This, too, has begun to be done clinically. As of 2009, over 400 babies around the world had been born as a result of in vitro maturation of eggs.[5] Follicles are surgically retrieved from ovaries, plunked into a hormone broth for a few days, and then used in IVF. This method is particularly attractive to women with polycystic ovaries, for whom the use of FSH and other hyperstimulation hormones can be dangerous. In 2007 a clinic at John Radcliffe Hospital in Oxford became the first clinic licensed by the Human Fertilisation and Embryology Authority (HFEA) to use the procedure; twelve clinics around the United Kingdom now provide the service.[6] In the United States, where no licenses are required, it is not clear how many of the roughly 500 fertility clinics are offering the procedure; one recent source says "only a handful" but a Google search quickly revealed more than ten.[7]

Is in vitro maturation currently safe? There seem to be no gross dangers to the babies born this way, though a 2010 study showed somewhat higher birth weights and more difficult deliveries than normal conceptions.[8] FDA approval has not been required so there has been no rigorous demonstration of either safety or efficacy.[9]

So the two technologies needed—freezing ovarian tissue slices and in vitro maturation of immature oocytes—are already in (limited) clinical use.[10] As far as I can tell, no one has tried combining them—using in vitro maturation on oocytes thawed from frozen ovarian slices. How attractive would that be? It might be attractive. It would still require a surgical procedure, at some cost and some, probably very low, risk. It would also require the prospective mother to have ovaries that can produce healthy eggs. But it has the advantage of being very likely to work—and to work soon.

Human Embryonic Stem Cells

As discussed in Chapter 6, the appeal of human embryonic stem cells (hESCs) is that they should be able to become any one of the hundreds of cell types in humans. Thus, hESCs should be able to make eggs and sperm because, in us, they do. It is crucial to remember, though, that our gametes have our genomic variations; those made from hESCs have the variations found in the embryo from which the hESCs were created. It would not have *our* genomic variations.[11]

The trick with hESCs is to get them to differentiate into the cell type you want. One approach is to just let them differentiate in an uncontrolled fashion, forming a wide range of cell types. The result of this kind of uncontrolled differentiation of hESCs is a teratoma, the equivalent of an unusual human tumor that contains a lot of different tissues types—and body parts. Naturally occurring teratomas are famous for including things like bits of teeth and various organs. Scientists test to see whether potential hESC lines truly are hESCs, and thus pluripotent, by transplanting them into mice and seeing if they create teratomas.

If you create enough teratomas, with enough cell types, you might create some gametes, or gamete precursors. If you can then successfully separate out those gametes or gamete precursors from the other stem cells, you have a possible source for gametes. It is a fairly random process, waiting for a teratoma that has created just the cell type you want, but germ cells were created this way as early as 2003 in mice and 2004 in humans.[12]

An alternative approach is to take embryonic stem cells and subject them to factors that will lead them to become gametes. By 2009 Renée Reijo Pera had managed to turn hESCs into sperm cell precursors that went through meiosis, though the process required adding various genes in a relatively uncontrolled fashion and the results were still short of being sperm, let alone participating in the development of an embryo.[13]

A variation on Reijo Pera's approach has now been shown to be successful. The experiments involved mice, not people, but they are powerful evidence of the plausibility of deriving human gametes from stem cells.

Mitinori Saitou at Kyoto University took mouse embryonic stem cells, let them become teratomas, and then plucked out cells that looked like primordial germ cells, cells that, in normal development, might become either sperm precursors (in males) or egg precursors (in females). He then took these cells, which he called, being very careful, "primordial germ cell-like cells," and implanted them into the testicles of newborn mice that had been genetically engineered not to produce their own sperm. While in the testicles those cells were exposed to the same hormones and other influences that normal primordial germ cells experience. And, like primordial germ cells, they became sperm.[14]

Then, in *Science* in October 2012, Saitou reported on a similar experiment that led to the production of mouse egg cells, and ultimately mouse pups.[15] Again, he turned mouse embryonic stem cells into "primordial germ cell-like cells" although these were from female embryos. Rather than put the cells into the testicles of living mice, he put them into "reconstituted ovaries," made of ovarian cells in vitro. After two days, he moved the entire "reconstituted ovaries" back into mice for further maturation of the oocytes. After four and a half weeks he removed the oocytes from the reconstituted ovaries and finished their maturation in vitro. Fertilization with normal mouse sperm followed, resulting in some embryos and, eventually, some mouse pups.

One could imagine taking the next step, going back to the Reijo Pera approach, and avoiding transplanting these primordial germ cell–like cells into living animals. Instead, you would expose hESCs (or mouse ESCs) directly to factors (probably proteins) to cause them to become first primordial germ cell–like cells and then, using other factors, to become eggs or sperm. This certainly would be more attractive for human use.

However it is done, using hESCs to make gametes has serious advantages. For one thing, the hESCs are basically the same cells that naturally become gametes. They have no genes or proteins added; they are just inner cell mass cells that have been extracted from a blastocyst rather than staying inside the blastocyst to become, eventually, germ cells.

One should be leery about putting too much faith into any one or two experiments. "Never believe any experiment until it has been replicated several times by several different labs" is a good rule of thumb. But it is at least interesting to look at Saitou's mouse sperm study, which showed great success with making sperm from mouse ESCs. Five out of six of the primordial germ cell–like cell colonies produced sperm and then healthy mice. Saitou's egg study also led to the birth of healthy mice, although only five mice out of 127 transferred embryos.

Of course, mice are not humans—if they were, we could cure just about every human disease because we can cure almost anything in mice. But we now know that hESCs can do at least some of what mouse ESCs can do in terms of making sperm and eggs. In late 2014, Jacob Hanna and M. Aziz Surani published results showing that they had made cells that look and act like human primordial germ cells—the precursors of eggs and sperm—from human stem cells.[16]

But remember two disadvantages of using hESCs. One is the continuing ethical and political controversy over destroying embryos to make them. Even some of those who might allow hESCs for use to treat deadly diseases might hesitate at allowing them to be used to enable Easy PGD.

But, secondly, hESCs carry the genes of the embryo that was destroyed to make them. And, most importantly, that is an embryo with a different genome from that of either of the hopeful prospective parents—effectively, it is egg donation from an embryo. Would a couple choose to use Easy PGD if it meant that one of the gametes that gave rise to the embryos did not have the same genes as one of the two prospective parents? And, if so, how often would they prefer hESC-derived gametes to those from living human donors? The cost difference might be a factor for eggs—living donor eggs are expensive—though not for sperm, which are relatively cheap. And with living donors, the prospective parents can get at least some idea of how some of the gamete's genes turned out by studying the donor. That isn't directly possible with hESC-derived eggs and sperm, though one might go back a step and study the people whose egg and sperm gave rise to the embryo in question.

On the other hand, it is possible that some prospective parents without their own gametes would prefer hESC-derived gametes because no living human is their source. If prospective parents are deeply concerned about possible interference, actual or psychological, from a living, breathing person who would have provided half of their child's genes, hESCs may seem attractive. The interference could be from the donor or could, in a sense, be a result of the desire of the child, when grown, to find his or her genetic parent. I suspect that will not be a common view, but I also suspect it will not be non-existent. Because they would be less expensive than donor eggs and could avoid the risk that a living donor would want to interfere in the family, hESC-derived eggs might attract a significant market among people without gametes. I suspect they would not be popular among people with gametes who just want to use IVF to do genetic selection, because they would not produce that parent's "genetic child."

Induced Pluripotent Stem Cells

Eggs and sperm safely and inexpensively derived from induced pluripotent stem cells (iPSCs) would likely be vastly more attractive than either living donor or hESC-derived gametes. Using iPSCs would allow the prospective parents to use "their own" eggs and sperm, derived from their own cells and carrying half of their own genetic variations, to create "a child of their own." This is likely to be such an important consideration that the widespread adoption of Easy PGD will likely hinge on the ability to use iPSCs or some other method of creating gametes from prospective parents' own genetic material. For the many parents who view a 50 percent genetic relationship as necessary in "a child of their own," iPSCs would be a wonderful solution—if they work.[17]

Remember that iPSCs use the patient's own cells, and hence own genes, to make pluripotent stem cells, which can then be differentiated into other cell types. Most of the work done so far with human iPSCs has started with human fibroblasts, a cell type from the skin. The researchers take a skin biopsy, using an elliptical knife to remove an oval patch of skin about three millimeters (one-tenth of an inch) in diameter and about one and a half millimeters (one-twentieth of an inch) deep. The skin is then minced and set to grow in a Petri dish. Some of the fibroblasts are then "induced" to become pluripotent.

Shinya Yamanaka earned the Nobel Prize for Medicine or Physiology or inventing iPSCs in 2012, only five years after his work on creating the first human iPSCs was published. That is practically a world speed record for a Nobel Prize. His iPSCs were exciting for several reasons.

For one thing, they did not involve embryos and so avoided much of the controversy around hESCs. But they also offered a way to get stem cells with particular genomes. If you are a scientist studying, say, Huntington disease, you could take a skin sample from a person with the genetic variation that leads to Huntington disease, turn it into iPSCs, then turn the iPSCs into brain cells and see what happens to them over time, thus, perhaps, learning more about the disease.

Or, for clinical use, if you wanted to replace damaged heart muscle cells in a patient whose heart was scarred by a heart attack, using hESCs turned into cardiomyocytes runs the risk that the patient's immune cells will attack the new heart tissue. If, on the other hand, you can take the patient's skin cells, turn them into iPSCs, differentiate the iPSCs into cardiomyocytes, and inject those into the patient, the patient's immune system should see those cells as its own—they will have its own genome—and so leave them alone.

Happily, Yamanaka's method was robust and easily repeated by other researchers—repeated and improved. Originally Yamanaka used a form of gene therapy to cause the skin cells to revert to an embryo-like plasticity. In some of the infected cells, all four genes began to work, making their proteins, and the cells reverted to an undifferentiated state. Although doing gene therapy in living people has proven frustratingly difficult, putting genes into cells in a Petri dish has long been possible.

The problem is that the genes involved made people nervous—for good cause. Some of those genes were known to cause tumors in humans. And by putting the genes into the cells, you cannot control how much protein they make, at what times, and for how long. This raised the possibility that iPSCs expressing these transplanted tumor genes would ultimately develop or become cancers themselves. It also raised the most subtle concern that these powerful genes might be changing which other genes are turned on and off, in ways that make the resulting cells abnormal and possibly affect their ability to perform their normal functions safely and effectively.

In subsequent years, these fears have abated somewhat. New techniques have injected the fibroblasts not with genes, but with proteins from genes. These proteins are then limited in number and timing in ways that gene transfer could not easily or reliably achieve. Perhaps more importantly, continuing research comparing hESCs and iPSCs has shown that the iPSCs generally react, in gross terms, like hESCs.

On the other hand, closer examination has shown that the iPSCs are expressing different genes—making different proteins—than otherwise similar hESCs. It is unclear how important these differences are, but until that is understood, therapeutic uses of iPSCs will be more questionable than using hESCs. On the other hand, the immune system advantages of iPSCs are so great that much work will go into exploring those differences, either to proving that they are not important or, if they are, finding ways to minimize or eliminate them. And note this is true for a broad range of tissues and medical applications, not just for making gametes.

We do know, however, that, at least in mice, iPSCs can make gametes, thanks again to Mitinori Saitou, but that evidence comes with a warning. The Saitou group made sperm and eggs not just from mouse ESCs but also from mouse iPSCs. Each was made the same way and each worked—but the mouse ESCs worked much better. Far more of the ESC colonies gave rise to sperm: five out of six in the best attempt compared with three out of twenty-eight for iPSCs. Even worse, although most of the mice born from ESC-derived sperm were healthy, two of the five mice born from the iPSC-derived sperm died young from odd cancers in the neck.[18] His results with eggs were similarly much better for hESCs than for iPSCs.[19] Using iPSCs may be the best hope for Easy PGD, but gamete production from iPSCs will have to be much safer and more effective than that before this process should be used in humans. The good news is that with several decades to work on it, it should improve greatly.

Gametes from Cloned Embryos

The best hope may be iPSCs, but there is another—cloned embryos. As discussed in Chapter 6, Dolly the sheep was produced by somatic cell nuclear transfer (SCNT), accomplished by moving the nucleus from one sheep's egg into another sheep's egg that had had its nucleus removed.

Using SCNT to make living animals works, at least in some species, but it does not work very well. So, nearly twenty years after Dolly shocked the world, there is only limited demand for using cloning to make new animals. Even pet cloning, in spite of predictions of great prospects, has not really caught on.

But, as noted in Chapter 6, cloning to make blastocysts, from which one can derive cloned embryonic stem cells, remains exciting. Doctors very much like the idea of having cells that are copies of the cells of their patients—transplanting cells from a clone of the patient should eliminate, or at least minimize, problems from the patient's immune system.

But it would also mean that, unlike hESCs but like iPSCs, SCNT would offer eggs and sperm that are the same as eggs and sperm from the person, presumably the prospective parent, whose somatic cell was used in the SCNT. Unlike iPSCs, but like hESCs, the cells that give rise to these gametes are not treated with extra, and possibly dangerous, genes or proteins but develop more naturally from inner cell mass cells. SCNT does require a supply of human eggs for use in making the SCNT embryos, but once the process has worked, SCNT eggs should themselves be usable to make more SCNT embryos.

There are two problems with this rosy scenario. One, as with hESCs, is the continuing controversy around destroying embryos. The second is that we are not sure how well SCNT works. Chapter 6 briefly mentioned the long but interesting story of fraud and disappointment in efforts to make SCNT work in humans, but in May 2013 Shoukhrat Mitalipov and his team at the Oregon Science and Health University finally succeeded in using SCNT to make human blastocysts, from which they derived stem cells, a feat that a different group replicated in April 2014.[20] Research with these kinds of cells is just starting.

At this stage, we just do not know how useful these SCNT-produced hESCs will prove, in general or for making gametes, but they are certainly promising. I suspect making iPSCs, which has already become routine, will always be easier (and cheaper) than using SCNT to make hESCs. It is also less troubling to people concerned about the destruction of human embryos. If both methods turn out to be able to make eggs and sperm with roughly equal safety and efficacy, hESCs seem unlikely to be used. The hESCs will only take over if iPSCs cannot perform as well.

How likely is this? Remember that the efforts to make iPSCs work as well as hESCs will not be driven by a desire for Easy PGD. They will be, and are being, pushed forward by the desire for a way to make all kinds of human cells for research and for cell replacement therapy. Heart disease, neurological disease, diabetes, and other similar plagues will drive the development of safe iPSCs (as well as safe hESCs). Eggs and sperm for Easy PGD will just be a secondary use, a "side benefit."

The rest of this book assumes that iPSCs will be used, but remember, for purposes of Easy PGD, it really does not matter (except possibly politically) which method is used. As long as either iPSCs or SCNT-derived hESCs can be used to make safe eggs (and, of less importance, sperm) for a reasonable price, Easy PGD can go forward.

We have now surveyed three different ways of making gametes—human embryonic stem cells, induced pluripotent stem cells, and somatic cell nuclear transfer. Not one of them is proven.

Will any of these methods work? We cannot be sure, but, given twenty to forty years to perfect them, I would bet that some, and perhaps all of them, will. For the last three, the various kinds of pluripotent cells (hESCs, iPSCs, and SCNT-derived hESCS), we know that pluripotent human cells can make eggs and sperm because they do, in all of us. The trick will be to find some way to make those cells safely and efficiently do the same thing outside of a living, developing human. Given the great medical and scientific interest in learning that trick for all kinds of human cells and tissues, it seems likely that eggs and sperm will not turn out to be exceptionally difficult.

Cross-Sex Gametes

So now assume that one of the last three methods for making eggs for prospective mothers (and sperm for prospective fathers) works. Consider the following variation—what if we could make eggs from cells taken from men and sperm from cells taken from women?

In theory getting at least part of the way there may not be too difficult. Pluripotent cells make different cell types, including eggs and sperm, depending on the environments they find themselves in. In nature, germ cells with an X and Y only develop in a male environment and those with two Xs only develop in a female environment. As with other pluripotent cells, the right environment may be extremely important. But we know, or think we know, from studying some genetic cases of male infertility that environment is not sufficient, at least to make sperm. Some genes on the Y chromosome must work for fully functional sperm to result. But could we figure out how to make eggs from men and sperm from women? The short answer is we don't know. The longer answer seems to be a definite maybe, bordering on probably.

Making eggs from men may not actually be very difficult. After all, men do have one X chromosome, as well as one copy of the Y chromosome. This raises two questions. If the environment is right and the appropriate (female) hormones are bathing the cells, do cells need two X chromosomes to become eggs or is one X chromosome enough? And, if one X chromosome is enough, does the possession of a functioning Y chromosome block the production of eggs?

Start with the X chromosome. Every diploid male cell will have one copy of the X chromosome and that chromosome will have every gene that a second X chromosome would have; are two copies of the chromosome necessary to make eggs? That question is particularly interesting when discussing the X chromosome, because, although women have two copies of the X chromosome in every cell, one of those copies is almost totally deactivated.[21]

Early in the development of female embryos, different cells randomly turn off different copies (maternal or paternal) of the X chromosome, leading to something called "X mosaicism." From that point on, all the cells derived from the cells that made the choice have the same copy of the X chromosome turned on and the same one turned off. Women, therefore, actually have, in effect, two genomes. Each has the same pairs of autosomes, but the genome in some of their cells uses the maternal copy of the X chromosome while the genome in other cells uses the paternal copy.

Therefore, all the cells that have produced eggs have done so with (mainly) one active X chromosome. But that "mainly" is a catch. Not all the genes on the Barr body (the inactivated copy of the X chromosome) are turned off. A few are used. But are second copies of those genes important to making eggs? We do not (yet) know. If those genes or, more properly, the molecules produced by those genes are important, making eggs from males would require either adding a second X chromosome or, at the least, adding the relevant genes or factors to substitute for a second X chromosome.

If the "second X" problem turns out not to be a problem, or is a problem with a solution, is the presence of a Y chromosome a problem? This would require that some of the molecules produced by the Y chromosome have somehow interfered with the development of eggs. That may not be the case. Or if the Y chromosome does interfere with egg production, maybe the Y chromosome can be turned off. Basically, at worst, one needs only to eliminate a chromosome or some genes. And eliminating chromosomes—or, more accurately, using cells from which chromosomes have been eliminated or have just "dropped out"—may not be that difficult. Whatever kinds of stem cells are used to make the gametes will be dividing in culture for some time, both as undifferentiated cells and as they are in the process of differentiating toward gametes.

Things happen to cells as they are being cultured. Among other things, they can sporadically add or subtract chromosomes.

This is usually a bad thing. As we have seen, aneuploid cells, those with the wrong number of chromosomes, are trouble. Some of those cells may drop out their Y chromosomes, becoming cells with 22 pairs of chromosomes and one, and only one, X chromosome. But, as discussed in Chapter 4, such a person will be a woman with a condition known as Turner syndrome. Women with this condition have some distinctive physical characteristics and health risks and conditions, including greatly reduced fertility, but the condition appears not to prevent the production of eggs but rather their proper maturation and functioning.[22] Neither would be necessary to derive eggs in vitro from iPSCs. Even if the absence of a Y chromosome were crucial, it might be possible to take a man's cells, turn them into iPSCs, culture the iPSCs, pick out daughter cells that have lost their Y chromosomes, and turn them into eggs.

Making sperm from women is likely to be more difficult. The Y chromosome does not have many functioning genes on it, but some of them are known to be important for making functional sperm. Women do not have Y chromosomes, so to make sperm from a women's cells it seems likely that, rather than something being subtracted from the cells, something would have to be added to them. And adding genes, let alone chromosomes, has proven difficult. The first human gene therapy experiments began thirty-five years ago and the successes have been few and far between. On the other hand, gene transfer is improving substantially, as are other methods of genome editing, discussed in Chapter 12, so this might not be a problem in twenty to forty years. Also, adding the proteins rather than the genes, as is done with some forms of iPSC to de-differentiate cells back to pluripotency, may make the process easier.

Also, remember that it may not be very important to make fully functional sperm. Thanks to ICSI these women-derived "sperm" need not swim long distances, survive in the difficult vaginal or uterine environment, or be capacitated. They don't even need flagella (tails). They just need to be injected. And although eggs are packed full of substances crucial for the subsequent development of the early embryo, sperm seem to provide only two things of importance in human reproduction—a haploid genome and, it seems, a structure called the first mitotic spindle, which is essential for the first division of the zygote.[23]

"Spindle" or "spindle apparatus" is the name for a complex structure in cell nuclei that is important when cells are dividing. The spindle apparatus guides the chromosomes through the complex dance they do in mitosis, separating carefully so that the 92 chromosomes (a doubled number before cell division) are properly packed into the two daughter cells, each cell getting identical copies of the original 46 chromosomes. It seems increasingly likely that this father-derived, sperm-delivered "first mitotic spindle" is essential to the proper carrying out of the earliest divisions of the fertilized egg. Would a female cell turned into a "quasi-sperm" make a functional mitotic spindle?

And one other potential problem remains: imprinting. Some genes, probably not many in humans, are "imprinted" differently depending on whether they come from the mother or the father, which means whether they are delivered in an egg or a sperm. Some of these imprinted genes may be necessary for successful embryonic development. As discussed in Chapter 2, in some vertebrate species, but no mammalian species, eggs will sometimes develop into organisms without the addition of sperm through parthenogenesis. No one has been able successfully to make mammals through parthenogenesis, or even get as far as establishing a pregnancy. One common explanation is that, in mammals, successful embryonic development may require some genes that have male imprinting.[24]

Assuming the hypotheses about the crucial roles of the sperm-delivered first mitotic spindle and paternally imprinted genes (and they are just hypotheses at this point) are correct, what would that that mean? It means that making sperm from female cells will require more than just making a female cell go through meiosis and hence go from having a diploid genome to having a haploid one. Those female cells will have to develop as sperm, at least to some extent. They will need to acquire male imprinting and a male-derived spindle apparatus.

Maybe both of those necessary steps will occur when a female cell is exposed to an environment that promotes the development of sperm. But maybe not. It is at least plausible that female cells would need some boost from Y chromosome–derived genes or proteins in order to accomplish those goals. They probably would not, however, need a full Y chromosome and all the proteins or RNAs it produces. They just need enough to produce cells with haploid genomes that are male imprinted and have a male spindle apparatus.

There is one other arguable problem with women making sperm, compared with men making eggs: their sperm won't (easily) be able to make boys. Eggs from male cells could lead to boys, most easily if the sperm that fertilizes them has a Y chromosome. But a cell from a woman does not have a Y chromosome to contribute to a sperm. And an egg from a woman does not have a Y chromosome. If two women want to have a son "of their own," they will have to find a way to add a Y chromosome, or its relevant parts, to their gametes. Whether this could be done by inserting someone else's entire Y chromosome, by adding specific Y chromosome genes to the X chromosome, or by synthesizing and adding the parts of the Y chromosome that are crucial to male development—or at all—is not clear. Of course, this may not be important for anyone interested in an all-female utopia!

To sum up, if we can make prospective parent gametes at all, then making eggs from men or, more accurately, male cells seems quite possible. Making sperm from women seems more complicated, particularly making sperm that will yield boys, but it is not clearly impossible. And, again, twenty to forty years is a long timespan—human embryonic stem cells were first isolated about fifteen years ago and the first IVF baby was born about thirty-seven years ago.

But cross-sex gametes do face one disadvantage, unique among the challenges of Easy PGD. Time, money, and effort would have to be expended solely to try to make eggs from male cells and sperm from female cells. Those expenditures might be done for nonhuman uses—consider the possible value of this technology in endangered species, particularly those with few, or no, reproductively competent individuals of one sex. But they would not be the result of research on something other than reproduction.

If men could produce eggs and women could produce sperm, what follows? Well, the implications for the gay and lesbian community are pretty clear. Right now, if a same-sex couple wants "a child of their own," the closest they can come is by using one gamete (sperm for men, eggs for women) from one partner and the other gamete from a first-degree relative of the other partner—a sibling or possibly a parent or child. That's close, but not the same as the kind of child that heterosexual couples are able to have. This technology could remove that barrier, to the likely delight of many same-sex couples. Of course, gay men who want a "child of their own" will still need to find a woman willing to carry the pregnancy, but lesbians could undertake the pregnancy without a third person, assuming either of the couple is willing and able to carry a pregnancy.

Uniparents and Unibabies

Cross-sex gametes do, however, raise another, more disturbing (even to me) prospect, one I had not considered at all until one of my law school colleagues raised it—a "uniparent." If cells from men can produce both sperm and eggs and cells from women can produce both eggs and sperm, any one person, male or female, could have both sperm and eggs produced from his or her cells and thus be both the genetic father *and* the genetic mother of any offspring. Note that if the person involved was a woman, she could also be the gestational mother of the child.

The child would not be a clone of the uniparent. Where the uniparent had the same allele in both copies of a given gene, the child would necessarily be identical to the parent (barring new mutations). But where the parent was heterozygous—had, let us say, one allele for type A blood and one for type O blood—the child could be AA, OO, or AO/OA. The child has a 50 percent chance of being, like the parent, heterozygous at that locus, but also has a 50 percent chance of being, unlike the parent, homozygous. It is not easy to find out at how many genes humans, on average, are heterozygous. If, say, we are homozygous at 50 percent of our genes and heterozygous for 50 percent, the child should be identical to the parent in about 75 percent of its genes (the 50 percent where the uniparent is homozygous and half the time when the uniparent is heterozygous), but different in 25 percent.

It never occurred to me that anyone would want to do this, but, on reflection, it is certainly imaginable—it's a big world with a lot of odd people in it. There would be risks. Each of us is estimated to carry about five to ten nasty autosomal recessive alleles, alleles that do not make us sick because they are counterbalanced by a normal allele. Each uniparental child would have a 25 percent chance of being homozygous for the "bad" allele and hence for having a bad genetic disease. If the parent carried eight alleles linked to bad autosomal recessive diseases, the average child (and embryo) would be a carrier for four, would have two normal alleles for two of the diseases, and would have two "bad" alleles for two of the diseases. But using Easy PGD, a uniparent (just like a parent couple) could screen out embryos with two copies of autosomal recessive "bad" alleles. A "uniparent" might have to make, and screen, more embryos than two genetic parents to be sure of getting one without a serious autosomal recessive disease, but Easy PGD does seem to make this possible.

RESEARCH INVESTMENT, INDUSTRY, MEDICAL PROFESSIONALS, AND HEALTH CARE FINANCING

Many things that are technically possible do not happen or, at least, do not become common. Consider flying cars. Or, more realistically, supersonic commercial flight. The first supersonic human flight was achieved nearly seventy years ago in 1947. A supersonic commercial jet, the Concorde, went into service in 1975, offering trips from New York to Paris in three and a half hours instead of eight. Only twenty Concordes were ever built and the last supersonic Concorde flight came twenty-seven years later, in the aftermath of a fatal crash in July 2000 and air traffic declines after 9/11. The service had never really caught on, stymied by a combination of environmental politics and high costs.[1] Will Easy PGD, which raises far greater political and emotional concerns, suffer the same fate?

No. Or, at least, probably not. The Concorde never had a pathway to acceptance. It appeared likely always to make too much noise (sonic booms), emit too many pollutants (especially at vulnerable high altitudes), and cost so much as to remain a scarce, vaguely frivolous luxury product. And even if further development and broader use might have solved those problems, supersonic flight did not remain a consumer product long enough to achieve those goals. Supersonic flight has not gone away, but it has been limited to places where the broad social environment is less forbidding—to military uses where arguable necessity can outweigh environmental damage and cost is little object.

Easy PGD, on the contrary, has a clear path to success, one that features intermediate steps with the biomedical, economic, legal, and political factors to make the technology succeed. This chapter explores the economic factors that will provide research investment and lead industry to become involved in gamete production, attracting the medical profession and making health care financing systems pay for it. Although my focus (and my knowledge) is primarily on the United States, I believe these factors will not just allow Easy PGD there but will also help it become part of the lives of hundreds of millions of people around the world.

Factors Driving Research Investment

For biomedical research possibilities to become reality, they need to be scientifically possible, but they also need funding for research and development and a significant potential use. These two points are, of course, related. The more significant the potential use, the greater the chances of getting the resources needed for research and development. And note that the resources involved are not just monetary. More significant, but sometimes overlooked, is the need for people—intelligent and skilled people who will devote their time, attention, and lives to the problem.

Easy PGD requires investment in advances in two areas: genomics and making or retrieving gametes. The genomics investment is as certain as anything in research; investment in gametes is less certain, but also highly likely.

Genomics

The necessary advances in genomics are better DNA analysis (probably whole genome sequencing but conceivably some other, less complete method) and better interpretation of the meaning of DNA variations—although, in the second case, even today's knowledge is probably good enough to make Easy PGD a success. But both of these technologies are well placed to improve even more, for one important reason—they are seen as crucial to better health care in every (or almost every) area of medicine. The same genomics technologies, both in DNA sequencing and in DNA interpretation, that will make Easy PGD possible are expected to (or, at least, are being hyped as certain to) revolutionize all of medicine. No investment in genomics specific to Easy PGD will be necessary—and the general investment is enormous and growing.

Faster and more accurate DNA analysis is important to bringing genomics to individual patients. Some people are already having their genomes sequenced, not really as "cases," but as early adopters—people with the money or influence to be sequenced who think the technology is cool and want to try it. Early adopters, even in Silicon Valley, are not going to be a large market, though as the cost of sequencing becomes lower and lower, the market will get bigger. Depending on the regulatory framework, we may see a repeat of the direct-to-consumer genomics business, where 23andMe, Navigenics, deCODEme, and other firms offered genotypes to individual consumers without the need to go through a physician. When they started, 23andMe and deCODEme were selling genotypes for just under $1,000. The main remaining firm in this market, 23andMe, is currently selling its service for $199.

But full genome sequencing is also already moving beyond the early adopters and being used for actual clinical cases. At medical centers around the country, the occasional mysterious case is being investigated with full genome sequencing. These are cases, usually involving children, where some kind of genetic condition seems likely but the child's signs and symptoms do not match any known genetic syndrome. Whole genome sequencing is being used there in a semi-clinical, semi-research way, looking both to explain the malady of a particular patient, but also to discover the cause of the problem. The distance between knowing a genetic cause and providing a therapeutic solution is, sadly, often vast, but hope for an eventual treatment, the parents' desire for knowledge useful in future reproductive decisions, and the often very strong desire for any explanation—whether or not that explanation is useful—are driving the huge investment in improving DNA analysis.

This is clearest in the investment in new methods of doing DNA sequencing. The NIH and other national research programs have committed large sums to support this work. But, perhaps more importantly, selling sequencing machines is a competitive business. Companies around the world are pushing to speed up whole genome sequencing, improve its accuracy, and lower its cost. In the United States alone, major players include Illumina (the market leader), Pacific Biosystems, Life Technologies, and Roche. China's BGI, which recently purchased the California firm Complete Genomics, is also now producing its own sequencing machines for sale.[2] That list is not exhaustive and certainly does not include start-up companies not (yet) publicly funded or widely known.

These market pressures extend beyond just research and development into sequencing machines, but also into how to use those machines most effectively. Some of the companies that make sequencing machines also provide sequencing services; Illumina is a market leader in both. But other companies specialize in taking someone else's machinery and using it most efficiently to provide sequence results. BGI, formerly known as Beijing Genomics Institute, is the other market leader with Illumina in providing sequencing services, especially after its acquisition of Complete Genomics. And, ultimately, if it wants to have a future, 23andMe will also move into providing sequencing services (directly or through subcontracting). Cheaper, faster, and more accurate sequencing, whether aimed at whole genomes (as I expect) or at smaller parts of the genome, such as whole exome sequencing, has been regularly arriving for several years. There is no reason to expect that to stop anytime soon.

This same easy pathway is available, of course, for genomic interpretation. Researchers, whether governmental, academic, or industrial, want to learn the connections between genetic variations and various traits or diseases for use not just for embryos but for adults, children, and fetuses. Efforts to associate the sequence of particular genomic variations with traits have been ongoing ever since we have been able to sequence DNA and they will continue. Although it has proven harder than expected to understand genetic contributions to common diseases, research on these associations continues. Any successes in this work will be directly useful in Easy PGD.

Some aspects of Easy PGD will not benefit as fully and directly from existing (and near certain future) research. Those will be uses of Easy PGD that predict results other than diseases, notably cosmetic traits. The focus of biomedical research, funded publicly or privately, has been on diseases and, ultimately, treatments for them. Funding for research into the genetics of skin color, for example, has suffered because skin color is rarely a disease. Researchers interested in this subject have difficulty interesting governmental or nonprofit funders; often they have to resort to connecting their research interests more closely to diseases. Thus, a researcher interested in the genetics of skin color may get funded to work on the genetics of inherited partial or total albinism.

Sometimes this tactic may be useful. Research on the genetics of some mental illnesses may provide insights on normal personality traits that are exaggerated in people with the illness. Research on muscle diseases may provide useful information about the likely physique and some aspects of the future athletic ability of embryos. But substantial private investment in genomics research into nondisease traits does not seem highly likely. There is little potential for a very profitable product or service at the research's end.

On the other hand, research on all genomic associations with traits should become much easier and cheaper. As more and more people are completely sequenced, those whole genome sequences, once ensconced in a databank, can be used as easily for research on musical ability as for research on pancreatic cancer. If the genome is available, the hardest part will be getting information on the traits or diseases themselves. This favors diseases. Medical records are full of information about diseases; the merging of electronic health records with genomic databases should make hunting for associations between diseases and genetic variations easier than ever. Those health records will not have much information about nondisease traits—perhaps height and weight, but that may be about all.

Information on nondisease phenotypes will have to be gathered and that is often expensive. Of course, one might try to gather the information directly from the research subjects. The direct-to-consumer genomics company 23andMe has done some of this. The problem with self-reported data is that it is self-reported—and people do not always tell the truth, to researchers or to themselves.

On the other hand, some nondisease traits of interest will be easy for an impartial third party to detect and record. Height and weight are fairly easy; eye, hair, and skin color, though capable of being manipulated to some extent (by colored contact lenses, hair dyes, and tanning) will be reasonably ascertainable. So will some other outward cosmetic traits, such as nose or eye shape, hair type, baldness, and so on. But for some of the traits of potentially greatest interest, such as personality traits or cognitive skills, getting good data may prove hard.

Hard is not impossible, but it is not clear who will pay for collecting these phenotypes. The interpretation of genetic associations with difficult-to-measure nondisease traits may be the area where financing the path forward is hardest, but even there the existence of databases with genome sequences of hundreds of thousands of research subjects (and the prospect of very cheap sequencing of the unsequenced) seems likely to lead at least some academic researchers into these fields.

Stem Cell–Derived Gametes

Most of the technologies for easy acquisition of gametes rely on starting with stem cells, of one kind or another, and turning them into eggs or sperm. These approaches will benefit from the strong interest in, and support for, stem cell research in general, whether hESCs, iPSCs, or SCNT-derived cells. The specific step of turning stem cells into gametes will require some more narrow support, but that should also be forthcoming.

The first human embryonic stem cell line was isolated in 1998, less than eighteen years ago. The first human induced pluripotent stem cell line was isolated in 2007, less than a decade ago. Since that time billions of dollars, public and private, have been invested in stem cell research. The state of California alone has borrowed $3 billion to fund stem cell research; several other states have created their own stem cell research programs, adding tens of millions of dollars to the sum. And although the federal government sharply limited its funding for human *embryonic* stem cell research during the George W. Bush administration, the overall federal research funding for stem cell research since 2000—both embryonic and nonembryonic—is several billion dollars. Governments in other countries have also funded substantial research, especially in the United Kingdom, Singapore, Australia, Sweden, and China.

Private corporate funding has not been nearly as robust, but given the early stage of the technology, that is not too surprising. A great deal of basic research needs to be done before anyone could expect to develop a profitable product from pluripotent stem cells. It is a common pattern for industry to let government fund the basic research, while paying for the latter-stage, more product-specific development itself.

It is true that funding, certainly public but also in some instances private, for research on human embryonic stem cells has been controversial and, in some cases, limited. That, in some ways, helps Easy PGD, which is most likely to use iPSCs, which do not require destroying embryos and result in children with their parents' own genes. To the extent controversies over hESCs drive research funding to iPSCs, that probably helps Easy PGD.

Research on all the broadly potent stem cells is aimed at learning to derive, preserve, and manipulate stem cell lines to make them into differentiated cells that can be used safely and effectively in medicine. All of that work will necessarily be useful in learning how to derive gametes from stem cells because the early stages are the same. It is only after the pluripotent cell lines are created and they need to be differentiated into specific cell types that stem cell research on heart cells, liver cells, and gametes should diverge.

But what about research on turning pluripotent cells into safe and effective gametes? Where will support, financial and otherwise, come from for *that* research? The U.S. federal government has been extremely reluctant to fund any research on fertility issues for political reasons; it is hard to imagine the NIH spending money to support the derivation of gametes from stem cells. Nonetheless, I believe there will be plenty of support, driven by the desire of some infertile couples to have children "of their own."

People can lack gametes, or a sufficient number of sufficiently effective gametes, for many reasons. Some are, for one reason or another, born without the ability to make gametes. It might be a genetic flaw or just bad luck—somewhere in their fetal or childhood development, something took a wrong turn and the result was a lack of useful gametes. Other people are born with the ability to make useful gametes, but lose it. Castration of boys is not common today, but one need not go back many centuries to find castrated slaves, singers, or even fabled leaders. Today, other common causes of acquired agametousness are accident and disease. Accidents are more likely to affect men, with their sperm factories hanging out in acutely vulnerable positions, than women, whose ovaries are more carefully hidden away, but devastating trauma can happen to both sexes.

Disease, though, is even more likely. Historically, a variety of infectious diseases could produce male infertility, including the otherwise fairly benign and common childhood disease of mumps. But other diseases can also render men infertile. High fevers, for example, can kill off the spermatogonia, leaving men without any sperm-producing cells. The "Father of His Country," George Washington, may well have been sterile from nonpulmonary tuberculosis, abetted by smallpox he contracted in Barbados when he was nineteen.[3] Today, cancer is more likely to ruin the reproductive futures of men and women, both cancers of the gonads and other cancers whose treatment destroys eggs and sperm.

But the most likely reason for being, in effect if not literally, agametous is, for one of our two sexes, age. Women's oocytes disappear with age, but even before they disappear, they become less and less effective at producing babies. This decline in the productivity of eggs with age produces the sharp decline in women's fertility through their thirties and into their forties. Although a few women give birth from their own eggs in their late forties and even early fifties, the chances of a successful pregnancy with forty-year-old eggs are lower—and they get lower with each additional year.

People who cannot produce effective gametes but who want to have "children of their own" will drive research into stem cell–derived gametes, either through their direct donations, through their organizations, or through their attractiveness to investors as a potential market. I explore this in more detail in the next section, but there are probably millions of them in the United States who have such a medical problem, and many of them will be willing to pay enough to provide a financial incentive for research on solutions.[4]

Factors Driving Industry Involvement

No special industry involvement is needed on the genomics side of Easy PGD—those DNA analysis and interpretation facilities and firms will arise for nonreproductive reasons. The same is true of the process for making iPSCs from skin cells. But making eggs out of iPSCs is likely to require special industry investment. Would the possibility of owning a process useful in Easy PGD be attractive to industry?

We know from existing experience that some people—enough to pay for about 160,000 IVF cycles a year in the United States right now—are willing to spend tens of thousands of dollars to have "children of their own." Most of them have gametes but have some fertility problem that prevents the gametes from becoming babies, whether it is low sperm counts, ineffective sperm, blocked fallopian tubes, or something unknown. Unfortunately, we do not know how many of those people have dysfunctional gametes.[5]

We do have some information on the number of IVF cycles that use donor eggs. In 2012, the latest year for which the CDC has data, of about 160,000 IVF cycles, over 10,000 used donor eggs in fresh cycles (and about the same used frozen embryos from donated eggs). That report does not keep track of the use of donor sperm and apparently neither does anyone else. The figure of 30,000 donor sperm births a year is widely used, but its source is unclear; some industry sources argue the number is closer to 5,000.[6] And, of course, some donor sperm is not used because of male infertility but to provide sperm for people without any male partner, single women or lesbian couples. For present purposes, let us estimate the annual use because of male infertility due to sperm quantity or quality at 5,000 cases.

Presumably, many of the roughly 20,000 women using donor eggs each year and the 5,000 men in couples using donor sperm would be interested in replacing donor gametes with those derived from their own cells. In addition, over 30,000 women in the United States aged forty or over used nondonor eggs in IVF cycles in 2012, with only about one in eight ending up with a live birth. Many of them would also be interested in their own stem cell–derived eggs, at least if those eggs were more successful than those that had already spent over forty years in their ovaries. That is already over 50,000 people who might want to use such eggs. And many other older women, facing problems with or concerns about their fertility but unwilling to use IVF, with or without donor eggs, might well be tempted by stem cell–derived eggs. The fact that the age of motherhood continues to rise, particularly among women with higher education and income levels, will help to encourage a powerful market for stem cell–derived eggs and, in many cases, broader use of Easy PGD.

A disease with over 50,000 new cases a year in the United States is not small. That is more than the number of new cases each year in the United States of multiple sclerosis (about 10,000), brain cancer (23,000), stomach cancer (about 24,000), and liver cancer (about 35,000). It is about the same as pancreatic cancer (49,000) and HIV infection (about 50,000), and not far below all leukemias (54,000), uterine cancers (55,000), kidney cancers (62,000), and thyroid cancers (62,000).[7] While infertility is not life threatening in the way cystic fibrosis or pancreatic cancer are, it is powerfully life limiting to some of those who have it—and a lot of people have it.

Assume, for the moment, that 50,000 people a year who are infertile because of problems with their gametes would like to be able to use stem cell–derived gametes to have children. If each one is willing to pay (or to have their insurer pay), say, just $1,000 for those gametes, that's a market of $50 million per year in the United States alone. Though not huge, it is surely enough to get the attention of physicians, drug and biotech companies, and venture capitalists. Now consider some other possible users.

As discussed in Chapter 8, we may be able to use stem cells to make eggs from men and sperm from women. That would create another market for this service, among gays and lesbians who want to have "children of their own" with their partners. How many gays and lesbians would be interested in having children through this method? That's also impossible to know, in part because we have so little good data about the number of gays and lesbians, let alone their parental desires, and what statistics exist are controversial. Plausible estimates of the U.S. gay, lesbian, and bisexual population range from 1.4 percent to 3.5 percent.[8] That would be about 1.1 to 2.8 million people between the ages of twenty and forty. How many want to be (genetic) parents? That is again unknown—but a figure of 200,000 to 1,000,000 couples seems a reasonable guess, or about 10,000 to 50,000 people for each birth year in that twenty-year cohort. (Of course, only one member of the couple would need cross-gender gametes.)

Another very large market may be women with ovaries that produce eggs but who, for other reasons, such as blocked fallopian tubes, need to use IVF and may well embrace stem cell–derived eggs as an easier and safer alternative to egg retrieval. Right now, about 100,000 American women under the age of forty undergo egg retrieval each year. Deriving eggs from stem cells will surely be less uncomfortable and risky than retrieving eggs. If the price were less, equal, or even a little greater, almost all of those women would likely prefer them.

So between people without gametes, gays and lesbians without the "right" gametes, and women who want to use IVF but would welcome an alternative to egg retrieval, the total U.S. market each year (after an initial surge) might be around 200,000 to 250,000 people per year. Even at just $1,000, that's a market of about a quarter billion dollars a year. In today's medical world, that is not chopped liver. And these are just the people who want it for fertility purposes—before adding anyone who decides to use IVF (and stem cell–derived gametes) in order to do Easy PGD! If half of the roughly four million babies born each year in the United States had gone through Easy PGD, the manufacturing of the gametes that produced them at $1,000 per baby would be worth two billion dollars.

Factors Driving Medical Involvement

"Market" is not always an appreciated term in medicine and sometimes it is not an appropriate one.[9] In the world of assisted fertility, however, it is very appropriate. Fertility clinics are a large and very profitable industry. The roughly 500 fertility clinics in the United States are, with a few exceptions, for-profit businesses. The exceptions are mainly clinics associated with nonprofit medical schools or health systems, but even there, although the overall enterprise may not be "for profit," the chance to make money in one department of a medical school in order to cross-subsidize unprofitable but otherwise important areas will often be quite strong.

The fertility industry is very profitable for the ironic reason that insurers typically do not pay for the services it provides. It is too simple to say that fertility services are not covered by insurance, but only a little too simple.[10] A few sources of infertility may be covered by most insurance, as when infertility is a result of a broader hormonal problem. And a few states require most insurers to cover fertility services, usually at a pretty minimal level. Even then, a state's ability to mandate coverage by insurers has some severe limits—under a federal law called the Employee Retirement and Income Security Act, states cannot regulate medical coverage that is paid for directly by the many self-insuring employers.

One might think that insurance coverage would provide a larger market and hence the chance for more profit. In many areas of the economy that would be true. But medicine is not a normal part of the economy. Patients are not very good at finding good bargains, at assessing the trade-offs in price and quality, and at checking the claims of doctors about their success rates. In much of American medicine, the insurance industry, including government-funded Medicare and Medicaid, plays the main role in constraining prices. They will not pay for "whatever the market will bear."

Fields without substantial insurance coverage, such as cosmetic surgery, laser eye surgery, and fertility, have largely unconstrained prices. Ironically, IVF prices are only low in states where insurance covers them. (And, if, as I suggest in the next section, payors cover Easy PGD, IVF clinics will follow that path and have a much bigger market with much smaller profit margins.)

Even if someone wanted to compete in the fertility clinic market on price, they may find it difficult because the supply of doctors is artificially limited by the limits on medical school graduates. Of more direct relevance, this kind of supply limitation applies not only to physicians but also to specialists. A fertility clinic needs a reproductive endocrinologist. Reproductive endocrinologists (or, to be more precise, specialists in reproductive endocrinology and infertility, or REI) are not trained as endocrinologists. Instead, the field is a subspecialty of obstetrics and gynecology. In the United States, the American College of Obstetrics and Gynecology (ACOG) sets the standards for recognition as a reproductive endocrinologist. Those interested must successfully complete four years of medical school, three years of an ob/gyn residency, and another three years of an REI fellowship. The Society for Reproductive Endocrinology and Infertility is the main organization for REI specialists. It has only about 700 members in the United States.

ACOG accredits fellowships in REI. For 2016, the forty-five American institutions offering REI fellowships were providing a total of about fifty first-year positions; the vast majority train only one new fellow each year, a few train two, one trains three, and several were training none.[11] The possibility for new competition in the field is not great.

The point, though, is not just that the fertility industry is and likely will remain profitable. It is also intensely commercial. Freestanding fertility clinics, usually not attached to staid medical schools with "respectable" reputations to maintain, seek to enlarge their markets. They advertise. They attract paying patients. And to do so they take advantage of an immensely strong human drive—the drive to procreate.

To call it an immensely strong drive is not to call it universal. There are people, even when involved in settled romantic heterosexual relationships, for whom parenthood is easiest, who choose not to have children. Nonetheless, in the United States as around the world, most adults have children. In fact, in 2013, 74 percent of adult Americans reported having had children. (Of respondents over forty-five years of age, the percentage was 86.) Another 16 percent of all adult respondents said they wanted children and 3 percent said they had not had children but wished they had. Only 5 percent said they did not want children.[12]

Even after the introduction of safe and effective birth control, allowing people to enjoy the pleasures of sex without the previously high risk of pregnancy and parenthood, people have continued to have children. Once a society has made the economic leap that eliminates the need for children to act as either laborers on the family farm or in the family business or, longer term, as the parents' retirement plan, the economic value of having children would seem to be negative. From one perspective, the combination of birth control and assured retirement incomes makes the continuing existence of children a mystery.

And yet, have children we do, although at below population replacement levels in some countries. Contraception and economic changes have meant that people have fewer children; such factors have not meant that (many) fewer people have children. For some combination of cultural and biological reasons, most people are eager—perhaps irrationally, sometimes desperately—to have children. And not just children but "children of their own"—children carrying their genes and not adopted children, who carry the genes of strangers, or stepchildren, carrying the genetic variations of one stranger.

Fertility clinics invoke this powerful urge in their efforts to grow the market. Right now, it is used to attract customers who have been unable to have the wanted "child of their own." But that same force might be harnessed for Easy PGD. There the selling point is not just, as it is for agametous patients, to have your own child, but to have your best possible child.

The first hook, I suspect, will be disease. Invest a little more to guarantee that your child will not have one of 4,000 terrible genetic diseases. The sales pitch will then move to decreasing the chances that your child will have a higher risk for various truly dread diseases. Already that list includes colon cancer, breast cancer, ovarian cancer, and Alzheimer disease. The list will only get longer in the coming years. The disease argument will be the most overt, probably for broadly political reasons, but tucked away in the advertisements will be the idea of having not just a child of your own, but the child you want—with the cosmetic results, the behavioral traits, and, perhaps most powerfully, the sex you want.

The slogans almost write themselves. "You want the best for your child; why not have the best child you can?" "We cannot guarantee that your child will be healthy, but here are 4,000 diseases we can guarantee he won't have." "The child of your dreams is available for only a few dollars more." "You spend $30,000 on getting the car you want; how much is the baby you want worth to you?" And in the United States at least, even commercial advertising is protected by the First Amendment.[13] As long as the advertisements are honest and are not for illegal products, the government cannot forbid them. The large and very profitable fertility clinic industry is quite likely to see Easy PGD as a way to become even larger and more profitable.

Health Care Financing and the Price of Easy PGD

Easy PGD would not just affect parents and their children, but could also have substantial effects on the health care financing system, which currently only rarely covers IVF and then poorly. But Easy PGD should change that, for reasons discussed below, in ways that are likely to make the price of Easy PGD, to the prospective parents, zero, or very close to it.

That its price is zero does not mean Easy PGD will have no cost. It will require various human and technical inputs and those will cost money. Predicting the cost twenty to forty years from now of procedures not yet invented seems almost the definition of a fool's errand. Nevertheless, I will provide some back-of-the-envelope guesses below.

One cost will be performing the skin biopsy in order to get cell samples to be transformed into iPSCs. A trained nurse could do that in a few minutes; let's cost that out, in today's dollars (not adjusting for inflation in the next twenty to forty years), at $20. The hardest cost to calculate is that of de-differentiating the donated cells into iPSCs and then redifferentiating them into a large supply of eggs or sperm. Progress in more general use of stem cells is likely to have made that process relatively mechanized and routine, but there will be costs. I will estimate, without a great deal of confidence, that this would cost about $1,000.

The PGD process is likely to require a skilled professional to extract cells from each embryo. Just the extraction seems likely to cost at least $10 per embryo. The actual genetic diagnosis and interpretation, however, should be very inexpensive that far into the future. The sequencing will be mechanized and the interpretation will be computerized. Call that another $50 per embryo. The most expensive part of the process will be the discussion of the results with the prospective parent. As explained in the Second Interlude below, that might be done in a variety of ways, including a completely automated process. A more likely alternative would include some automated return of results (and counseling), probably through Internet-delivered videos, and some interaction, preferably face to face, with a trained professional. Call that another $500.

All in all, this adds up to $1,520 per attempt, plus $60 per embryo. At 100 embryos, the total cost would be about $7,520; at ten embryos it would be about $2,120. Now add another 50 percent for overhead (including liability insurance) and the bottom line cost should be somewhere around $11,000 for 100 embryos or $3,200 for ten embryos. If the technologies develop as I expect, the cost seems unlikely to be more than twice as high as estimated and could easily be half as much. (The cost estimates are particularly responsive to the estimated costs per embryo of extracting one cell and then of performing PGD on it.)

Assume, then, a cost around $11,000 for 100 embryos. Why would the price be effectively zero? Because at that cost the procedure should pay for itself for health reasons. One end result of Easy PGD should be fewer sick children and, ultimately, healthier adults. At $11,000 per cycle, it would cost $1.1 million to use Easy PGD in 100 pregnancy attempts. Assume that those 100 Easy PGD cycles yield 100 births. This would not, of course, be the result of 100 percent efficiency in Easy PGD pregnancies, but some of those pregnancies will produce twins or (perhaps) other multiples and in other cases the prospective parents will use embryos from one Easy PGD attempt in several different pregnancies through embryo freezing.

Out of 100 live births, one would expect about two babies to have, or rapidly to develop, serious health problems that could have been predicted with genetic testing. Most of these children will have rare diseases, but 5,000 rare diseases spread over four million births a year produce a lot of tragedy, and a lot of costs. Neonatal intensive care units (NICUs) have an average cost of over $3,000 per day and that's before any surgeries or other expensive procedures; about 65 percent of babies admitted to those units stay for an average of twenty days.[14] Time in an NICU might be just the beginning of years and years of expensive treatment. If the net present value of the cost of the health care that will be provided to those children were to be $550,000 each (a very low estimate for serious diseases), Easy PGD is, in effect, "free" to the health care system. And that is before counting the costs of the higher risks of later onset diseases that might be avoided, from cases of breast and colon cancer to cases of sudden cardiac death to cases of Alzheimer disease. The later the onset of the disease, the lower the net present value, but $1 million spent in long-term care for an Alzheimer patient at age seventy and beyond would still have a nontrivial net present value. (It would work out to over $100,000 using a 3 percent discount rate.)

Of course, the "health care system" is not necessarily the same as any one insurer or government program that pays for health care. In a system with multiple health care payors, the one that pays for PGD would not necessarily be the one that benefits from the lower health care costs. Predicting health care financing systems twenty to forty years in the future seems, to me, even crazier than predicting the costs of not-yet-invented technologies, but whatever that system is, if all the payors provide the service, all will benefit from it. It seems more than plausible that overall health care system costs will, in one way or another, be considered in deciding whether to cover Easy PGD.

And, of course, health coverage is not intended to make money or even to break even. We spend money to prevent and treat human suffering. Otherwise, we would pay for no health care and have no health care costs. People "buy" health care, through insurance or otherwise, because health care and the anticipated decreases in suffering and increases in lifespan have value to them. In addition to the financial costs avoided by Easy PGD would be the costs in human suffering for the people born with the diseases, as well as for the families into which they are born. Those values should dwarf the financial costs, but if, in addition, the financial costs are negative—if Easy PGD in the long run saves the health coverage system more money than it costs—that will surely be a strong incentive to adopt it. And this holds true whether the health coverage is provided by private, for-profit insurers; private, nonprofit insurers; or government programs. It may not be a perfect incentive—there may well have been cost-effective disease prevention efforts that insurers have *not* covered—but it will be a push toward subsidizing Easy PGD.

How does this square with the current reluctance of insurers to pay for IVF? Today they are not paying for a procedure that will reduce health costs in the future but one that will, for the most part, merely increase births of babies across the whole range of health outcomes (though, thanks to the problem of multiple births, skewed a bit toward the unhealthy end). PGD can sometimes offer healthier children when it is used by parents with a strongly familial disease that using PGD could avoid—and insurers sometimes cover PGD, including IVF, for that purpose. But Easy PGD, by using whole genome sequencing, greatly expands the range of diseases potentially prevented and makes the economics much more favorable for those paying for health care. (Note that this argument does not necessarily work for the "roll out" of stem cell–derived gametes for those infertile because of missing gametes; the economic argument only kicks in with the use of PGD for whole genome—or other broad—sequencing that will be used to avoid diseases in the children born.) Easy PGD may still not be a "medical necessity," the usual, though vague, criterion for health insurance coverage, but if it saves the insurers money it is likely to be included (like, for example, influenza vaccinations).

So combine a population primed to have children with increasing maternal age, a thriving for-profit industry looking to expand its market, a legally protected culture of saturation marketing, and a procedure paid for by health coverage. It is hard to see the result as anything other than the broad uptake of any reasonably priced Easy PGD.

10

LEGAL FACTORS

The current American legal framework contains no insurmountable obstacles to Easy PGD—at least if the science works. This chapter discusses how existing law is likely to allow Easy PGD in the United States, with a few comments on other countries. In the United States, the procedure will face some significant challenges with the FDA about the safety and efficacy of the process, but at least under current law will be largely free from other limitations.

Direct Regulation of Assisted Reproduction

The United States is an odd place in many ways. The most relevant for present purposes is that it has almost no direct regulation of assisted reproduction. Most other countries regulate assisted reproduction in many ways, from assuring safety to making moral choices.[1] For example, in the United Kingdom, a clinic can only offer any particular assisted reproduction service when a government regulator, the Human Fertilisation and Embryology Authority, approves of the procedure, the clinic, *and* that clinic's use of that procedure.[2] In France, IVF is limited by law to heterosexual couples who are either married (to each other) or in at least a two-year relationship.[3] In Italy, the statute provides that no more than three eggs may be fertilized at a time and all the resulting embryos have to be transferred.[4] Austria bans the use of donor eggs and only allows sperm donors for married couples and artificial insemination, not IVF.[5] Sometimes the rules are different even within the same country: some Australian states allow single women to receive IVF; others do not unless they are medically infertile or at serious risk for genetic disease.[6]

In the United States, assisted reproduction is governed by almost no specific legal regulation. We know this as a result of the Octomom. In January 2009, Nadya Denise Doud-Suleman became known as the "Octomom." She gave birth to octuplets, only the second set of octuplets ever to be born alive in the United States, as a result of IVF. It turned out that she already had six children, all conceived through IVF while Doud-Suleman was single, unemployed, and on public assistance. Her octuplets created a public uproar and a great deal of pressure on the fertility industry. Many people, from many different perspectives, thought something was deeply wrong with the story. And, in fact, in July 2011 the doctor who supervised all of her IVF cycles and births, Dr. Michael Kamrava, had his license to practice medicine revoked by the Medical Board of California, after an investigation by the board found that he had transferred not eight but twelve embryos into Ms. Doud-Suleman's uterus.[7]

The Octomom scandal put pressure on the American Society for Reproductive Medicine (ASRM), an organization created to advance reproductive medicine (along with, no doubt, the interests of the fertility professionals who are its members), about the lack of regulation in the United States compared with the situation in other countries. The ASRM responded by convening "a meeting of professionals, patient advocates, government representatives and legal experts in December 2009 to examine the oversight of assisted reproductive technology."[8] The ASRM report on that meeting says, "After examination of the complex network of state and federal regulation as well as professional self-regulation governing ART practice, we conclude that Assisted Reproductive Technologies are among the most regulated medical procedures in the United States." The ASRM's statement might be technically true because few specific medical procedures are significantly regulated. It is deeply wrong in its implication that IVF is actually substantially regulated.

At the federal level, assisted reproduction services, of course, must comply with FDA regulation of drugs, biological products, and medical devices and, to the extent they involve clinical laboratory services, with the Clinical Laboratory Improvements Amendments Act (CLIA), both discussed below. But that's about it. Congress has passed exactly one statute dealing specifically with assisted reproduction, the Fertility Clinic Success Rate and Certification Act, back in 1992. The act sets some definitions and "requires" fertility clinics to report information about the number of cycles of IVF they perform each year and their success rates.

The Centers for Disease Control and Prevention (CDC) administer this law and each year since 1997 the CDC has published annual reports showing success rates for each participating clinic. The act requires clinics to provide the relevant data to the CDC; it does not provide any sanctions beyond requiring the CDC to publish the names of the scofflaws. According to its latest report, published in November 2014, thirty clinics did not report their results for 2012, while 456 did. The annual report contains a great deal of valuable information and makes interesting and potentially useful reading—but it scarcely counts as "regulation."

Of course, the United States is a federal country with many important responsibilities left to the states. Yet the states have not directly regulated assisted reproduction substantially either. States regulate who can practice medicine and they (and their localities) license businesses, including fertility clinics. Some states regulate the use of gestational surrogates; others regulate the parenthood status of those adults who use assisted reproduction. And Louisiana forbids the disposition of "excess" (nontransferred) embryos. But, in contrast to foreign countries, no U.S. state regulates how many embryos may be created or transferred at one time, the acquisition or use of donor eggs or sperm, the marital or relationship status of those who can use IVF, the sexuality of those who can use IVF, the age of women receiving IVF-produced embryos, the use of PGD to select embryos in order to avoid disease, the use of PGD to select embryos for sex or other traits, or, in fact, almost anything else. (A handful of states have recently started regulating abortions performed on the basis of the fetus's sex, race, or, in North Dakota, disability, but these laws are not only of questionable constitutionality but govern abortion, not PGD.)

The ASRM report relies substantially on professional self-regulation. Some of that regulation is done through ACOG's certification of ob/gyns. But most of the ASRM's emphasis is on regulation of practice by the ethics and practice guidelines of the specialty society for physicians who focus on infertility—the ASRM. The ASRM has an affiliate specifically for fertility clinics, the Society for Assisted Reproduction Technology (SART). About 90 percent of fertility clinics in the United States are members of SART, who are supposed to adhere to the ASRM ethics and practice guidelines. But the guidelines are not significantly enforced.[9] In spite of its efforts, the report fails miserably in demonstrating the significant specific regulation of assisted reproduction and hence of Easy PGD.

The FDA

The FDA, which does not regulate assisted reproduction or any medical procedures, provides the main legal hurdles for Easy PGD. Relevant to Easy PGD, the FDA regulates drugs, biologics, and medical devices. For FDA purposes, drugs are

> (B) Articles intended for use in the diagnosis, cure, mitigation, treatment, or prevention of disease in man or other animals; and
>
> Articles (other than food) intended to affect the structure or any function of the body of man or other animals.[10]

Devices have a similar definition.[11] The definition of biological products takes a different approach (and appears in a different statute), but also applies only to a list of items that are "applicable to the prevention, treatment, or cure of a disease or condition of human beings."[12] Whether a new product is a drug, device, or biological, it cannot legally be released into interstate commerce without a prior decision by the FDA that it is safe and effective (except of course, for various exceptions).

The FDA regulates these products but it does not regulate medical procedures or, more broadly, "the practice of medicine." Lawmakers say so, the agency frequently says so, and the American Medical Association regularly reinforces that position by stringently opposing any such intrusion. This is, of course, nonsense.[13] Determining which drugs, biologics, and devices doctors can use or prescribe is clearly some "regulation of the practice of medicine."

The FDA, along with the Federal Trade Commission for some products, regulates how covered products can be marketed—what their manufacturers (or sponsors) can say about them and how (subject to a few intensely fought-over exceptions)—but almost never, at least without the somewhat voluntary agreement of the sponsor, how the products are used. The "off label use" doctrine is one result.[14] The FDA examines products for their safety and efficacy—that is, their use in a particular way, at a particular dosage, for a particular "indication." If the FDA decides a drug or biological is safe and effective for that use, it approves the product for that use, which goes "on the label." But generally any physician is free, subject only to his or her own conscience (and malpractice coverage) to use an approved drug, device, or biological product in different ways, for different patients, and for different diseases. The drug maker cannot promote the drug for any unapproved purpose, but the FDA cannot stop the doctor from so using it.

So, to what extent will Easy PGD require FDA approval?

Under current law, at least, "Easy PGD" will *not* require FDA approval. It is a procedure or, more accurately, a set of procedures and that is not what the FDA regulates. But some of the sequencing methods will require FDA-reviewed devices, and the derived gametes used in Easy PGD will most likely have to be approved by the FDA as drugs or biologicals.

Genome Sequencing and Interpretation

Various "things" used for genetic testing, when done "for use in the diagnosis of disease or other conditions, or in the cure, mitigation, treatment, or prevention of disease" are medical devices. The sequencing machines used to analyze cells from embryos are clearly medical devices when employed for that purpose. If someone only wanted to do PGD for nondisease traits, such as eye color or musical ability, those genetic tests might not involve regulated medical devices, but Easy PGD will no doubt use sequencing machines that do medical as well as nonmedical tests and hence involve regulated devices.

The interpretation of the DNA sequence may also involve regulated devices. If a whole genome sequence were interpreted directly by a human being, that human would not be a device but would be both incredibly busy and bored. Analysis of genomes may probably always involve some human oversight, but the bulk of the work will necessarily be computerized. Software programs will compare the results of the whole genome sequence of a cell from an embryo with a database that correlates genetic variations with traits. That software may also be a regulated device. Software may not seem much like a "device," but the FDA has already successfully regulated some medical software as devices.[15]

The substance of the tests themselves, whether the DNA variations actually predict what the test says they predict, is not currently regulated by the FDA, at least when the results are not being provided directly to consumers without a medical intermediary. The FDA insists they are technically medical devices, but it has used its discretion not to regulate what it calls "laboratory developed tests" (LDTs). In October 2014 the FDA released a draft guidance that would have brought far more genetic (and other laboratory conducted) tests under review.[16] At this point it is not clear what will happen with that effort.

Just how much proof of safety and efficacy the FDA will require before authorizing the use of sequencing machines, genetic interpretation software, and possibly genetic tests will depend, but this approval is unlikely to prove a substantial barrier to Easy PGD. In fact, in 2013 the FDA already authorized at least one sequencing machine for clinical medical use, one made by Illumina.[17] And under Section 510(k), a provision of the FDA's device regulatory statute, once one of a certain type of product has been allowed, other "substantially similar" products will be "cleared" by the FDA, usually without requiring much evidence of safety and efficacy.

The exact same sequencing machinery, genetic interpretation software, and genetic tests will be used in Easy PGD as in any other human genetic testing using whole genome information. And thanks to the off-label use provision, even if those were to be approved only for use in, say, adults, they can legally be used for other purposes, such as testing embryos. FDA regulation on the genomics side will not prove a significant barrier to Easy PGD.

Stem Cell–Derived Gametes

The FDA's real role is likely to be with respect to stem cell–derived gametes, but that story is complicated. Eggs and sperm do not seem to be obvious drugs or biological products, let alone medical devices. As far as I can tell, they are not medical devices, but, under the FDA's current view, they will be both drugs and biological products and will require premarket approval as either.

First, the gametes would be "articles" and would almost certainly be drugs when "intended for use in the diagnosis, cure, mitigation, treatment, or prevention of disease in man or other animals" or "to affect the structure or any function of the body of man." Similarly, if we grant that the gametes fall within one of the list of nouns in the biologicals definition, they are likely biologicals as long as they are "applicable to the prevention, treatment, or cure of a disease or condition of human beings." If the gametes were used for the purposes of providing gametes for an otherwise infertile couple, they would meet both definitions because they would be "intended for use" or "applicable to" the treatment of infertility, which is both treating a disease and affecting the "function of the body of [wo]man."

The only way to avoid that conclusion would be to say that gametes, by their nature, are not the kind of "articles" or one of the list of nouns in the drug and biological definitions. As discussed below, that runs against the FDA's practice with human cells (and even human embryos) and the (admittedly very few) judicial interpretations of FDA's jurisdiction over human cells.

That's the only really clear part. But what if the stem cell–derived gametes are just intended to allow otherwise fertile parents to select their offspring's genetic traits? It is not, in that case, for the diagnosis, treatment, or prevention of any disease in the parents, but maybe the embryos, in their role as possible offspring, are the relevant group. If parents are using stem cell–derived gametes to be able to choose embryos to avoid some genetic diseases, perhaps it is the "diagnosis" of disease in the embryos, even though in most cases the embryos will not have symptoms of these diseases, which will not appear until birth and sometimes many years after.

Alternatively, it might be considered "prevention of disease" in the children. Typically, though, one thinks of preventing disease in a particular person, not by "selecting" which embryo might become a person in a way that "prevents" the creation of a person with the disease. That seems more like "avoiding" disease than "preventing" it. On the other hand, every embryo, and hence any embryo chosen that becomes a child, will have some disease risks in its genomes, which Easy PGD is diagnosing, in advance, for any children born after testing.

These arguments both seem plausible, if not definitive, but they still have a problem. Either way, PGD, not the gametes, provides the diagnosis. Only by taking a broad view could the egg and sperm that created the embryo be diagnosed as articles intended for "use" in the diagnosis. Such a view is not crazy—after all, in these scenarios the derived gametes were only used (instead of old-fashioned gametes and conception) as part of a plan for diagnosis. But neither is it clearly right.

And even if that problem were overcome and either the argument from diagnosis or prevention of disease were accepted, what if parents were to claim that they were *only* selecting for nondisease traits. This may not seem very plausible, but it might if the parents said they only wanted to use Easy PGD to determine the child's sex, which is not a disease. Would their use of gametes that had been generated specifically for use when no disease was to be examined therefore not be regulated?

If your head is spinning and you are wondering why we should care about the number of angels dancing on this particular pinhead, I sympathize. And, frankly, I think it is not likely to be important in any event because I strongly suspect that stem cell–derived gametes will first be produced and used in order to treat infertility. These would clearly be drugs or biologicals and would require the FDA to approve the gamete-production process before it could be used and, after that approval, off-label uses other than infertility would be allowed.

We do have some evidence that the FDA thinks it has jurisdiction in somewhat similar cases, starting with cloning. After Dolly, at least four different groups or individuals announced plans to use the SCNT process that produced Dolly to clone human babies: the Raëlians,[18] the ironically named Dr. Richard Seed,[19] and more credibly Dr. Severino Antinori, a well-respected fertility doctor in Rome, and Dr. Panayiotis Zavos, a fertility specialist at the University of Kentucky.[20] The efforts of the Raëlians, Seed, Antinori, and Zavos did not result in any cloned babies (although some births have been asserted), but they did lead to an October 1998 letter from Stuart Nightingale, Associate Commissioner of the FDA, asserting that the FDA had jurisdiction over human reproductive cloning. The letter made only a vague claim as to the basis for its jurisdiction: "Clinical research using cloning technology to create a human being is subject to FDA regulation under the Public Health Service Act and the Federal Food, Drug, and Cosmetic Act."

This claim of jurisdiction was the subject of several law review articles, with some supporting the claim and some rejecting it.[21] The arguments against jurisdiction focused on whether the embryos were really "articles" or otherwise "things" for purposes of regulation, on issues of whether cloning would be used to diagnose or treat disease, and on the inconsistency between this regulation and the FDA's previous avoidance of regulation of assisted reproductive technologies. We have never received a clear answer on the FDA's power over cloning; no court case has tested it. Nonetheless, the FDA's 1998 letter provides evidence that it would consider stem cell–derived gametes as coming under its jurisdiction. It was willing to call an actual human embryo a "drug" or "biological product" in spite of similar questions about medical purpose (the would-be cloners did not have disease-related ends in mind).

Since then the FDA has asserted jurisdiction over reproductive technologies in another similar situation: mitochondrial transfer. Women with diseases caused by the DNA of their mitochondria will necessarily pass that disease-causing DNA to any of their children. In addition, some researchers believe the decline in fertility of aging eggs has to do with problems in their mitochondria (or, perhaps, more broadly in the "cytoplasm," the material between the egg's outer membrane and its nucleus). These problems might be solved for the woman concerned either by transferring the nuclei of her eggs into the enucleated eggs of a healthy young woman or by transferring the cytoplasm (including the mitochondria) from a healthy young woman into some of her eggs. Both of these approaches were tried nearly twenty years ago and with some success. And then the FDA stepped in.

One group, at NYU, created embryos using these reconstructed eggs for at least two patients, though no pregnancies resulted. In October 1998 they presented their results at an ASRM meeting and were rewarded shortly thereafter by a letter from the FDA, telling them that they were using biological products (presumably the SCNT eggs) without prior FDA approval.[22] Another fertility doctor in New Jersey tried the other approach, moving not the nucleus but the cytoplasm from one egg to another. By 2001 thirty children had reportedly been born around the world using this technique. In July 2001 the FDA intervened again, claiming that, like the eggs resulting from nuclear transfer, these eggs were "more than minimally manipulated" cells and, as such, were biological products subject to its jurisdiction.[23] Rather than pursue FDA approval, both groups stopped performing the procedure (at least in the United States). It should also be noted that both of these procedures produce "three-parent children," as noted in the headlines from the 2015 debate (and positive parliamentary vote) on the procedure in the United Kingdom.[24] Whether this process is safe or not remains to be shown; it clearly has raised some strong, and sometimes silly, public responses.[25]

As with the claim of jurisdiction over reproductive human cloning, no court case followed so we do not know whether the FDA's claim of jurisdiction over these eggs is valid. But it is further evidence that the FDA would consider stem cell–derived gametes as subject to its jurisdiction.

What does all this mean? The FDA is likely to assert jurisdiction over all stem cell–derived gametes, whatever their intended use (as long as reproduction was involved). It will require that they be proven safe and effective, through appropriate nonclinical and clinical trials before being approved under a New Drug Application (NDA) or a Biologics License Application (BLA). It is possible, though I think unlikely, that the FDA might take a different position. It is also possible, and perhaps somewhat less unlikely, that a court would take the position that such gametes are not subject to FDA regulation. In that case, given the importance of making babies in a safe way, I suspect Congress would step in to give the FDA such jurisdiction.

I must briefly describe one more FDA issue before moving on. The FDA has had a difficult time figuring out how to deal with living tissues. Blood was transplanted for decades before the Federal Food, Drug, and Cosmetics Act was passed. And when solid organs began to be transplanted regularly in the late 1960s, the FDA did not assume jurisdiction over them. But in the 1990s the FDA decided it did need to regulate some aspects of living tissue used medically and it ultimately, in 2007, adopted regulations to govern "human cell, tissue, and cell- and tissue-based products" (HCT/P). Many of these HCT/Ps do not require premarketing approval by the FDA, but are subject to regulation only to avoid passing infections from the donor to the recipient. Some HCT/Ps, including some gametes (donated eggs and sperm), do not need to be proven safe and effective before they can be used clinically, but that only applies to cells and tissues that are "not more than minimally manipulated." The exact contours of "not more than minimally manipulated" are not clear, but gametes derived from hESCs, iPSCs, or SCNT cells will certainly not qualify. In 2014 the FDA's general approach to HCT/Ps and the minimal manipulation exception was generally upheld by the federal appellate court most expert in the laws dealing with administrative agencies (including the FDA), the Court of Appeals for the District of Columbia Circuit.[26]

So the FDA will, probably successfully, try to require stem cell–derived gametes to be proven safe and effective before they can be used clinically. What does that mean?

It is useful to break the derivation process down into two steps. The first would be the derivation of pluripotent cells. Cells from hESCs, iPSCs, or SCNT cells will clearly be subject to the FDA premarket approval requirements as long as they are intended to be used for an appropriate medical purpose. Those kinds of processes will have to be judged, and approved, by the FDA before any therapies with cells derived from pluripotent cells can be adopted. Gametes will not be the first successful stem cell product; those first steps will have already been approved in other products, some of which are in clinical trials today. FDA approval for medical use of such cells will almost certainly arrive long before Easy PGD.

The second is the process of turning those stem cells into gametes. Here the FDA would require evidence for the safety and efficacy of that process, judged, presumably, by how well it leads to healthy babies. This will require first work in laboratories and on nonhuman animals before moving eventually to human clinical trials. This FDA process will add greatly to the time and expense of developing this technology. Typically, FDA approval for drugs or drug-like biological products takes eight to twelve years and hundreds of millions of dollars. I suspect the FDA would be extremely careful (and slow) before approving a stem cell–based method for making babies.

In Chapter 13 I argue that such caution would be appropriate and propose some steps the FDA should take to assess the safety of stem cell–derived gametes for the children they produce. I think FDA approval should, as a matter of good policy, be required. It seems likely the FDA and the courts will take the position that it already is required, although that is a bit uncertain, at least where the gametes are not intended to treat infertility.

Non-FDA Regulation

The FDA will not pose the only barriers to Easy PGD under existing law, but it will pose almost the only serious barriers. The DNA sequencing part of Easy PGD will also be subject to some non-FDA regulation, as will, in a few states, at least one of the ways in which gametes could be derived.

The DNA sequencing will presumably be done by a clinical laboratory, perhaps one that is part of the fertility clinic or possibly (as is done now in PGD) a separate entity. As noted above, many of the devices that laboratory would use, notably the sequencers, are products subject to FDA regulation, but in addition the clinical laboratories themselves will be subject to other federal and state regulation.

CLIA—the Clinical Laboratory Improvements Amendments Act of 1988—is the most important regulation of clinical labs. This federal statute, administered in different parts by the FDA, the CDC, and the Centers for Medicare and Medicaid Services, sets quality standards that clinical laboratories have to meet. (Laboratories can usually either meet the CLIA standards directly or meet them indirectly through accreditation by the College of American Pathologists or some states.) The quality standards involve the training and compensation of the personnel, the quality controls in the laboratory, and a number of other steps aimed at ensuring a good "process." CLIA does not regulate what the laboratory measures or its medical value, but just how, and how well, it measures it. Becoming, and staying, CLIA-certified is somewhat expensive and time-consuming, but not difficult. Laboratories doing the DNA analysis for PGD should have no trouble becoming CLIA-certified.

Clinical laboratories, though, can also be regulated by states. Most states do a fairly desultory job of that regulation. New York has taken a more aggressive position with respect to clinical laboratories and, unlike CLIA, requires them to demonstrate that the tests they do actually measure something useful. It is possible that various state health departments could use their regulatory authority over clinical laboratories to limit genetic sequencing as part of PGD, though presumably they would need to make an argument that the safety and efficacy of those tests had not been proven.

Although no state's laws raise concerns about the legality of deriving gametes from iPSCs, using gametes made through either hESCs or SCNT may be impossible in a few American states. A few states in banning reproductive cloning explicitly banned the use of SCNT to start a pregnancy, either by name or by requiring that the pregnancy start with an embryo from combining an egg and a sperm. (Interestingly, Illinois went further and banned any "transfer to a uterus or attempt to transfer to a uterus anything other than the product of fertilization of an egg of a human female by a sperm of a human male"—probably not foreseeing cross-sex gametes but with the effect of foreclosing their use.)

Other states reach the same result in different ways. In Louisiana, for example, destroying embryos is illegal so making stem cell–derived gametes by making hESCs or SCNT cells in Louisiana would be illegal. (Of course, if the hESCs or SCNT cells are made outside Louisiana and then shipped into the state for gamete derivation—or the gametes are made outside the state and then shipped in—this law would seem not to apply.) Other states have banned research involving the destruction of embryos, either in the early 1980s as an early political move by the pro-life movement or around 2000 as part of the political fight over hESCs. Neither would seem to affect production of gametes from stem cells for FDA-approved clinical use.

A newer pro-life approach, through "personhood" laws, might make it illegal to use hESCs or SCNT in Easy PGD. In 2011 Mississippi voted on an initiative that would have defined an embryo as a living human person from the moment of fertilization and would have protected—or tried to protect—its right to life. The Mississippi initiative failed (in large part for fears about its effect on conventional IVF), and even if it had passed, to the extent its protection of embryonic life conflicted with the U.S. Supreme Court's protection of a right to abortion, it would have been unenforceable. The Supreme Court, however, has not found a right to make gametes; the use of hESC (and probably SCNT) may well have been banned under that initiative. Groups in other states have tried to pass similar laws, so far without success.

In addition to embryo protections (and hESC prohibitions), the small, early burst of would-be human cloners led several American states, starting with California, to pass laws banning human reproductive cloning. Some of them defined cloning broadly enough (usually intentionally) to include SCNT. In those states, using SCNT as part of the process to make gametes would be illegal.

So, in the United States, legal challenges do exist to Easy PGD, but, except in a few states in cases where hESCs or SCNT cells are used, the only serious one is FDA approval. This is not true in other countries; many have adopted more restrictive direct regulation of reproduction.

Sometime in the next twenty to forty years, Easy PGD will be approved by the FDA and, where specific laws do not make methods illegal, in equivalent foreign bodies. Under existing law, in the United States and in many other countries, those regulatory agencies are required to examine safety and efficacy and only safety and efficacy—not moral or ethical concerns. But the law can always change, either expressly or as applied, which leads us to the next chapter, and politics.

11

POLITICS

Current American laws may not prevent Easy PGD but laws can be changed. Will political forces prevent the rise of Easy PGD? That question cannot be answered with confidence, even in United States, the one country I know something about. Nevertheless, my unconfident guess is that Easy PGD will not be subject to significant regulation in the United States or in very many of its individual states. For an explanation, we need to start with some discussion of the American politics of IVF.

The Politics of IVF

As discussed above, most developed countries have significant substantive regulation of IVF and other assisted reproductive technologies. They regulate some combination of the marital status, sexual orientation, and age of the people who can use it as well as the procurement and use of donated eggs and sperm, the number of eggs that can be fertilized, the number of zygotes that can be transferred, and some of the procedures that can be used. The United States has almost none of that.

On its face, this is a mystery. Pro-life groups should detest almost all forms of IVF. The Catholic Church certainly does; it classifies IVF, along with almost all other interventions in reproduction, as sinful. In addition to the sin of letting artificial techniques interfere with "the unitive nature of reproductive sex within marriage," IVF almost always produces embryos that are not transferred. These embryos are treated, for at least prudential purposes, as living, "ensouled" persons in Catholic doctrine, in part because of uncertainty and in part because of the certainty that, if not destroyed, they, and only they, would eventually receive their individual souls even if they have not yet.[1] To many conservative Protestants, they are "babies." Unless every embryo is transferred (as the law used to require in Italy until overturned in court), some of these persons will never get the chance to live and will, in fact, usually be disposed of as medical waste.

The United States has a vociferous pro-life movement, but it has no laws banning IVF or even requiring the transfer of all IVF-created embryos. The nearest approach is Louisiana's statute prohibiting the intentional destruction of human embryos, resulting in clinics with freezers full of embryos that will never be used but can never be destroyed.[2]

A few aspects of IVF that particularly trouble some feminists are lightly regulated. The paid "donation" of eggs may exploit women, particularly poor women, by offering them substantial sums for undergoing an unpleasant and somewhat risky medical procedure, typically for the benefit of women (and men) who are *not* poor. A handful of states, as part of laws generally encouraging human embryonic stem cell research, have prohibited the payment of women who donate eggs for research. California, for example, in the same initiative that authorized the spending of $3 billion on stem cell research, prohibited any research funded by that money from using eggs whose donors had been paid anything beyond reimbursement for their out-of-pocket expenses. Even paying for any wages they lost as a result of missing work for the donation process is impermissible.[3]

Yet no U.S. jurisdiction forbids the payment of women who are "donating" eggs for reproductive purposes. The average payment to such a donor is between $5,000 and $10,000 per cycle. Nor does any U.S. jurisdiction forbid paying men for sperm donation, where the average payment is $50 to $200 per "donation." Similarly, paying gestational surrogates to carry other people's babies poses the same risks of exploitation. Both liberals and conservatives worry about this. Yet only a few American states have banned paid surrogacy or surrogacy contracts and a few others have limited payments to surrogates to reimbursement for their expenses. Most have not taken any position.[4]

What is the answer to this mystery? I think it has several reinforcing answers. One is the recognition in the United States of the strength of people's desires to have "children of their own." Even, or perhaps especially, the generally more conservative people who make up the pro-life movement are acutely aware of the "blessings" of children. While the official Catholic position sees IVF as impermissible, the powerful fundamentalist Protestant part of the pro-life coalition makes no such argument. Pro-life advocates tend to idealize the innocence and desirability of a child. That general posture makes it difficult for many to condemn people—especially married, heterosexual, middle class people—who desperately want a child.

A second is an American unwillingness to take no for an answer, especially from medicine. The same resistance to fatalism that impels families to push for additional medical treatments to and beyond the point of futility fuels an expectation that medicine should, both scientifically and legally, be able to help people have their own children.

The relatively libertarian nature of U.S. society also plays a role. Although many things are banned in the United States, the arguments for bans may have to be better, or at least more time honored, than in other countries. There are, of course, exceptions, yet, in general, aided and abetted by a popular belief in the existence of constitutional rights (a belief widely shared by people who disagree vehemently on what those rights might be), Americans seem less likely than many peoples to accept government bans.

It is also the case that, at least at the federal level, the structure of American government, and particularly of the Congress, makes passing any regulation over the opposition of a well-organized group difficult. State legislation is often not quite as subject to legislative gridlock.

This opposition to government limitations on private choices may be particularly strong when it comes to the family. Restrictions on how parents can raise their children exist, but they are exceptional. Parents can give their children any names they wish, a right Americans are amazed to find is not universal. Although American parents must educate their children, they are free to educate them in private schools or at home. Parental rights even extend against the members of the extended family; in 1980 the U.S. Supreme Court held unconstitutional a Washington State law that gave grandparents visitation rights against the parents' wishes.[5]

But one other factor needs to be added. In the United States fertility services make up a large, profitable industry. It advertises, it lobbies, and it can call on the good will of hundreds of thousands of Americans who are grateful to IVF clinics for giving them "children of their own" (as well as, increasingly, those adult IVF children themselves).

The relative immunity of assisted reproduction to regulation in the United States is a function of all of these factors. Their relative strength can be debated. The result cannot. In spite of the self-serving protests of the ASRM, assisted reproduction is almost totally unregulated—at least by governments—in the United States. Even the FDA's regulation of the safety and efficacy of assisted reproduction has been late, limited, and half-hearted.

The Likely American Politics of Easy PGD

Will that limited regulation continue in the face of Easy PGD? Although predicting politics even a year in advance is hazardous and predicting it twenty to forty years in advance is certifiably insane, nonetheless, I think it will, at least through the first, crucial stages of the technology. And that is because the critical technical steps that will make Easy PGD possible will be perfected and widely adopted without being about Easy PGD at all.

Infertility as the Entering Wedge

At least when it comes to human health (as contrasted with, say, genetically modified food), genomics technologies face almost no political opposition. Quite the contrary; they have strong political support because of the hopes invested in them for preventing, treating, and curing disease. Even hESC research, a flash point for opposition to bioscience research, has been pulled along enough by medical hopes that it has not been banned at the federal level or in most states. Differentiating eggs and sperm from stem cells could provide a moment for political action, but it will not, because it, too, will be viewed as a politically admirable health measure.

Stem cell–derived gametes will first be approved and introduced to help people who are infertile because they lack functional gametes—a group that will likely include many older potential mothers whose own eggs are no longer effective for making babies. Efforts to prevent help for people who "just want to have children of their own" will not gain political support. Consider the following advertisement. Start with an attractive and articulate married (heterosexual and most likely white) couple, in their late twenties or early thirties. They explain that cannot have children of their own because one of them had a youthful cancer. They talk of their grief at not being able to have children and how this application of stem cell technology could reverse the way that this lingering effect of cancer has blighted their lives. In the American context, at least, that's a compelling story—and one that the ASRM, or others, would be happy to put on television, either in an effort to influence legislators or as part of an initiative campaign. It is hard to believe that any American state would forbid this technology, at least for these people.

One might imagine that politics might draw the line at married, heterosexual couples, but that also seems implausible in the American context. Over a third of babies born in the United States are now born to women who are not married, including an increasing percentage of births to well-educated, economically successful women. And any efforts to limit the use of this baby-making technology to heterosexuals (if cross-sex gametes proved possible and safe) would face, in many states, well-organized and well-funded opposition from same-sex organizations, as well as triggering the same kinds of litigation that have confronted bans on gay marriage. And, particularly after the recent *Obergefell* decision finding a federal constitutional right to same-sex marriage, it would raise very substantial constitutional questions.[6] Even people who do not like homosexuality may find it hard to turn a deaf ear to the pleas of people, particularly, now legally married people, for "children of their own." The use of cross-sex gametes will, and should, face hard regulatory questions about safety and efficacy, but I doubt that it will face effective political opposition.

Health-Related Expansions beyond Infertility

Once the procedure has been approved for use in infertility, what will stop it from being used to make eggs for IVF even for parents who actually make their own eggs? Here the issue is not the child's health but the mother's. Stem cell–derived eggs will offer safer (and cheaper) IVF. It is, again, hard to see political support for legislation that, while allowing the use of such eggs (or sperm) for people without their own gametes (or without gametes of the "right" kind), would forbid it to the vast majority of the adult population—and electorate.

Similarly, the use of Easy PGD to avoid the birth of children (or the abortion of fetuses) with serious diseases is unlikely to be (very) politically contentious. Prenatal genetic testing has been legal and practiced in the United States for over forty years. PGD has been in clinical use for over twenty-five years. A few conservative states, in recent years, carrying out the current incremental strategy of the pro-life movement, have banned abortion for sex or race, and in one case, North Dakota, for disabilities.[7] It is unclear whether those bans are constitutional, but it is interesting that, as noted in Chapter 10, they have banned the abortions, not the tests necessary to provide information on which to base the abortions. And, of course, for most people, not transferring an embryo is less morally worrisome than aborting a fetus so PGD is less worrisome than fetal testing. Easy PGD to avoid health problems (or, at least, nontrivial health problems) should not face much political opposition.

Nonhealth Reasons

Once we move beyond health, at least four uses of Easy PGD might generate enough political opposition to lead to bans or substantive regulation: nondisease traits, sex selection, selecting for a disability, and uniparents.

The nonhealth uses of Easy PGD, such as for cosmetic traits or (nondisease) behavioral characteristics, might be a palatable target. Certainly some prospective parents will want to use Easy PGD for reasons other than avoiding diseases; at the same time, many in the population may want to ban anyone from using it for such purposes. But the argument for such regulation is simultaneously an argument against it. These traits may not be important enough to justify intervention by Easy PGD, but, on the other hand, is a parental choice to have a brown-eyed or tall child important enough to justify legal intervention?

And note that thanks to whole genome sequencing, Easy PGD will yield all the genetic information about prospective parents' embryos. If they are given that information, who can say whether parents are making a decision based on medical risks, on nondisease traits, or (most likely) on some combination of the two? As every embryo will have some genetic health risks, any regulatory scheme that wanted actually to stop selection based on nonhealth characteristics (as opposed to making a merely symbolic statement against such selection) would have to forbid parents from getting genomic information relevant to nonhealth traits.

It is hard to see a legislature allowing prospective parents to be shown some parts of their embryos' genomes but not others, particularly when the actual consequences of different genetic variations remain at all complicated and uncertain. We have already seen, in the direct-to-consumer genetic testing debates swirling around 23andMe, resistance to the idea that the government can control personal access to one's genome.[8]

Telling prospective parents that they cannot have access to information about their embryos' genetic variations that their doctors know, for fear the parents will make morally bad choices, seems a political loser (as well as suspect as a matter of constitutional law, as will be discussed in Chapter 16).

It does seem possible that some jurisdictions could forbid, or regulate, one set of nondisease traits: sex. As noted above, a few U.S. states have already banned sex-selective abortions and a few have banned selective abortion based on the fetus's race.[9] These statutes seem to be more of a ploy in the ongoing abortion wars—picking an issue where public, and judicial, sentiment against aborting a healthy fetus might be at its peak—rather than a genuine expression of concern about sex-selection abortion. There are, however, good reasons to be concerned about the long-term social consequences of unbalanced sex ratios, reasons that might be seen as rationally justifying restrictive legislation.[10] One could see a coalition of the bioconservatives and feminists, worried about selection against female embryos, pushing for restrictions on sex selection.

Sex selection, though, does have some support, at least when used for "family balancing." A 1999 ASRM committee had discouraged PGD solely for gender selection, but then in 2001 the committee opined that it was permissible for "family balancing." The ASRM's most recent position on sex selection, adopted in spring 2015, takes no position on nonmedical sex selection, other than that it is controversial and clinics should develop their own thoughtful positions.[11] And although we like to think of sex selection and, specifically, a strong preference for sons as alien and exotic cultural habits, found mainly in Asia, some people even in the United States would like to choose the sex of their children.[12] My own guess is that even strict regulation of sex selection through PGD is unlikely to be adopted in the United States, certainly at the federal level and probably not in most states. And, of course, if a few states do ban it, their ability to enforce it when their residents can get the same service right across a state line would be very limited.

A third potentially attractive target for regulation of Easy PGD is to forbid people from choosing embryos with particular genetic diseases or disabilities. Those parental choices may seem bizarre and unlikely, but at least in some contexts, they are plausible.

Some deaf parents want deaf children. Some have already used lower-tech assisted reproductive technologies, such as artificial insemination using sperm from a man with a hereditary form of deafness, to increase their chances of having a "child of their own"—or, more specifically, a "child like themselves."[13]

Others who might be interested in this kind of selection are "little people," those with dwarfism or other forms of very short stature. Couples where each partner has the most common form of dwarfism, achondroplasia (the condition of most of the actors who played the Munchkins in the movie *The Wizard of Oz*), already often use prenatal genetic testing. They have a one-quarter chance of conceiving an embryo that will inevitably die, either before birth or just after it. When couples with achondroplasia use fetal testing or, even more so, PGD, they have the possibility not just of avoiding futile pregnancies but of choosing babies who will have achondroplasia—who will be "like them."[14]

Achondroplasia is a fairly benign genetic condition. Although it is associated with a few health conditions, people with achondroplasia have normal intelligence and close to normal lifespans. Should people with achondroplasia be allowed to have "children like them"? Serious deafness is actually a surprisingly powerful disability; people born with profound hearing impairments have education and economic levels far below average. But their life expectancy is close to normal and they are otherwise healthy. We would not allow deaf parents to deafen any normal children they have, but should we forbid them from selecting deaf embryos?

Would the desire of people with these disabilities to have children "like them" be any worse than the desire of most of us to have "children of our own"? Would it be any different? And, of course, does that mean we should allow people with a particular disability to select children with that disability but forbid it to people without that disability? What about people with a beloved relative, a sister or brother, a mother or father, with that disability?

It would be tempting (it would tempt me) to forbid parents from selecting embryos in order to have children with a nasty, inevitably fatal genetic disease, like Tay-Sachs. But, as discussed below in Chapter 16, this would lead to some very difficult problems in implementation, including how to have the government say that lives with some diseases are worth living, but others are not?

The fourth argument for banning the use of stem cell–derived gametes is politically the strongest—its use by "uniparents." The idea of a person who is both the mother and father of a child will strike many people as profoundly wrong; for one thing, many people will view it as incest or confuse it, inaccurately, with reproductive cloning, a thus far science fiction procedure with an enormously negative public image. For another, unlike the other uses of stem cell–derived gametes, it is hard to come up with any compelling story about why someone needs to be a uniparent. It is more likely to appear an unattractively egocentric, even egomaniacal, step. Whether any legislature would find such a ban politically appealing is not clear, although it could easily be a sop to those who want some kind of action taken against "all this" new reproductive technology. (Whether such a distinction, singling out one and only one kind of use of this technology, would be constitutional in the United States is another interesting question, discussed in Chapter 18.)

The Politics of Public Financing

One last political issue must be mentioned. Even when it is politically or legally too hard to ban something, it may be possible to ban the use of public funds for it. Abortion is legal (at least in theory) throughout the United States but no federal funds can be used to pay for abortions, even when the woman involved has federally funded health care. In fact, physicians receiving federal Medicaid funds to treat a pregnant woman are not even allowed to discuss the option of abortion.[15]

Fertility treatments are not currently covered by Medicaid, the huge federal-state program for the poor, or by Medicare (probably because the vast majority of Medicare recipients are over sixty-five). That is true even before adding the likely political opposition to Easy PGD, either for creating and destroying so many embryos or for its uses for genetic selection. It is certainly plausible that some states (or perhaps the federal government) would ban the use of governmental funds to support these procedures.

It will be interesting to watch how those political fights turn out. On the one hand, conservative legislatures may not want to encourage Easy PGD, but they will also want to keep their Medicaid costs down, which Easy PGD could help do. Some of them may also be uncomfortable, for some of the same reasons that drove eugenics, with the fact that only the poor will be "breeding recklessly." At the same time, others will be arguing that it is peculiarly unfair, and perhaps unconstitutional, effectively to deny such an important family-creating tool only to the poor. Of course, in most countries with a unified national health coverage requirement, this would not be an issue. Whether Medicaid will exist in the United States in twenty to forty years—or what its health care financing system will look like at all—is unknowable.

Summing Up the Likely American Politics of Easy PGD

Two anonymous reviewers of a draft of this book said that the political fate of Easy PGD would depend on whether people viewed it as more like IVF or more like reproductive cloning. I think they were right. Several states (and many countries) quickly banned human reproductive cloning after Dolly.[16] But that procedure seemed deeply unnatural in ways that had no precedent. Easy PGD is just another application of two longstanding (and entirely legal) medical procedures, IVF and PGD.

Reproductive cloning also had few, if any, sympathetic possible uses. Two might be asserted. One is parental grief at the loss of a child and the desire to have another child as similar as possible. The other (in a world without the possibility of generating new gametes) is infertility—a cloned baby might be the only kind of genetically related baby a person without gametes could have. Easy PGD fights not just infertility (and in a closer to "natural" way than cloning) but also fights disease.

Reproductive cloning was still distant and speculative (and remained so for at least fifteen years). Easy PGD is much less speculative and its near relative, "hard" PGD, already exists. It may also be significant that cloning has much scarier fiction (*The Boys from Brazil, Never Let Me Go, The Island,* the two *Star Wars* films set during the Clone Wars, and, in part, *Brave New World*) than Easy PGD (*Gattaca* and again, in part, *Brave New World*).[17]

The same reviewers also noted the growing political strength of fundamentalist Christians in the United States (or, at least, some parts of it). I am not convinced the political strength of that movement is increasing today. More importantly, I suspect, in part from past cycles of political involvement interspaced with long periods of relative political quietism, that any increase today is not likely to be sustained over the next several decades. But, of course, if such a major shift in the American political landscape occurs, it could certainly have negative consequences for Easy PGD, at least beyond its uses in fertility and possibly in avoiding serious health conditions.

All things considered, I expect that there will be little likelihood of banning or heavily regulating Easy PGD in the United States. Two more general factors make this true. First, inertia is the most important force in human affairs; it is almost always easier to do nothing than to do something and that is particularly true in the federal government with its checks and balances. And second, Easy PGD will have substantial political support, some of it with money and influence. The IVF industry will surely support it. So will the millions of Americans—whatever their sexual orientations—who need it for fertility. So will many prospective parents who just like the idea of having healthier children, or children with traits they want. So might even people who think the government should, by and large, keep out of parental choices. (The transhumanist movement would also support Easy PGD, as at least a small step, though its political importance seems to me unlikely to be great.)

I do think that the four particular uses noted above (under "Non-health Reasons") might be subject to some regulation, either in a few states (sex selection or trait selection) or by many states and the federal government (uniparenthood). It is also possible that a few states will limit public funding of Easy PGD for patients. But, subject to all the caveats about the uncertainty of political predictions, most uses of Easy PGD by most people seem likely to be permitted in a future United States—if and when the procedure is shown to be safe.

Outside the United States

However limited my expertise in American politics, I have enormously less for any other country. And lifespans being what they are, no realistic prospect of obtaining much more. All I can say is that just as different states may take somewhat different positions on Easy PGD, other countries will also take different positions, only more so. But I think I can offer a few generalizations.

Countries with a politically powerful religion or religions opposed to Easy PGD are less likely to allow it. The Catholic Church is the example that comes first to mind, though some fundamentalist Protestant groups may be at least as strongly opposed (if less likely to be politically dominant at a national level). Even there, it is easy to overestimate or mistake the effects of religion. Some very Catholic countries, like Malta and El Salvador, ban all abortion; others, like France, Spain, and Uruguay do not.[18] And within the Muslim world, some (mainly Sunni) countries limit assisted reproduction, particularly using donor gametes; other (mainly Shia) countries, including Iran, allow it more broadly.[19] And religion in

Israel, at least Orthodox Judaism, strongly supports assisted reproduction.[20] Finally, both religious membership and the political power of religion within a country can change substantially over a few decades.

Some countries have histories or cultures that make them particularly skittish about human biotechnologies. Germany, with its Nazi legacy, is the prime example.[21] More broadly, much of Western Europe is a lot more concerned about genetic interventions, particularly in food, than the United States. The human side of this can already been seen in the greater restraints on IVF and other forms of assisted reproduction in those countries than in the United States.

On the other hand, some countries have histories and cultures that are likely to make them even more accepting of Easy PGD than the United States. In most East Asian countries, embryos and fetuses are much less important than in many countries with Christian backgrounds. Sex selection, through prenatal testing and abortion, has been quite common in some of these countries even though illegal, notably in China.[22] It is easy to imagine East Asian countries adopting Easy PGD enthusiastically—perhaps disconcertingly so.

As Tip O'Neill, the former Speaker of the U.S. House of Representatives, famously liked to say, "All politics is local."[23] Ultimately, I suspect the world will be a patchwork in terms of its regulation of, and use of, Easy PGD. It will be widely permitted, or even encouraged, in some, banned in a few, and limited in its uses in some in between. And many countries, with more immediate life and death concerns, may not regulate it at all. But I will be quite surprised if, within forty years, it is not available to many—probably most—prospective parents around the world who have access to good health care.

12

SOME OTHER POSSIBLE USES OF NEW TECHNOLOGIES IN REPRODUCTION

This part has, so far, laid out a case that Easy PGD is likely to be possible, accepted, and common sometime in the next few decades. But the revolutions the biosciences may bring to human reproduction will not necessarily stop with Easy PGD. Four other possible technological advances lying farther down the path deserve mention. Each could profoundly affect human reproduction and at least the first three seem plausible within something close to our twenty- to forty-year time window: human reproductive cloning, genome-edited embryos, synthetic chromosomes, and artificial wombs.

Human Reproductive Cloning

Remember the uniparents and the "unibabies" from Chapter 8? They would not be clones, but it now seems plausible that someone who wanted a clone, of himself (somehow, with no real evidence, it seems to me men are more likely to want a clone than women) or of someone else, might be able to do it sometime in the next twenty to forty years. I would not have said so five years ago. The lack of progress in making cloned human embryos had been quite impressive and the progress made in nonhuman primate cloning was no better. But the eventual success in 2013 of using SCNT to make cloned human blastocysts makes me think baby clones may well be possible.

"May well be possible" is not a ringing endorsement, but then, so far, no one has made any primates, human or other, through SCNT. Cell lines have now been created from cloned blastocysts, but cell lines are not babies. Still, blastocysts are a large and essential step toward babies. Those human cloned blastocysts are a first step toward reproductive cloning.

Good reasons exist to worry that "possible" will not be the same as safe. The safety of using SCNT to clone individuals in other species is quite mixed. About twenty species of mammals have been cloned.[1] For some, like mice, it has become fairly routine. For larger mammals, the incidence of miscarriages, still births, and deformed offspring remains higher than in "normal" reproduction.[2] We might be willing to put up with unusually high numbers of dead or disabled livestock, but we care much more about human babies. It is unclear whether human reproductive cloning would ever become safe enough for the resulting babies (which should probably be about as safe as other assisted reproduction methods in common use) for it to be permitted. And it is unclear just how one should go about proving the safety of human clones—it will be a brave or foolish (or both) person who, however good the results in, say, nonhuman primates, first tries to make a human baby through cloning.

In the longer run, though, it would be interesting to see just how popular human reproductive cloning would be even if it were "safe enough." A clone would not be the same person as its "progenitor," any more than identical twins are the same people.[3] And, in spite of almost identical DNA, these clones would not be as similar as identical twins. They would not have developed inside the same uterus at the same time, exposed to the same conditions. They would not be raised in the same world, with the same physical and cultural environment. How many people would want to make a baby that was nearly the same as an identical twin of someone else, but not quite? My guess is that even if safe and approved (and human reproductive cloning has been banned in many countries and states), it would be a minority taste and probably quite a small minority. (At least, for some reason, I hope so.) But that, necessarily, is only a guess.

Genome Editing and "Designer Babies"

So far, I have assumed that parents will want to make their children using eggs and sperm with the prospective parents' genetic variations—that they will want "a child of their own" or, for a clone, a "child of a particular someone." But that may not be the case. Consider a parent who, by very bad luck, has two alleles associated with an autosomal dominant disease or disease risk—say, two copies of the Huntington disease allele or two copies of a dangerously mutated *BRCA1* gene. Any egg or sperm from that parent will necessarily pass the disease-associated allele on to any child. Similarly, if both parents have two copies of the same allele for any gene, the child will necessarily get two copies of that allele. If the allele is the one for cystic fibrosis, the two parents, each with cystic fibrosis, will necessarily have a child with cystic fibrosis.

Perhaps parental devotion to having "a child of their own" will not extend that far. That seems plausible where diseases are concerned, though the disease examples I have given are likely to be rare. It might also be the case, though, for other traits, such as cosmetic traits or behavioral traits, where the parents would like children who would receive a "better" genetic inheritance than the parents' own gametes can provide. What then? New technologies seem likely to make it possible for parents to edit their prospective children's genomes to change alleles they do not want into alleles they do. None is more exciting, and more concerning, than CRISPR/Cas9.

In earlier drafts of this book, I did not pay as much attention to genome editing. It was plausible, but expensive, unreliable, and time-consuming. That has changed in the last few years thanks to the development of a technique called CRISPR (which stands for "Clustered Regularly Interspaced Short Palindromic Repeats")/Cas9 (the name of a particular protein).[4] The invention of this technique is usually traced to 2012 and Jennifer Doudna at UC Berkeley and Emmanuelle Charpentier, then working in Europe—but there is controversy over the "true" inventors, as well as a patent fight specifically over CRISPR/Cas9.[5] It has been widely and enthusiastically adopted by laboratories around the world.

CRISPR/Cas9 seems to be the Model T of genome engineering. Like cars before the Model T, other genome editing methods existed earlier, but they were expensive, difficult, and, as a result, not commonly used. With CRISPR/Cas9, cheap, easy, and fast genome editing is now available, like the Model T was, to everyman—or at least everyone with molecular biology training and a few thousand dollars. And, like the Model T, CRISPR/Cas9 will almost certainly be overtaken by some improvements (CRISPR/Cas10? 20?)—but it will have been the advance that changed the world.

The broad potential of CRISPR/Cas9 has drawn concern, not least from co-inventor Jennifer Doudna. Although the technique has many potential uses, such as improving the potential for gene therapy to treat diseases of children and adults or greatly easing the creation of all kinds of genetically modified nonhuman organisms, attention has focused instead largely on the possible use of CRISPR/Cas9 to make inheritable modifications in the genomes of people—so-called "germline modifications." These are changes taken up by all the person's cells, including her eggs or his sperm, and so passed on to future generations.

In January 2015 Doudna was instrumental in convening a workshop to discuss the ethical and social implications of CRISPR/Cas9, with an emphasis on human germline genome editing. The workshop resulted in an article published in *Science* that spring calling for a moratorium on using this or similar methods to edit the human germline genome, along with extensive public discussion and further research.[6] (I was one of the participants in the January meeting and one of the eighteen co-authors on the *Science* piece.) The upshot of that article, bolstered by concerns expressed by other parties, was a commitment by the United States National Academies both to hold a "summit" on the implications of CRISPR/Cas9 and to convene a special committee to study and report on the issues it raises.[7] The Summit, which was cosponsored by the United States National Academy of Sciences and National Academy of Medicine, the British Royal Society, and the Chinese Academy of Sciences, was held in early December 2015; the study committee has just begun its work and is expected to report sometime in 2016.[8]

Although many will no doubt misunderstand this point, Easy PGD is not about "designer" babies, but about "selected babies." CRISPR/Cas9 and similar genome editing methods offer "designer babies." Prospective parents could choose which DNA variations their children would have, whether or not the parents carry those alleles. As noted above, there are only a few times where this would be necessary to avoid a genetic disease or trait, but parents might want to use genomic engineering to add "enhancing" variations to their children that they themselves do not possess. Right now, few such enhancing variations are known, but that is likely to change in the next twenty to forty years.[9]

How likely are CRISPR-edited babies? That remains unknown. First, it is not clear exactly when and how the genomes would be edited. Each cell needs to be modified to take out the old and add the new DNA. If some cells were edited successfully and some were not, the result would, unfortunately, be a mosaic, whose actual phenotype would be some mix of the results of the two genetic components.

One might try to edit the zygote, where only one cell has to be modified, though both the sperm's and the egg's pronuclei would have to be modified. Two-, four-, or eight-cell embryos could also be tried, but the more cells, the greater possibility of incomplete editing and mosaicism. The embryos would have to be tested for such mosaicism and, presumably, the mosaics discarded. This method could use a lot of embryos, which would be a logistical problem (unless stem cell–derived gametes become possible). But the unsuccessful editing of otherwise healthy embryos and the destruction of those improperly edited would also be a moral problem for many people, and not just those at the pro-life extreme.

Probably the better course would be to edit the germ cells used to make the embryos. Those eggs and sperm, or, more likely, the cells that produce them, could be grown and edited in vitro. The cell lines could be tested to see that they had been edited correctly before using them to make embryos. Making cell lines from human germ cells is not yet possible, but it is essential to the gamete derivation part of Easy PGD and I assume it will be possible in the near future.

Of course, it is not clear whether genome editing will ever be made to work safely and effectively enough for making babies, where the stakes are very high. The possible use of CRISPR/Cas9 to treat an adult volunteer for a serious illness is quite different from its use to create an unconsenting baby who might have a long and painful life.

Thus far, one known effort has been made to use CRISPR/Cas9 on human embryos. In March 2015 a Chinese team published the results of their editing of nonviable human embryos, embryos that had been double fertilized and thus had three copies of every chromosome.[10] The group found that CRISPR/Cas9 did work but not very effectively—lots of mosaicism—or very accurately—lots of potentially dangerous unintended editing of other ("off target") DNA sequences. And beyond these questions of the short-term effectiveness of the genome editing lie concerns about the longer-term effects of the process on embryonic and fetal development.

Before any such editing could be tried for making human babies without being criminally reckless, extensive safety work would be necessary, perhaps including even trials on other great apes (probably chimps). And if, instead of using alleles already present in other humans, the process were used to edit DNA to give humans genomic variations found only in other organisms or not found in nature at all, the safety testing would have be much more stringent.

Will parents want to use genome editing? Should we allow genome editing, either in a world with Easy PGD as an alternative or in one without that option? And consider that without Easy PGD as an option, or at least its egg creation ability, genome editing (if done to embryos and not to gametes) may well fail because of a need for too many harvested eggs to make the embryos. Much will depend on its safety, but not everything. What kinds of enhancing gene sequences will be found and, more importantly, just how significant will they be? A 1 percent edge in, say, math ability is not very exciting, but a 50 percent improvement might be. Also, how expensive will the process be? And in addition, how many parents will want, for all but the most serious medical reasons, a child that is "theirs" and not one that has genes added for nonmedical purposes by genomic editing?

The safety issues of Easy PGD revolve around the safety of deriving gametes from iPSCs; we already accept PGD as safe enough. Germline genome engineering would have those safety issues if, as seems almost inevitable, it uses derived gametes. It would also have the safety issues of the editing intervention and, for "enhancing genes," the safety and efficacy issues of their use. As a result, I would expect safe Easy PGD to be available sooner than safe germline genome editing, as well as more acceptable to prospective parents, and so probably to be used much more, at least in the next few decades. But things come and go very quickly in bioscience—after all, two years ago, in writing an earlier draft of this chapter, I did not know CRISPR/Cas9 existed. So when I try to look more than a few decades ahead, my crystal ball remains foggy.

I need to make one last point on genome editing. We have been trying for thirty-five years to do a kind of genome editing, one where, rather than replacing a particular sequence, we just want to add a particular sequence. This has been called gene therapy. It was first tried in humans in 1981, but it is just beginning to prove successful as a therapy. In fact, its fascinating history is, so far, largely a tale of woe, of unexpected failures, few successes, and, in some cases, failures in the midst of success.[11] That may be changing; in 2012 the European Medicines Agency approved the first human gene therapy for use in the European Union.[12]

Gene therapy could be used to add alleles to an embryo for a particular trait that its genetic parents did not have and thus confer those traits. This would only work where the trait can be changed by adding a new dominant allele without subtracting an existing one and it would have its own safety risks. And, with the very likely rise of CRISPR/Cas9 for editing genomes, it is hard to see much of a future for this pure "allele addition" in modifying embryos (or the gametes that create them).

Similarly, another technique, called "gene silencing," which may eventually prove useful for "turning off" dangerous or otherwise unwanted alleles, also seems less attractive in light of CRISPR/Cas9.[13] Why not just edit the unwelcome alleles instead of silencing them for an unknown period of time? Gene silencing may well be useful in treating some conditions in people who have been born; it seems unlikely to be a valuable technology in making babies.

Wholesale Chromosome Construction

We have been talking, so far, about small-scale genome editing, modifying a few alleles at a time. But some researchers have much greater ambitions. The synthetic biologists want to be able to construct, from scratch, whole genomes, base pair by base pair. And, in fact, they have already done so.

In May 2010, a team led by Craig Venter, the driving force behind the private sector's version of the Human Genome Project, announced that it had created the genome of a living organism from scratch.[14] Using small pieces of DNA it built a variant version of the genome of a bacterial species, *Mycoplasma mycoides,* and transferred that genome into the cell of a related bacterium, *Mycoplasma capricolum,* whose own DNA had been removed. And the new construct was alive.

The Venter team constructed the *M. mycoides* genome by ordering from a DNA synthesizing company 1,078 pieces of DNA, each 1,080 base pairs long. The ends of each of these pieces of DNA overlapped by eighty base pairs the piece of DNA that was next in the *M. mycoides* genome. The 1,078 pieces of DNA, which they called "cassettes," added up, absent the overlaps, to the entire 1.08 million base pair genome of *M. mycoides.*

In spite of occasional press claims to the contrary, the Venter group did *not* "create life" in a test tube. The *M. capricolum* bacteria into which they transferred the synthetic *M. mycoides* genome had been alive. And the *M. mycoides* genome, by itself, would not be alive. It required the membrane, the cytoplasm, all the proteins, and the various other bells and whistles that the *M. capricolum* host provided. All the Venter group really did was transform one species into another, a trick, as we saw in Chapter 1, Oswald Avery and his associates had managed in the early 1940s in their proof that DNA and not protein was the stuff of inheritance.

What the Venter group did that was new and different was to construct the whole genome of one species, which, when put into the shell of the other species, transformed it utterly. The result was a paper in *Science,* a lot of publicity, and an order from President Obama to his then-new bioethics commission to report to him on synthetic biology.

The Venter accomplishment, or stunt, or both, is just the most famous achievement of the synthetic biology movement, which can be described as an effort to turn biology into engineering, using standardized biological parts to recreate existing biological organisms and tissues or to create brand new ones. Why is this relevant to the end of sex? Because it may be possible to synthesize the whole genomes of gametes, rather than to use prospective parents' genomes as a starting point to make those gametes. This would allow the prospective parents to select exactly which alleles they wish their children to have at every point of the genome.

It is a long way from the Venter publication to a whole human genome. The whole genome for *M. mycoides* is only 1,080,000 base pairs long and, like most bacterial genomes, it is contained entirely on one circular chromosome. The human genome is more than 6,000 times larger than that of *M. mycoides* and is contained on 46 complicated chromosomes. The very smallest of the human chromosomes, chromosome 21, has 47,000,000 million base pairs, nearly fifty times as many as the genome Venter synthesized. Humans have about 23,000 genes; some of those individual genes are longer than the entire genome of *M. mycoides.*

On the other hand, twenty to forty years might be a long enough time to go this great distance. I would not bet against the possibility that we could synthesize a complete human genome in the next twenty to forty years, but I do think there are reasons to doubt how important synthetic genomes or chromosomes will be in human reproduction. Most of those revolve not around the DNA itself, but its packaging.

In what form would synthesized DNA be packaged to be introduced into a gamete or a zygote? One simple approach would be to make the DNA into one or more circles, like a bacterial genome. One million base pairs, roughly the size of the bacterial genome constructed by Venter, could easily contain five to fifteen human genes of average length; if we learned enough to "edit" those genes and only include the important parts of them—say, all exons and regulatory regions, but no introns—that number could go to more than a hundred. If, instead, we built a circular chromosome as large as the largest bacteria genomes, it might hold up to a thousand intron-less human alleles. It would take about twenty-three such circles to hold one haploid human genome and forty-six—by coincidence, the same number as our normal complement of chromosomes—to hold two haploid genomes (one from each parent).

The problem is what happens to these DNA circles in humans. They might function in the cell in which they are introduced, but would they be duplicated, with very high accuracy, and passed down evenly to the two daughter cells each time a parent cell divided? Mitosis in humans is set up to duplicate pairs of chromosomes, which undergo a complex dance to ensure that each daughter cell gets the same pairs of chromosomes the parent cell had. There is no reason to think that the machinery that makes mitosis work would also work with the equivalent of human plasmids. And meiosis, the cell division that turns the diploid ancestors of gametes into haploid eggs and sperm, is even more complicated.

One answer to this problem is not to put synthesized alleles into gametes or zygotes as circular bacterial DNA, but as one or more actual humanlike chromosomes. This would certainly be more complicated. Chromosomes are DNA molecules wrapped around a long protein backbone. Human chromosomes are between about 250 million (chromosome 1) and 50 million (chromosome 21) base pairs long and have special regions—centromeres in the middle, telomeres at the end—and other physical features that may, or may not, be of vital importance. On top of this, the DNA in human chromosomes can have various other molecules attached to it that affect when and how it is expressed. These include methyl groups, which seem to prevent use of the gene; acetyl groups, which encourage gene expression; and phosphate groups, whose function is unclear. Constructing a functional human chromosome would be difficult. Being confident that it would function properly indefinitely—for at least seventy or eighty years—would be even more challenging.

The work needed to make perfect (or, at least, "good enough" for eighty years) chromosomes would, no doubt, be substantial. The safety of the processes by which one, ten, or all forty-six old chromosomes were removed from cells and the new chromosomes added would also require much study. Ultimately, one might want to build not just chromosomes but an entire cell nucleus in order to limit the disruptive effects of the transfer. Each move up in complexity, however, adds time, cost, and, mostly, uncertainty.

On the other hand, a widely distributed team has already synthesized an entire chromosome of a yeast species.[15] And although yeasts are not humans, unlike bacteria they are eukaryotic organisms, the kinds of complex organisms that include all "big" creatures, including us. And although their chromosomes are smaller and have fewer introns, they are much more similar to human chromosomes than bacterial chromosomes are.

So creating whole new custom chromosomes for embryos seems (somewhat) plausible. Eventually. It does not seem easy. And I suspect that for most parents (and for companies seeking to profit from the decisions of parents), it will not seem worthwhile when compared with either Easy PGD or germline genomic editing. Synthetic chromosomes for making babies may (or may not) become significant in the next century, but probably not in this one.

Artificial Wombs

Artificial wombs are the least plausible of the technologies discussed in this book—and the least directly related to its subject. Such a technology would not affect the genes that children carried or the choices that parents made about their children's genes, but it would have major effects on the future of human reproduction. And, as the least implausible version of such an invention stems from one of the technologies discussed here, it seems worth mentioning.

The uterus, plus all the pregnancy-supporting tissues it hosts, is a remarkable organ. It produces, or allows the passage of, signals that help guide embryonic and fetal development. Through the placenta, it provides life support to the fetus, conveying oxygen, glucose, and water to it and removing its waste products. It allows those transfers while generally isolating the fetus from all outside influences. It grows enormously in size and complexity in the course of nine months and then returns to its entirely different nonpregnant state. Exactly how it does all of this is quite mysterious.

From time to time, newspaper articles appear talking about exciting research being performed with artificial wombs, usually in Japan and usually involving goats.[16] By the later stages of a pregnancy, when a fetus has a fighting chance of surviving outside the womb, it may be possible to use a mechanical device (or combination of devices) to provide the fetus with its needs for oxygen, food, fluid, and waste removal. Machines already exist for extracorporeal oxygenation (providing necessary oxygen directly into the blood of humans, thus bypassing the lungs), parenteral nutrition can provide food directly into a person's bloodstream, and kidney dialysis cleans waste products from the blood. Combine these methods with a direct link into the fetus's circulatory system—and a pump if necessary to help the fetal heart keep the fetal blood moving—and you may have a form of life support for a well-developed fetus that might be viewed as a kind of artificial womb.

It seems highly unlikely, however, that a mechanical device, or combination of mechanical devices, will be able to replace the womb throughout development, for the nine months from the implantation of the blastocyst to the moments just before birth. We have little idea of what is actually happening inside the human womb in the early months of pregnancy and it is hard to imagine how to conduct ethically acceptable studies that would help us find out. A mechanical artificial womb, for use from the earliest moments of pregnancy, seems highly unlikely, perhaps forever.

But why do artificial wombs have to be mechanical? The excitement about stem cells comes from the idea that they can be used to produce replacement tissues for humans—new blood, new heart cells, new insulin-producing cells for the pancreas. And there are hopes for even more complex constructions. Researchers have already used stem cells to create, and transplant into human patients, semi-artificial bladders, built with human cells over a synthetic scaffolding. Similar progress has been made in constructing stem cell–based trachea, the "tube" connecting the mouth with the lungs.[17] Other human organs would also be of great interest, from skin to livers to kidneys to hearts.

The uterus is an organ. It is the product, ultimately, of stem cells, as are all other human body parts. Why not use stem cells to construct a human uterus? Organ construction work to date has focused on transplantation, growing trachea or bladders for use as transplants into humans. This makes sense. Apart, perhaps, from some research uses, it is hard to see how a human bladder would be valuable outside the body of a human who needed one. But the uterus might be useful outside a human body.

Consider the following scenario. A box is built that is able to provide human blood with appropriate control over its oxygen, glucose, waste products, and hormone levels. Inside that box, and fed by the blood supplied and maintained by the mechanical side of the box, a human uterus is grown. The uterus receives, through the blood supply, the appropriate hormonal signals needed, first to mature and then to prepare it for beginning a pregnancy. At the appropriate time, a blastocyst is transferred to the inside of the uterus where, if all goes well, it implants in the uterine tissue, begins growing a placenta and the other support apparatus of a pregnancy, and starts on its path to becoming a baby.

This is not a vision for the next ten years or even twenty. I suspect it would only be tried after long experience with other kinds of stem cell–grown organs used for transplants.[18] Its adoption would be slowed further by the fact that the technology would never be medically essential. Even for women without functional wombs, there is already a workable replacement—gestational surrogates, living women who carry another woman's pregnancy. We know that works.

But we also know that it does not work perfectly. The gestational surrogates sometimes misbehave: drinking, smoking, using drugs, or doing other things that endanger the fetus. Sometimes they become emotionally bonded to the eventual baby and cause difficulties in the child raising. And they are always subject to the risks of accident and disease that afflict anyone. Artificial wombs would also be at risk for some problems, like power outages, earthquakes, or fires, but parents might consider those more easily avoided than the risks that come with another human being carrying their child.

How would making babies without requiring women to be pregnant change the world?[19] Certainly it could change the behaviors of prospective mothers, allowing them to continue their normal work, lives, and habits without the interruption of nine months of pregnancy. Would it change the emotions between mother and child? Perhaps. There seems to be little evidence that surrogacy has had that effect, although, to be honest, there is little good evidence about the effects of surrogacy at all. Would children be different if they didn't experience their mother's heartbeats, food choices, and daily motions (and emotions) for the nine months before birth? And, if so, could those not be easily simulated by, for example, just building a heartbeat into the artificial womb?

Some women love being pregnant and claim they never felt better in their lives. Others endure it, but not with pleasure. I suppose it is possible that some women love labor and delivery, though I have not yet heard anyone say so. Would there be a market for safe and effective artificial wombs? I think so. What would the implications be? Beyond "vast," I do not know—but unlike the implications of Easy PGD, I believe we will have a long, long time to think about them.

I want to end this chapter with a brief note on one more reproductive possibility. In December 2015, while I was reviewing the copy edits for this book (which was my last chance to make substantive changes in it), I read the advance, on-line publication of Sonia Suter's article, (mentioned briefly in the notes to Chapter 8) on the implications of deriving gametes from stem cells.[20] The article describes many of the uses and implications of such a technology along lines similar to those in this book. But she proposes one use that I had not imagined—multiplex parenting. In this scenario, three people want to mix their genes to make a baby. Two of them make an embryo, have hESCs created from it, turn those hESCs into a gamete, and combine it with a complementary gamete from a third person. (More than three people can be involved; it just requires making more embryos for use as gamete sources.) In effect, two people want their child to mate with someone else, without the wait and bother of actually having a child and raising him or her to puberty.

It is not clear to me how many people would be interested in making such children (few, I suspect) or how serious would be the implications. (On the latter point, I recommend those of you interested read Professor Suter's article.) But here is what was most striking to me about the idea. I have been working on this book for five years, immersing myself in the subject—and this idea had never occurred to me. That might be evidence for my own lack of imagination. I prefer to take it as evidence of just how wide-ranging and non-intuitive the implications of new biological technologies may be for human reproduction. And how far short of reality Easy PGD, as well as the four main examples discussed in this chapter, may fall.

SECOND INTERLUDE

EASY PGD: THE FUTURE

The last six chapters have laid out one likely pathway to the near future, as well as a few more distant possibilities. Where will that path have led us in, say, forty years? I believe to a world where most pregnancies, among people with good health coverage, will be started not in bed but in vitro and where most children have been selected by their parents from several embryonic possibilities based on the genomic variations of those embryos and hence the genetically influenced traits of the eventual children.

This will not—and should not—happen overnight. Even if all the scientific developments necessary happened tomorrow, the FDA process (or its equivalent in other countries) would likely take a decade or more. And the scientific developments will not happen tomorrow. Even once the technologies are approved, their widespread adoption will take more time. Twenty years is a realistic lower bound for this future; forty years is, I think, a realistic upper bound. This brief section sets out what reproduction is most likely to look like in that future world and how widely new methods will be used, as well as the most foreseeable obstacles to, and variations in, that future.

The Most Likely Future

Assume the FDA or an equivalent regulatory authority approves Easy PGD, or its constituent parts, and that Easy PGD has not, in whole or significant part, been banned on moral grounds by legislation. What happens?

First, IVF clinics will offer Easy PGD. They will form alliances with genetic testing laboratories, just as they have done already for PGD, unless the technologies become easy enough to bring in-house. Similarly they will ally with physicians or laboratories to derive gametes from stem cells, unless, again, it becomes sufficiently easy for them to do so in-house.

Prospective parents—married or unmarried, gay or straight—will make an appointment with the local clinic when they decide they want to have children. They will in advance have authorized the clinic to have electronic access to their whole genomes from their medical records. The prospective parents' first conversation at the clinic will include some discussion of what diseases or traits their embryos might carry based on their own genomes. If they decide to proceed, at that meeting or shortly thereafter, the clinic will take a skin biopsy, or use possibly other, even less intrusive cell selection methods, from the person (or people) involved and begin the process of deriving the relevant gametes.

Sometime later—days or weeks?—the derived gametes will be ready. If both eggs and sperm were derived, no further visits will have been necessary for the gametes. If fresh sperm are to be used, the man will go to the clinic to provide a sample.

At this point, the key question—presumably answered earlier—will be how many eggs to fertilize and, effectively, how many embryos to create. This will not be limited by the supply of eggs, effectively infinite, or of sperm, millions of which can be found in a normal man's ejaculate. It is more likely to be a decision made based on the costs of analyzing each embryo's genome, as well as some statistical analysis of the number of embryos necessary in order to have a great likelihood of getting some without serious disease risks. Both of these factors may well vary from couple to couple; some couples may be willing to pay more to check more embryos, and some couples at particularly high risk of genetic diseases may need to make more embryos in order to be confident of having some that avoid their high risk.

Probably five days after fertilization, all the thriving embryos will have several cells removed for genetic analysis. The embryos will, most likely, be frozen to provide time for the next steps. The whole genome sequencing should happen rapidly, as will the computerized analysis of the resulting variations. This will lead to the part of the process that will require the most skilled labor and human judgment. The prospective parents will have to choose, based on expert information provided to them by someone interpreting the embryos' genomes, which one embryo to transfer (or to transfer first). (As IVF has become easier and cheaper, it is likely that only one embryo will be transferred at a time, avoiding the risks of multiple fetuses.)

Let's take a moment to think about the process of this choice. The prospective parents will be presented with genomic information on, say, a hundred embryos. They would first be told which of the embryos are unlikely to be viable, through chromosomal errors or other genetic variations that are lethal during fetal development.

For the remaining, say, eighty embryos, most parents will be told that their embryos would get none of the five to ten thousand dangerous, highly penetrant genetic diseases. In ninety to ninety-five percent of the cases, the parents will not both be carriers for the same autosomal recessive disease (and the woman, or women, will not be a carrier for an X-linked disease) and neither prospective parent will have an autosomal disease allele (and the woman or women will not have a genetic mitochondrial disease). For the five to ten percent of prospective parents who do carry such dangerous alleles, about a quarter (if recessive or X-linked), or half (if dominant) of the embryos will be affected. If those couples know their genetic status in advance, they could make enough more embryos to counterbalance that loss. (Women with genetic mitochondrial diseases will need to use genome editing or mitochondrial transfer—some remedy other than Easy PGD as all of their embryos, and hence children, would be affected.)

These first steps—rejecting embryos that would not be viable at all as well as those with powerful genetic variations that would cause severe disease—will be, effectively, determinative. We will assume for now that no one would want to transfer nonviable embryos or embryos doomed to a serious genetic disease. For the remaining embryos, let's say eighty, the prospective parents will have a chance to get more information.

First, they might learn about each embryo's less powerful health risks—say, a percentage chance, based on the genes (and not on any environmental factors) that the embryo would grow up to develop each of, say, fifty diseases. Then they could get information on which autosomal recessive disease alleles each embryo carries—alleles that could not cause disease in the person that embryo might become, but that could cause disease to its offspring.

They could also get cosmetic information for each of those eighty embryos—hair, eye, and skin color, as well as nose shape, hair type, likely height, probability of male pattern baldness or early gray hair, and so on. They would be given the chance to receive some, probably weak, information about likely behavioral characteristics—this embryo will have a 60 percent chance of having above-average intelligence, a 35 percent chance of having above-average musical ability, a 45 percent chance of having above-average mathematical ability, and a 75 percent chance of being better at sports requiring endurance rather than sports requiring power or quickness. Finally (or perhaps first), they will be asked whether they want to know "boy" or "girl."

If they have asked to get all that information, the results may look something like this—but sixteen times larger and more complicated.

EMBRYO 1

- No serious early onset diseases, carrier for Tay-Sachs, PKU
- Higher than average risk of coronary artery disease, colon cancer, type 1 diabetes
- Lower than average risk of schizophrenia, breast and ovarian cancer, type 2 diabetes, asthma
- Dark eyes and hair, graying early in life; moderately tall, straight hair, thin build
- 55 percent chance of top half in SAT tests, much lower chance than average of being an athlete, good chance of above-average musical ability
- Girl

EMBRYO 2

- No serious early onset diseases, carrier for PKU
- Higher than average risk of type 2 diabetes, cataracts, colon cancer, prostate cancer
- Lower than average risk of asthma, autism, pancreatic cancer, gout
- Dark eyes and light brown hair; male pattern baldness; medium height, straight hair, medium build
- 40 percent chance of top half in SAT tests, likely to be introverted, good chance of above-average musical ability
- Boy

EMBRYO 3

- No serious early onset diseases, carrier for PKU
- Higher than average risk of bipolar disorder, rheumatoid arthritis, lupus, colon cancer

- Lower than average risk of leukemia, autism, gout, Alzheimer disease
- Blue eyes and light brown hair; medium height, curly hair, heavy build
- 65 percent chance of top half in SAT tests, good chance of above-average athletic ability, likely to be anxious
- Girl

Embryo 4

- No serious early onset diseases, carrier for Tay-Sachs
- Higher than average risk of bipolar disorder, cataracts, autism, prostate cancer
- Lower than average risk of schizophrenia, Alzheimer disease, asthma, pancreatic cancer
- Dark eyes and hair; early graying; above-average height, straight hair, medium build
- 50 percent chance of top half in SAT tests, average athletic ability, above-average chance of exceptional musical ability
- Boy

Embryo 5

- No serious early onset diseases, carrier for Tay-Sachs, PKU
- Higher than average risk of coronary artery disease, type 1 diabetes, lupus, colon cancer
- Lower than average risk of schizophrenia, leukemia, autism, pancreatic cancer
- Blue eyes and dark hair; average height, curly hair, heavy build
- 45 percent chance of top half in SAT tests, above-average chance of exceptional athletic ability, likely to be extraverted
- Boy

Even looking at just five embryos, how would you decide? How could you decide? Each is a mix of advantages and risks, your views of which are likely to be affected by your own life and family.

The prospective parents will get a printout like this and be asked to select the embryo they want to transfer (and possibly a second or third choice in case their preferred embryo does not thaw well or for some other reason the pregnancy does not take). Then, when the couple and the woman who will carry the pregnancy are ready, the lucky embryo will be transferred into a womb. Within a few days testing will be able to tell whether or not the pregnancy has started, at which point Easy PGD may well be over and regular prenatal care begun. The "may well be over" comes from the possibility PGD's error rate may still be high enough to require subsequent follow-up, probably through noninvasive prenatal testing (NIPT) at about the eighth week of pregnancy.

About forty weeks after the embryonic transfer, if all has gone well, the prospective parents will no longer be "prospective." At that point, they will begin the disconcerting but exciting challenge of learning the near-infinite number of things about their new baby that Easy PGD did not warn them about.

Likely Amount of Use of Easy PGD

That's what Easy PGD would look like for prospective parents. What would it look like for society? That depends largely on how widely it is adopted. As one anonymous reviewer of this book pointed out, we see very few "discretionary" instances of PGD today; it is only used about 8,000 times a year in the United States and largely to try to improve the chances of pregnancy, to avoid a known family genetic disease, or to try to create a "savior sib." (The main discretionary use occurs when it is used for non-medical reasons to determine the embryo's sex, either as one part of PGD also done for one of the three medical reasons above or on its own.) Just how much use can we expect?

That is another difficult question. Assume, as suggested in Chapter 9, that thanks to health coverage, the cost of Easy PGD to parents will be effectively zero. For the last few decades about four million children have been born each year in the United States as a result of about 3.9 million pregnancies that end in "live births." (Multiple births account for the difference.) Roughly another million pregnancies have ended in abortions each year.

One often reads that only about 50 percent of births are the result of planned pregnancies. In fact, the best survey source shows that between 2006 and 2010 in the United States, 62.9 percent of births were. This figure could be seen as an upper bound for the market for Easy PGD. After all, the 37 percent of pregnancies that were unplanned could not use this process. On closer examination, though, that number is not entirely fair.

Only 13.8 percent of the total number of pregnancies were unwanted. Of the remaining roughly 23 percent, 9.2 percent were wanted but came two years or less early while 14 percent came more than two years early. Those women may not have been affirmatively trying to have a child at that point, but were thinking about it. And, on the other end, about 1.5 percent of births are the result of IVF, the ultimate planned pregnancies.[1]

We have only a little data to help us guess the likely choices of prospective parents who have the option of Easy PGD based on their current use of various ways to predict fetal health and other characteristics. We know that, in 2012, about 5 percent of 158,000 IVF cycles included PGD. We also know that somewhere under 1 percent of all U.S. pregnancies undergo amniocentesis or CVS, but several hundred thousand per year are undergoing NIPT even though it has only been available for three years. And most pregnancies now, perhaps almost all at least in those where the pregnant woman receives any prenatal care, will use ultrasound, often several times, to look for problems in fetal development.[2]

Between these methods lies prenatal screening, mainly for Down syndrome and neural tube defects, discussed in Chapter 5. This kind of screening has been available, with increasing accuracy, for over thirty years. Beginning in 1986, California required every obstetrician to offer prenatal screening to any woman first presenting to the doctor at less than twenty weeks of pregnancy.[3] Across the country in 2011 and 2012, about 72 percent of U.S. pregnancies received prenatal screening.[4]

One could think that women who accept this prenatal screening would be willing to use Easy PGD. But there are at least two barriers to that conclusion.

First, it is not clear that all the women who agree to prenatal screening actually know what they are accepting. The procedure requires them to sign a consent form whose contents are mandated by the state, but pregnant women, like other patients, sign many forms and, on occasion, do not understand or even read them. The procedure just involves, in the first trimester, one more tube of blood to be drawn in one of the many blood draws that are part of prenatal care. In the second trimester, it requires another tube of blood, plus an ultrasound that the woman was almost certainly going to have anyway. Genetic counselors, who often have the task of explaining a high-risk screening test result to a woman, report anecdotally that many of the women do not realize they have had such a test, deny authorizing it, and say that had they known what it was, they would not have had it. We have no data on how many women respond that way; it seems to be a minority but not a tiny minority.

Second, the women who choose prenatal screening are already pregnant. Women need to choose Easy PGD before they become pregnant. And, unlike the prenatal screening test, Easy PGD will involve additional procedures (at least the skin biopsy) and additional doctor visits (both for the biopsy and to discuss the results).

But there is more in play than just procedures. Becoming pregnant is important, and primal, and, I think even a man can confidently say, not always a matter of logic. Naturalness, romance, mystery, fate, and a variety of other emotional and cultural responses can be bound up in the process of becoming pregnant, or even of deciding to become pregnant—or deciding to be open to the possibility of becoming pregnant. Making a decision about tests for a pregnancy that has already started may well be quite different from making a decision definitely to become pregnant.

So, how to add this all up? Presumably almost all parents who use IVF would be willing to use Easy PGD, if only to improve their chances of successful IVF. It also seems likely that most planned pregnancies would choose to use free, safe Easy PGD. Some may do it to avoid the most serious health risks, some may do it to select their baby's sex, and some may do it for other traits. If Easy PGD is free, presumably many, though not all, of those planned pregnancies would begin with Easy PGD. Of the "not accidental but not planned" pregnancies, some prospective parents will see enough advantages to PGD to push them over into the "planned category," but others will not. Some prospective parents will choose not to use the technology as a result of religious, philosophical, or ethical concerns—or less easily described personal and emotional responses. And some accidental pregnancies will still happen. Babies will continue to be conceived in bed, in the back seats of cars, and under "Keep off the Grass" signs, whether by people too young and impulsive to think about alternatives or by those who could have, but didn't, think about them.

Even in a country where it is safe, free, and legal, how many pregnancies will start with Easy PGD? It depends on, among other things, just how deeply these new technologies affect how people live. If, in fact, you could make gametes safely, easily, and cheaply from iPSCs, people might make some very different choices about reproduction. They might even choose as contraception what we currently use for sterilization, knowing that thanks to stem cell–derived gametes, they can always get gametes.[5]

Finally, if it works well, the use of Easy PGD will undoubtedly increase with time. Some people love to be early adopters—of new computers, of new cell phones, of new medical technologies, and of new sources of genetic information. Not everyone stood in line to get the first generation iPhone or iPad, just as not everyone signs up for direct-to-consumer genetic testing. When it comes to medical technologies, most people—and most doctors—prefer to be neither the first nor the last to adopt something new. And some people resist novelty as long as possible.

My own guess is that once Easy PGD has been available clinically for ten years or so, somewhere between 50 and 70 percent of pregnancies in the U.S. will have been started using it. Most of the people who currently plan their pregnancies will want it, about half of those who have desired but unplanned pregnancies will switch to planning in order to use Easy PGD, and a few people with undesired pregnancies will instead have desired Easy PGD pregnancies. If the technology continues to be, and to seem, effective, that percentage should rise over time, but, without coercive measures or major social changes in controlling reproductive capability, it is unlikely ever to reach 100 percent. Some people will always refuse the technology, for reasons of principle or personality, and some pregnancies will always be unplanned accidents. But, in the long run, I could imagine 90 percent of U.S. pregnancies being the result of Easy PGD.

These percentages, of course, will doubtless vary in other countries, with other cultures, health care systems, and economies. Some will use it less; some may well use it more. These differences in the use of Easy PGD have important social implications that will be discussed in Chapter 15.

Alternative Futures

In spite of the logic, to me at least, of the spread of Easy PGD, science, and its translation into clinical practice, could lead to several variations in this future.

The biggest showstopper would be the discovery of some irremediable safety problem with IVF. For all we know, every person conceived through IVF could drop dead at age forty. The oldest such person, Louise Brown, was born in 1978. I will discuss the safety issues around IVF, with or without Easy PGD, in Chapter 13. Major issues seem unlikely to arise, but they cannot be considered impossible. And certainly the discovery of smaller but nontrivial risks is entirely possible. In a world where IVF had serious and unpreventable risks, not only would Easy PGD not be used but IVF would be used much less frequently, if at all.

In another one of those futures, the safe and effective derivation of oocytes from stem cells proves to be impossible. In that case, Easy PGD might still exist for some people, but in a much less popular form, one that is not nearly as "easy." Women who want to use PGD but who want to avoid the arduous process of preparing for egg retrieval may choose extraction of immature oocytes followed by in vitro maturation. Both of these technologies are currently possible and presumably will improve in the future.

But these immature oocytes will still need to be extracted. The extraction will not require the hormone injections, with all their side effects, but will still require an invasive surgical procedure, albeit a relatively minor one (similar to the actual egg retrieval procedure today). Women might be able to choose to do it only once in their (reproductive) lives, extracting and freezing a slice of ovarian tissue, to be thawed and used as necessary. If tissue freezing has unexpected problems, they might require a new ovarian tissue retrieval with every attempt at pregnancy. These procedures would probably be more expensive than egg derivation, if only because they require more skilled labor for the surgery. And they will necessarily be more invasive and uncomfortable for the woman, though not as uncomfortable as current egg harvest methods. I would expect, in this future, for PGD to be used more widely than it is today, but still probably a minority decision and definitely less common than in an Easy PGD world.

Note, though, that in vitro maturation would not allow all of the interesting applications of Easy PGD. It would not be available to provide gametes for people who do not have their own, the most medically, and politically, compelling use of Easy PGD. They would have to continue to use donor gametes, to adopt, or to be childless. This future also rules out making sperm from women or eggs from men (or both from one person). Avoiding the possibility of "uniparents" may cause little concern, but preventing prospective parents who are gay or lesbian from having "children of their own" would be a substantial loss to them.

Another possibility is that deriving eggs proves to be possible, but not through iPSCs. It might still be the case that the process for turning skin cells (or other cells) into embryonic-like pluripotent cells will never be made safe. This seems unlikely, in part because of the great interest throughout medicine (and not just in reproduction) in making iPSCs. But there may turn out to be insurmountable problems. Then what?

If eggs (and sperm) can only be derived from embryonic stem cells, the process may still be used by people who do not make their own gametes. It would be much less widely popular than Easy PGD because it will not give prospective parents a "child of their own" though they may be able to choose among a wide range of different hESC lines to find one that has a genome "closest" to their own.

Stem cells from SCNT cells could provide yet another option. It is not clear how popular this would prove, if effective. It is hard to see that SCNT would ever have any advantages over a safe and effective version of using iPSCs to make gametes—at the very least it requires an additional step of replacing the donor egg's embryo—but, like iPSC and unlike hESC, it would give people gametes made from their own genomes.

Alternatively, germline genomic editing, through CRISPR/Cas9 or its successor, may prove a viable alternative. If coupled with iPSC-derived gametes, it could do everything Easy PGD could do, and more. It could give parents the chance to have children with traits their own DNA does not make possible. I suspect some parents will find that attractive while other parents, except in the unlikely case of serious disease issues, will want a baby made from only "their own" DNA, but in what ratio? I do think the political pathway to acceptance of "designer babies" when the medical need is low will be much more challenging. But I suspect safety questions will pose the greatest obstacle to germline genome editing—whether and how quickly those questions can be resolved remains to be seen.

The other uncertainties are economic and political. It is possible that Easy PGD will not be as inexpensive as I expect. In that case, its use may be limited, especially if political concerns about the procedure prevent it from being subsidized, either by governments or by insurers. The cost of the PGD will almost certainly be quite low, but the cost of making safe iPSCs from individuals is unknown at this time. Even more unknown is the cost of deriving gametes from stem cells. With IVF currently costing $15,000 and more, even if Easy PGD costs $15,000 instead of $1,000, it will have some market, but that market is not likely to be big enough to be revolutionary. Of course, that conclusion depends not only on issues of cost, insurance, and subsidy, but also of economic growth and income distribution. If average family income were to double in the next forty years (an annual average growth rate of about 1.8 percent per year), even $15,000 out of pocket would be affordable for many more people.

Finally, political actions may change the future. I laid out in Chapter 11 some of the reasons I believe that political factors are not likely to limit Easy PGD substantially, at least in the United States, though they may lead to regulations or bans of some extreme uses. But political sentiments are volatile. The United States may end up enacting, on a federal level or, perhaps more likely, in individual states, prohibitions or restrictions on Easy PGD based not on safety and efficacy, but on ethical and moral concerns. The next section of the book explores those ethical and moral concerns, as well as a few pragmatic issues Easy PGD would raise.

You can, and should, judge for yourself whether those concerns will, or should, lead to significant restrictions on Easy PGD. Empowering my readers to make an intelligent and informed decision about Easy PGD is the reason I wrote this book. So let's turn to Part III and look at some of the implications of Easy PGD.

PART III

THE IMPLICATIONS

This part of the book explores the implications of Easy PGD. It mainly looks at problems with Easy PGD. Some of these problems are speculative and will never come to pass, and others might be avoided by wise policy, but some will be real and inevitable. I have grouped the risks into six chapters, looking at safety, family relationships, equality, coercion, naturalness, and a last category, which I call "implementation," but which might be called "other."

Many of these issues have been explored at great length and depth since well before the technologies for even primitive prenatal genetic testing were available. Aldous Huxley's *Brave New World,* published in 1932, H. G. Wells's *The Time Machine,* published in 1895, and Mary Shelley's *Frankenstein, or the Modern Prometheus,* published in 1816, each touched on some of them.[1] More recently, many nonfiction books and articles have been devoted more specifically to the hazards of human genetic selection, motivated not just by earlier fiction but by too real atrocities committed by Germany during the Nazi regime in the name of eugenics.[2]

I will discuss very little of that preexisting literature. Life is short and this book is too long. Much of what I have to say about the implications though is, at best, my variations on the insights of earlier authors. My contribution lies in applying their somewhat abstract concerns to a concrete setting, and a setting that I hope I have convinced you is at the very least quite plausible. American lawyers like to apply principles to concrete cases, rich with facts and context. That approach has strengths and weaknesses, but, for better or for worse, it is mine.

First, though, I must say a few words about benefits. It may seem a bit unfair to have six chapters on risks, disadvantages, and problems, and only the next three paragraphs discussing benefits, but the cost/benefit scales do not weigh chapters, pages, or words. The benefits of Easy PGD are simple to state, if hard to quantify, and many of them are implicit in the earlier chapters of this work. I see three.

The first is a decrease in the amount of human suffering caused by genetic disease. Fewer babies will be born with disabling and fatal genetic conditions. If we set that amount at about 2 percent of current births and assume that easy PGD prevents even only half of those births, that would "prevent" the suffering of about 40,000 children—and their parents and other family members—each year in the United States alone. And it would prevent the suffering of some people later in life from diseases with a strong genetic component. It would prevent that suffering by preventing the children, not by preventing the disease in those children, but each child not born with a genetic disease would be replaced by a child born without that disease.

The second benefit is a (slightly) closer match between the children parents want and the children parents get. This must not be overestimated. I am a parent of two children who regularly surprise me. Easy PGD could have prevented very few of those surprises because few if any of the surprising details of their lives could have been predicted by their DNA. If parents think they want children who will grow up to be tall, with dark eyes, and a greater than average chance of being good musicians, getting such children should be counted as a benefit. One might question whether parents (and their children) really are better off if the surprises from children are somewhat diminished—this will be a major topic of discussion in Chapter 14—but if at least *some* parents think they will be, it is hard to second-guess those perceptions.

The third benefit is more abstract but very real. If Easy PGD can be developed as a safe and effective technology, the freedom to use it—or, more accurately, the freedom from prohibitions on its use—should count as a benefit, at least by people who prefer freedom. Freedom to choose, particularly in health care, is regularly infringed and, I believe, often for good reasons. Freedom to parent is not infringed as often. But infringing freedom is always a cost, even when that cost is far outweighed by the benefits.

One could spin those three benefits out in more detail—a decline in genetic disease, for example, is also a decline in health care expenditures for those genetic diseases—but, basically, those are it. They may seem frail reeds against the storms of the next six chapters, but, at the end, we will weigh those benefits against the costs. I will do that weighing in the concluding chapter and suggest what I think should be done. But each of you must work the scales for yourself and decide what you think should be done. To do that, though, you need to learn about the costs. So let us begin.

13

SAFETY

"Is it safe?" sounds like such a simple question. It is certainly an important one, to me the single most important question Easy PGD raises. But answering that question is anything but simple. We need not only scientific evidence about safety but also a conceptual understanding of what we mean by safety—and, ultimately, a process that can assure us about safety. This chapter will start by discussing what we should mean by the safety of Easy PGD. It will then lay out the questions (and evidence) about safety—the safety of current uses of PGD and IVF, the safety of making babies from gametes derived from iPSCs, and the safety (narrowly and broadly) of genetic selection. It will then suggest a process to assess, regulate, and, one hopes, assure the safety of Easy PGD, before ending by discussing "rogue" clinics.

The Meanings of Safety

Nothing is entirely safe. Risks can never be completely avoided and all safety is necessarily relative—relative to the alternatives, both actions and inactions. Our society's response to risks varies. Sometimes we pay them no attention, sometimes we try to provide information about them, and sometimes we directly regulate them.

Assume that you live in Chicago and want to visit Yosemite National Park in California. To do that, you will have to travel halfway across the country. You are likely either to drive the whole way—2,171 miles if you use Interstate 80—or take one or two flights to get near the park and then drive the rest of the way. Both flying and driving have risks (though the risks of driving will vary more with who you are, and who your driver is, than the risks of flying will). Either way, those risks are relatively low and we do not, as a society, try to persuade you to use one method or the other (although airlines and car companies may try to sway your decision).

Of course, those risks are so low in part because of thousands of social interventions, from highway and automobile safety laws to the Federal Aviation Administration's air traffic control network. We do not regulate which choice you make, but we do regulate the risks of the choices you are offered. Some of that regulation will involve direct rules, such as crash safety requirements for cars or training requirements for pilots. Other regulations will be indirect, such as the potential civil (or even criminal) liability that car manufacturers or airlines would face from disastrous mistakes.

But once society has imposed regulations, directly or indirectly, to improve the safety of a product or process, the question remains whether the product or process is "safe enough." Either driving or flying from Chicago to Yosemite will impose some risks, risks that for most people (though perhaps not for all) will be greater than the risks they face by staying home instead. And that is even before the risks posed by the steep trails, rockslides, black bears, and swift streams of Yosemite. The park averages about twelve to fifteen traumatic deaths each year.[1] A dozen or so deaths from among four million visitors is not a high death rate, but how does it compare with Chicago's? Or your part of Chicago's? Is it "unsafe"? We generally leave that decision up to the individuals concerned, although even at Yosemite particular parts of the park may be off-limits because they are considered too dangerous.

In medicine, our society takes a different path. We regulate medicine in order to improve its safety, directly through requirements like licensure rules for doctors and hospitals as well as indirectly through medical malpractice liability. But we also directly decide whether some kinds of medical interventions, notably drugs, biological products, or medical devices, are safe enough to be made available at all. Those products may only be sold for medical uses if the FDA has found that they are "safe and effective" for a particular use. That safety and efficacy, though, is necessarily judged in the context of the product and the disease it is intended to treat.

Metastatic pancreatic cancer is one of the worst diagnoses a person can receive. The vast majority of its victims are dead within a year of diagnosis and most suffer great pain before dying. A drug that cured half of the people with that disease instantly while quickly and painlessly killing the other half would be a wonder drug, astonishingly safe and effective in that context. If a treatment with the same results were offered for typical teenage acne, it would definitely not be safe or effective. The alternatives, both of treatment and of nontreatment, dominate such an intervention in that disease.

No drug has to be, or can be, perfectly safe or perfectly effective; it just has to be relatively safe and effective in its context. What is "safe enough" to try to avoid a terrible and inevitably fatal early childhood disease like Tay-Sachs may be quite different from what is "safe enough" for choosing a child's eye color.

One must also ask "safe for whom?" Are we worried about the safety to the embryo and (possible) eventual child, the safety to the family unit, the safety to society, the safety to the human species, or some combination of the above? If it is a combination, how do we weigh the safety of the different parties?

And, finally, we must ask, "safe as perceived by whom"? Who gets to make a decision after weighing the risks and the benefits? For many of the decisions that adults make in their day-to-day lives, they get to weigh the risks and make decisions, but for some the FDA has the final power. Even if a mentally competent, well-informed adult suffering from an incurable disease is willing to take risks the FDA finds unacceptable, the FDA has power to block that decision.

For Easy PGD, the FDA, and its equivalent bodies, will make an overall assessment of whether it is sufficiently safe for the fetus and potential child, as well, perhaps, as for the mother, to be allowed for particular specific uses. If the FDA approves Easy PGD, prospective parents and their physicians will have to decide whether Easy PGD is safe enough, in their view, for the parents' planned uses. And each society, through its government, will have to decide whether or when Easy PGD is safe enough for that society, and for the entire species, to be allowed for some ends—and, if so, for what ends? This chapter examines the health risks of Easy PGD to the prospective child and family, with some discussion of species safety. We will come back to issues of society's safety in later chapters.

Health Risks of Easy PGD

The safety risks of PGD, for present purposes, can be put into three groups: the risks of PGD (including those of IVF, its necessary accompaniment), the risks of making babies from gametes derived from stem cells (probably iPSCs), and the long-term risks to the safety of both the child and of the species from widespread genetic selection. Each is discussed below.

The Health Risks of Current PGD

The first baby born after the introduction of PGD arrived in 1990. In 2012, the latest year for which the CDC has published data, about 5 percent of 158,000 IVF cycles employed PGD, or roughly 8,000 cycles. Given the average success rate of IVF, about 3,000 children were therefore born in the United States–a little less than one-tenth of 1 percent of all U.S. births–after having one or a few cells ripped from their three- or five-day-old embryos. One can fairly say that PGD babies, though not common, are not rare. What does this experience tell us about the safety of PGD?

Not as much as we would like. No registry allows all PGD (or all IVF babies) to be followed to see how they do. This is, in some respects, not shocking. PGD is a clinical procedure, not a research trial. Patients use PGD to have babies and do not necessarily want, or want their child, to participate in research. The embryo certainly cannot choose whether or not to be followed after birth as a research subject. It would be wonderful, from a research perspective, if we had required every child, or every fifth child, randomly chosen, who was born after PGD to be followed indefinitely. We did not, and it is not entirely clear that we could, ethically, legally, or practically.

So we do not have great data about health results in children born after PGD. Whatever observational data we do have will be of limited value in forecasting the safety of the PGD part of Easy PGD. The children whose parents let them be observed are not necessarily representative of all PGD children and parents who feel forced to try PGD today are certainly not representative of the people who, in the future, may use Easy PGD.

We can say that, based on casual observation, PGD is not grossly unsafe. It has not been observed to lead to a huge number of miscarriages, stillbirths, or neonatal deaths. Neither has it been observed to lead to the births (or stillbirths) of babies missing large parts of their bodies because of the biopsied cells; when PGD is successful, the other cells make up the difference. As far as can be told from the studies that have been done, the process of PGD does not seem to add any discernible additional risks to the risks of IVF—though it must be noted that the data is so weak that small risks coming directly from PGD would not be discerned.

So how safe is IVF? Pretty safe—at least, safe enough to be used each year by about 160,000 American women. It has some known risks for the women who provide the eggs for IVF, but we can ignore those, as PGD does not use egg harvest. For fetuses the miscarriage and stillbirth rates for IVF are about the same as those for similarly situated women who have conceived naturally. (Note that these may be higher than the average rates; women who use IVF are statistically different from other pregnant women—older and often with a history of unsuccessful pregnancies.)

But children born as a result of IVF are somewhat more likely to suffer from health problems than similar children conceived the old-fashioned way, both immediately after birth and in the long run. The overwhelming cause of those problems is the disproportionate share of IVF babies born as multiples—twins, triplets, and more. Multiple pregnancies lead to more risks, both for the children and for their mothers. Over 30 percent of IVF live births are multiples, as are about 45 percent of IVF babies.[2] (Remember, one multiple "live birth" produces two or more babies so those percentages are not the same.) Children born as multiples have, on average, lower birth weights, longer hospital stays, and more long-term health problems than those born as singletons.

It is important to note, though, that, adjusting for maternal age, the risks to children born as multiples after IVF are not significantly higher than the risks to children born as multiples after natural conception. The problem is one of multiple births, not IVF, but IVF is not innocent of involvement in this risk. IVF as currently practiced encourages parents to accept, and possibly to seek, multiple births. Transferring multiple embryos increases the risk of multiple births, but decreases the parents' chances that an IVF cycle will not produce a baby. For emotional, physical, and possibly financial reasons, prospective parents may well err on the side of multiple births; moreover, some parents may think that getting two babies out of one pregnancy is an efficient use of the time, discomfort, and risk of pregnancy.

What about IVF risks other than from multiple births? Those risks, if they exist, are not large.[3] But there is some disquieting evidence that neither are they zero. Specifically, children born after IVF seem to have higher rates of some diseases, as well as of two rare genetic conditions.

The CDC reported in November 2008 that babies conceived with IVF have slightly higher rates of several birth defects that cause bodily malformations, such as cleft lip and palate, certain gastrointestinal system malformations, and some heart conditions. They looked at about 9,500 babies with birth defects and about 4,800 babies without; about 2.4 percent of the mothers of the babies with birth defects had used IVF, but only about 1.1 percent of the mothers of normal babies had done so.[4] Another study found somewhat increased cardiovascular disease risks in children conceived with IVF; as it points out, given the relatively young ages of all people born from IVF (none over forty), it will be several decades before we see the full effect of these risks.[5] A different kind of study, a "meta-analysis" that combined thirty-eight earlier studies, found that children conceived by IVF might be at some increased risk for childhood illnesses, though the evidence was neither entirely consistent nor strong.[6]

In addition to these structural birth defects, some evidence links IVF to two genetic diseases: Beckwith-Weidemann syndrome and Angelman syndrome.[7] Both are rare; the first strikes about one child in 13,000 and the second about one child in 10,000. But, if one looks at children with these syndromes, they are much more likely than average children to have been conceived through IVF, about ten times likelier in many cases. That this would be the case in "genetic" diseases is puzzling. If the diseases were merely caused by the particular variants in the children's inherited DNA sequences, there would be no reason for higher rates in children born after IVF (unless somehow the fact that one or both parents had that variant made it more likely that they would have to use IVF). Beckwith-Weidemann and Angelman syndromes, however, appear to be diseases caused, at least sometimes, not by sequence variations in the child's genome, but in how the genes involved are expressed.

If these studies are right (note they have only a small number of subjects), this does not mean the risks of IVF are high. If an IVF baby had ten times the risk of these syndromes as a naturally conceived baby, the risks would still be one in 1,300 and one in 1,000, compared to a normal rate of all birth defects or serious genetic diseases of about 3 to 4 percent. Still, those risks deserve some attention, especially to what may be causing them.

On balance, IVF does not have much greater risks than those of natural conception, but it has some, mainly from multiple births, which are more socially than medically caused. Other risks are, at most, small, but deserve attention.

Health Risks of Making Gametes from iPSCs

We at least have data, albeit sketchy, for the risks of IVF. As no pregnancies have even been attempted using gametes artificially derived from stem cells, we have absolutely no data about those—or, at least, no human data.

We do have a small amount of data from Saitou's mouse experiments. In his experiments with stem cell–derived sperm, the mice born from sperm made from mouse embryonic stem cells were, as far as he could tell, normal. He tried the same thing with three different iPSC lines; he could only make sperm from one of the three lines and although those sperm did lead to normally fertile offspring, he notes that "some" of them died prematurely from neck tumors. In his subsequent work making mouse egg cells from both mouse embryonic stem cells and mouse iPSCs, he reports no differences in the health of the offspring. So his work is evidence that embryonic stem cells might be a bit safer and more effective than iPSCs, but the sample is very small.[8]

It makes sense that embryonic stem cells are more likely to be safe and effective because they are more similar to inner cell mass cells than are iPSCs. But in humans we have absolutely no evidence whether either hESCs or iPSCs can make safe and effective gametes and one thing biomedical research has amply demonstrated is that humans are not mice.

One must also note that human gametes from stem cells will have to go through the long and complex development process that Chapter 2 described for naturally arising eggs and sperm. The process of deriving gametes from stem cells must not only be safe in itself but it must also allow a safe process of moving not just from stem cells to eggs and sperm, but moving from primary oogonia and spermatogonia all the way to mature, ready-for-fertilization eggs and sperm.

Long-Term Health Risks from Easy PGD

Two very different issues fall into this category—long-term health risks from Easy PGD to the children born as a result of its use and the long-term risks to the health of both those children and of the entire human species from the genetic selection that Easy PGD will enable.

The undeniable fact is that we do not know what the effects of existing uses of PGD or even, with great confidence, of IVF are later in life. Louise Joy Brown, the first IVF baby, was born in July 1978. As I write this book, it is possible that every IVF child will die on his or her thirty-eighth birthday. We have no reason to think that will be true, but we simply have no examples of healthy thirty-eight-year-olds born as a result of IVF. Of course, the relative good health of many IVF babies now in their thirties, and the hundreds of thousands of now-adult IVF babies, is some evidence that they won't fall off an unforeseen cliff. But what about elderly people born from IVF? Could there be an earlier aging process or heightened risks of the diseases of old age? There certainly could be, and although we have seen no evidence of problems, the possibilities that the manipulations entailed by IVF, especially the effects of "unnatural" circumstances of cell culture on gene expression, could have some long-delayed consequences cannot be ignored.

Similarly, effects on the descendants of IVF babies are plausible. Many IVF children, including Louise Brown, have subsequently had healthy children of their own "the old-fashioned way." But those children have not had time to have their own children. There is no reason to think that Easy PGD, current PGD, or IVF have these kinds of very long term or transgenerational effects—but we cannot be certain that they do not.

The second issue does not involve the processes of Easy PGD, but the process that it enables—the selection of particular genetic traits, and genetic variations, to be passed along to the next generations. We know that many genes play different roles in different tissues, as well as roles in many different gene networks. We are not close to understanding how all of those roles and networks interact.

Therefore, we do not know, with any certainty, what effects selecting for one variation in one particular gene might have when combined with the selection of another variation in a different gene, or in conjunction with a particular environment. It could be, when looking at two particular functions controlled by a different gene, that particular alleles of those genes might be superior for each of those functions, but that their combination would cause serious problems in some third, or fourth, or fifth function. Picking an embryo that has genetic variations that predict both unusual height and unusually good ability at mathematics might lead to a child with a high risk for a nasty disease.

Or it might not. Right now, we are in very speculative territory. Our knowledge of gene interactions is primitive. We know that scores, or hundreds, of genes may be involved in characteristics like height or intelligence, but we have, as yet, effectively no knowledge of the effects of various combinations of these variations on other traits. The results might be disastrous for any particular embryo. Of course, the same is and has been true of the random combinations of alleles that have produced all humans. Somehow it seems different when it is intentional, perhaps because that provides someone, beyond chance, fate, or a deity, to blame.

More broadly, those same risks afflict the whole species. With over seven billion humans, the risks of short-term bad combinations of alleles do not seem great. Even if, in the first decade of fully implemented Easy PGD, 10 percent of parents conceive using the process and 20 percent of those parents select combinations that turn out to be unexpectedly harmful, 2 percent of the world's births would be affected, scarcely enough to threaten the species.

But in the longer term, the consequences could be worrisome. Humanity could end up a monoculture, like the potatoes whose susceptibility to one strain of *Phytophthora infestans* caused the Irish potato famine—and my Greely ancestors' "leave or starve" emigration to the United States. If too many people chose the same variations, variations that may be crucial to humanity's survival in different circumstances might be selected against, particularly if they have some negative effects.

Sickle cell disease makes a useful example. An autosomal recessive disease, it is caused by an allele that makes a variant form of hemoglobin, the protein that carries oxygen in the blood cells. The allele that causes disease makes hemoglobin S instead of the normal hemoglobin A. People who inherit a copy of the hemoglobin S allele from each of their parents are unable to make normal hemoglobin and will have sickle cell disease. The interesting thing about the sickle cell allele is that it actually promotes human survival—in some environments.[9]

People with one hemoglobin S allele and one hemoglobin A allele are normal, or almost normal, except that they resist the most common form of malaria much better than people with two copies of the "normal" A allele. In places where malaria is quite common, as in the sub-Saharan African and Mediterranean locations where sickle cell disease is common, this autosomal recessive "disease" gene promotes survival by a population. It protects half of a carrier couple's children from malaria at the cost of killing one-quarter from sickle cell anemia. But if people decided to weed out hemoglobin S alleles using Easy PGD and conditions changed—if malaria spreads widely because of climate change or mutates in ways we cannot treat—our species could be at risk because of decisions by parents, using Easy PGD, that made sense in the short run—eliminate hemoglobin S—but not necessarily in the long run.[10]

Of course, the chances that Easy PGD might cause such a terrible thing to happen are very unclear. But if parents (or governments) use Easy PGD to homogenize our genomes, the loss of genetic diversity could come back to harm us, as it did the Irish potatoes, and the Irish who depended on them. Those problems might be mitigated by the use of somatic cell genome editing to put the missing alleles back into human cells, but only if that technology is available at the time. Loss of genetic diversity should be, in the very long term at least, a real concern for Easy PGD.

Regulating the Safety Risks of Easy PGD

So we know something about some of the risks of Easy PGD and nothing about some of the other plausible risks. What conclusion should we reach about the safety of this as yet uninvented process? The only sensible answer is "We don't know. Yet." And that answer comes with an obvious corollary—we should find out as much as we can before allowing this technology to move full speed ahead.

To me, that means regulation and, at least in the United States, FDA regulation. The FDA is in the business of determining whether medical interventions are safe and effective enough to be allowed into general use. Easy PGD should not be permitted in the United States without FDA approval. But that simple statement hides a variety of complications.

To start, as discussed in Chapter 10, it is not entirely clear that the FDA currently has the power to regulate Easy PGD or, at least, all uses of Easy PGD. As set out in Chapter 10, I think stem cell–derived human gametes are "drugs" or "biological products" subject to regulation, but that conclusion is not certain—and will not be unless and until a court case upholds it. And even then the FDA could only regulate their introduction into medicine, not how doctors chose to use them.

But let's assume, as I think likely, that the FDA does have authority over at least the first steps of Easy PGD—the creation of iPSCs and the derivation of human gametes from them. How should the FDA then try to ensure that Easy PGD is safe?

The FDA should apply its usual procedures for demanding proof of safety and efficacy for the stem cell–derived gametes, but with special rigor, as the effects of interventions that create gametes might be unpredictable and long delayed. The usual procedures first involve preclinical work, typically done both with human cells in laboratories and with nonhuman animals, followed by Phase 1, 2, and 3 trials in humans. IVF did not go through these stages because the FDA did not, and still does not, view most IVF procedures as falling within its jurisdiction. Easy PGD—or, at least, the use of stem cell–derived gametes—will.

In exercising its approval authority over those gametes, the FDA should first require three different types of preclinical research: nonhuman animal trials, human gamete studies, and human embryo studies.

Typically, before it approves trials in humans the FDA wants to see preclinical work in two different nonhuman species, often one rodent species (mouse or rat) and one nonrodent species. Work on both IVF and, later, on cloning have shown that different species have very different reproductive processes. In particular, many of the "usual suspects" for preclinical animal work—mice, rats, rabbits, pigs, cats, and dogs—have been cloned, giving rise to live animals, but not a single primate has been successfully cloned to the point of creating a pregnancy, let alone a live birth. Given both evidence that primate reproduction is different and the importance of getting this technology right, the FDA should require preclinical work in at least one nonhuman primate species as well as in another species. But working with nonhuman primates is complicated, for many reasons.

One complication is choosing the primate species. Chimpanzees are most closely related to humans and might seem a good choice, but working with chimpanzees (or the other great apes) raises serious ethical, practical, and financial issues. The smaller nonhuman primates, like marmosets or tarsiers, are easier and cheaper to work with, but also more distant from humans. Among the most common research primates are macaques, three species of old world monkey, and macaque reproduction has been closely studied. I suspect the FDA will require preclinical testing of gamete derivation in both a rodent species and in a macaque species, though, given the importance of its decisions, in spite of the difficulties, it should at least consider trials with chimpanzees.

Whatever species are chosen, substantial preliminary work will be necessary to create stem cell–derived gametes in those species and, depending on the species, to learn how to do in vitro fertilization in them. (Mice and macaques would be particularly useful as their reproduction has been deeply studied.)[11] The animal work should include not only an assessment of the efficacy and safety of the process in making offspring in the chosen species, but also in observing the stem cell–derived gametes and resulting embryos in that species.

A second type of research would use human stem cell–derived gametes, but only in the laboratory, not in humans. Human gametes derived from iPSCs (or hESCs or SCNT cells) would first be created and tested to see how similar they are to normal human gametes. To the extent they are different from normal human gametes, their differences should be compared to the ways that the research animal iPSC-derived gametes differed from their naturally produced eggs and sperm. Only if the human gametes were sufficiently similar in their characteristics to either normal human gametes or to nonhuman iPSC-derived gametes that had been shown to be relatively safe and effective when used in preclinical trials should the third kind of research follow—the creation of embryos for research.

Most human embryo research, in the United States and around the world, takes place on "leftover" embryos, initially created as part of IVF in the expectation that they might be used to make babies, but then not chosen for that purpose. The creation of human embryos solely for research is deeply controversial. Some would say that it is inappropriate—or unethical, or immoral, or murder—to create embryos without any intent for them to have a chance to be born. Some countries, and some U.S. states, allow creation of research embryos and others ban it. The federal government refuses to fund it. Yet the creation (and sometimes destruction) of embryos from iPSC-derived gametes could play a hugely important role in ensuring the safety and efficacy of Easy PGD, while transferring them for potential birth would short-circuit the safety testing process.

If the gamete work justifies taking this next step, the FDA should require embryos to be created from iPSC-derived gametes so that their early development could be analyzed. This would examine how normal that early embryonic development is, even though that analysis might involve damaging or destroying the embryo. Without that kind of testing, we could not know how similar those embryos were, or were not, to normal embryos. For the federal government to require such testing, even if it is funded privately, would today be extremely politically controversial and would, I suspect, be blocked by Congress.[12] I can only hope that, in the event, Congress will value the importance of ensuring the safety of babies over the safety of embryos.

These three stages of preclinical research would tell us how safe and effective reproduction using stem cell–derived gametes was in two nonhuman species, including a nonhuman primate. And they would tell us how similar human stem cell–derived gametes, and embryos made from them, are to normal human gametes and embryos or to abnormal but safe nonhuman gametes and embryos. That should allow the sponsor, and the FDA, to determine whether the results are sufficiently promising to move to the required human clinical trials.

If the sponsor decides to move forward, it would file with the FDA a request for an "Investigational New Drug Exemption" (IND). The FDA will block that request unless it is convinced that the intervention is sufficiently safe and plausibly effective enough to justify using humans as guinea pigs. The decision to move an intervention for the first time from nonhuman animals to people should always be frightening. You can never know how humans will react to a drug until you try it in humans. Even after successful preclinical trials, trials in humans sometimes go terribly wrong; in the famous case of TeGenero, six healthy volunteers ended up in intensive care after getting injections of an immune system–modulating monoclonal antibody that had passed all of its preclinical trials easily.[13] When the intervention could have unforeseen effects on babies, "terrifying" might be a better word.

Such trials typically take place in Phase 1, Phase 2, and Phase 3. Typically, Phase 1 is a small trial, involving about ten to fifty people, often healthy volunteers. This phase is intended mainly to prove the gross safety of the procedure and to provide useful information on how humans process the drugs or biological products, but not to show its effectiveness. If the Phase 1 trials go well, the product moves on to Phase 2 trials involving hundreds of subjects. They are intended to show some efficacy, as well as to provide more evidence of safety. And if the Phase 2 results are satisfactory—to the product sponsor and to the FDA—Phase 3 trials follow. These trials enroll thousands of subjects and test, in more depth, safety and efficacy, while also looking at different sizes, numbers, and timings of doses.

But stem cell–derived gametes would not be normal drugs or biological products. For one thing, you cannot test safety apart from efficacy—the safety involved is only partly the safety of the person into whom they are transferred but mainly the safety of the babies that might be eventually created. Their safety can only be ascertained if there is some effectiveness; no babies, no evidence about safety.

Also, you cannot test them on volunteers. The true research subjects are the embryos and ultimately the babies created from those gametes. This raises a serious ethical problem: the research subjects cannot give consent—they do not exist. But without resolving this problem, no research would be possible on this technology. We allow parents to consent (sometimes) to research on their children; we allow (some) research on the effects of products on pregnant women and their fetuses and eventual children, although it does not happen often. We will have to do the same to test the safety of human stem cell–derived gametes. Consent would have to be obtained from the people whose cells were used to derive the gametes, as well as the women who would try to carry the pregnancies.

Although, as noted above, the first trials cannot be classic Phase 1 trials, where efficacy is unimportant, it does make sense to start, as in normal Phase 1 trials, with a small number of subjects. The first efforts to do this in humans should be done in a way that would only affect a handful of babies, not hundreds or thousands.

If the Phase 1 trials do not reveal serious problems, moving to much larger combined Phase 2/3 trials will make sense. It seems likely it will not be necessary to separate out Phase 3 trials—the efficacy issues for which Phase 3 trials are often used, like the timing or dosing of a drug, are likely to be irrelevant here. The size of these subsequent trials will have to be calculated based, at least in part, on a decision about acceptable risks. If you want to be able to have confidence that you can detect risks that happen one time in a thousand, for example, you will need more than a thousand babies born from these trials.

These clinical trials, at each phase, pose some special questions. How long should the trials run—until the babies are born, until they are one year old, five years old, ten years old? Until they are adults who have, themselves, safely reproduced? Until they are dead at a ripe old age?

And who will bear the liability if something goes wrong during the trials? If the trials produce a disproportionate number of disabled or harmed babies, the costs to the parents could be staggering, but if the trial sponsor bears them, the costs could lead to bankruptcy—and the prospects of those costs could lead to refusal to proceed with trials, and the technology. What if the trials only produce, more or less, the number of disabled or harmed babies expected from the same number of normal pregnancies—will those babies (and their parents) be compensated? It is likely that most of the time no one would be able to prove whether the baby was harmed because of the use of iPSC-derived gametes or just by bad luck. And consider the enormous political risks to the FDA—or the presidential administration whose FDA allowed these trials to go forward. If the trials led to many harmed babies, the political fallout could be excruciating, making agencies (and administrations) leery of allowing them. The sponsor of the trial could also expect substantial political and public relations damage if things go wrong.

Finally, and importantly, who will pay for these trials and accept these risks? Clinical trials for drugs and biological products cost hundreds of millions of dollars. They are usually undertaken by companies that, if successful, will have some period of monopoly control over the products—through patents, through various FDA regulatory exclusivity provisions, or both. Will someone have an effective patent, one that cannot be easily circumvented, on a process for making and using iPSC-induced gametes? And, if not, will the government actually be able, politically, to award a monopoly period of regulatory exclusivity over a method of making human babies? Would either kind of monopoly provide an expectation of sufficient returns to make the costs and risks of clinical trials worthwhile?

We have no precedent for this kind of procedure. IVF never went through an FDA-like regulatory process. Although some drugs were essential to IVF as it came to be practiced, notably those that induce hyperstimulation of the ovaries, those drugs had been approved previously, not for use in IVF but as fertility-enhancing drugs. Neither have any of the current refinements of IVF, such as ICSI or in vitro maturation, gone through a regulatory process. The FDA's imposition of its clinical trial requirements on two mitochondrial transfer technologies in the early 2000s was sufficient to stop that technology for at least fifteen years.

But let's assume that, somehow, trials do get funded and approved and the trial data is ultimately submitted to the FDA, as part of a new drug approval application or biological license approval application. How should the FDA evaluate those data to determine whether making babies from iPSC-derived gametes is safe and effective?

Presumably, unless things change in the next twenty to forty years, the FDA will use its longstanding process. Its staff will receive the clinical trial results, analyze the data, ask questions of the sponsor, and eventually prepare information for review by an advisory committee, made up largely of non-FDA experts. The advisory committee will meet to listen to a presentation from the sponsor, hear comments from the public, ask questions of both, and deliberate before voting on a recommendation. The FDA is not required to follow an advisory committee's recommendation, but it usually does.

The harder "how" question is one that is part of every FDA decision but may take on special importance here—how should one define safe and effective? If the procedure appeared as safe as natural conception and childbirth with similar parents, at least within the age ranges for which data were available, presumably it would be safe and effective. What if it appeared as safe as IVF, with similar parents, but not as safe as natural conception? What if it appeared slightly less safe than IVF or if, quite possibly, one could not confidently say whether it was more or less safe than IVF?

Part of the answer in the usual case depends on "safe for what?" A drug for a terrible and fatal disease need not be as safe as a drug for a minor ailment. What category do iPSC-derived gametes fall into? If they are to be used by parents who cannot have "children of their own"—their own genetic children—in any other way, and hence cannot "treat" their infertility in any other way, does that matter? How much additional risk to their children should we let those parents take in their search for their "own" children? If they are to be used by parents for PGD in order to avoid serious genetic diseases, how much risk does that justify? Would it be justified if the net risks to the children were lower with Easy PGD than without it—somewhat lower risks of genetic diseases outweighing somewhat higher risks of the procedure? And what if the parents do not want to use iPSC-derived gametes (and Easy PGD) to avoid disease, but to pick their child's sex, or hair color, or likely personality? How much risk does that justify?

What if at least some of the worst outcomes from using iPSC-derived gametes could be easily determined by testing the embryo before making the decision to transfer it? (Remember, Easy PGD is not just using stem cell–derived gametes but it's also then doing broad genetic testing on the resulting embryos.) If it were known that 90 percent of all embryos created this way would be predictably too damaged for use, but that the 10 percent of "good" embryos could be easily picked out for transfer, should the creation of all those unused embryos be considered a harm to be weighed against the benefits? Or what if problems caused by using iPSC-derived embryos could be detected in utero, in plenty of time for a therapeutic abortion? How much harm does that count as—just the harms, physical and emotional, to the parents or also any harms to the fetus? If predictably a certain percentage of the parents would choose to continue the pregnancy to the birth of a damaged baby, should that count against the safety of the procedure?

The last three paragraphs have far more questions than answers, because those will be real questions, and hard ones. They might be made a little easier to answer if the FDA were able to satisfy three conditions: impose some post-approval requirements to limit uses to those with high benefits, mitigate some of the risks of the procedure, and understand better some of the other risks.

The FDA might limit the procedure's use to the people—infertile parents or otherwise genetically diseased children—who would benefit the most. Or it could mitigate the risks by banning the transfer of more than one embryo in this process, creating safety "savings" from avoiding multiple births to set off against the procedure's higher risks. Or it could require long-term follow-up of the children born from the procedure as a way of getting better information on long-term risk to reduce uncertainty about risks, if not the risks themselves.

Each of these ideas is attractive, but each would require a change in the FDA's power. The FDA does not have the power to regulate how a regulated product is used—either in terms of limiting off-label uses or blocking the transfer of multiple embryos. Neither can it require the children to be followed for years. The FDA can, in some circumstances, encourage or even require companies to do post-approval trials (sometimes called "Phase 4") and it can, sometimes, require that the companies create Risk Evaluation and Management Strategies (REMS), but neither requires patients to become research subjects if they want to use an approved product—especially if those "patients" are babies who never agreed to use (or to be) the approved product.

Rogue Clinics

This chapter has laid out a process for regulating the safety of making babies from iPSC-derived gametes, and thus (for the most part) the safety of Easy PGD. But that process is only useful if it is followed and that is hard to guarantee. There could be truly illegal and surreptitious efforts to perform the procedures without (or before) FDA approval. More likely, though, are activities undertaken outside the reach of the FDA (or an equally rigorous foreign agency). Most stem cell remedies are regulated by the FDA and are not yet approved, but stem cell clinics have nevertheless sprung up throughout the world (including in parts of the United States) to provide patients, including many Americans, with access to unproven therapies.[14] Similarly, various unapproved cancer remedies, often pure quackery, are offered at high prices to desperate people outside the reach of the FDA.

Without truly effective, and universal, international harmonization and enforcement, these kinds of unapproved (because unproven) clinics will spring up. A country could try to regulate "reproductive" tourism, but those actions can raise domestic and international legal problems as well as huge practical problems. Countries have not had much success in keeping their citizens from engaging in such tourism, whether it is for unapproved cancer treatments, domestically illegal abortions, or locally banned uses for IVF.[15] There is no reason to think that regulation of Easy PGD would be any more successful—probably less so as the regulated procedures just require the one-time removal of a skin biopsy and the one-time transfer of an embryo.

There is no perfect safety. There is no perfect way to assure safety. But there are good ways to try to get reasonable assurance of acceptable levels of safety. With the health of potentially huge numbers of children at stake, countries need to work very hard to apply those good methods to determine the safety of Easy PGD.

14

FAMILY RELATIONSHIPS

"Happy families are all alike; every unhappy family is unhappy in its own way." Tolstoy's start to *Anna Karenina* is one of the most famous opening sentences of any novel.[1] But would it be different for families built with Easy PGD?

The day-in, day-out happiness of families is rarely a concern for law and policy except at the most dysfunctional extremes. And yet it is, for many of us, the single most important issue in our lives. How would Easy PGD affect families, and the joy, pride, and comfort—or the pain, embarrassment, and hatred—we take from them? This chapter explores those issues in three different contexts. First, it considers how Easy PGD might affect the feelings and commitments of people in a traditional parent-child family built through that technique. Second, it discusses the argument sometimes advanced that genetic selection violates a child's "right to an open future." And third, it considers some of the more unusual family structures that Easy PGD makes possible and speculates on the happiness, and external consequences, of those families.

The Effects of Easy PGD on Traditional Families

Would the fact that the parents selected their children from among many embryonic choices based on their particular genetic predispositions affect an otherwise traditional family and, if so, how?

We do not, and cannot, know. We do not have any significant experience with the kind of genetic selection involved in Easy PGD. Although PGD has been practiced for nearly twenty-five years, it has not been used for the kind of detailed genetic choices that Easy PGD makes possible.

Instead, it has been employed for four specific purposes: to avoid serious genetic diseases known to run in the family, to select an embryo with immune system genes that make it a possible cord blood donor for a sick relative, to choose boy or girl, and, most commonly, to try to pick embryos that are more likely to survive the implantation and development process and become babies. We have no data about the happiness of the parents and children in families created by PGD and, even if we did, their reasons for using PGD, and their options with PGD, have been so different from the future Easy PGD as to make any comparisons apparently useless.

There is a bit of data about families constructed from IVF. Happily, they seem to be at least as happy as families constructed through natural conception—but that provides little comfort about Easy PGD.[2] IVF children are not selected for their particular genetic traits, other than being genetically related, usually, to one or both parents. Perhaps more importantly, they are the results of what is often a very long and trying process of overcoming infertility. A closer and more useful comparison with Easy PGD might be the happiness of adoptive parents and their families when the parents were able to make very detailed selections of the child to be adopted rather than having less choice, but I do not know of any data on the comparative happiness of those families.

So, in the absence of data, we do what we can—we speculate. But we can speculate, at least, in an organized way, considering first the possible effects on the parents and then the possible effects on the children.

The biggest issue with the parents seems likely to involve a greater investment in a particular future for the baby. Parents often (always?) build hopes and dreams around their pregnancies and eventual children. Will they feel more confident in, or entitled to, certain outcomes if they have selected the child in advance based on its genetic variations?

At some level, that answer must be yes. Parents who selected their embryo so that their baby would be a girl would almost certainly be more upset at a boy than parents who, conceiving the old-fashioned way, knew that the baby's sex was a coin flip. How long-lasting that disappointment would be and what would be its likely consequences for the parent-child relationship can only be guessed at.

But, of course, Easy PGD should be able to get right the question "boy or girl?" as well as questions about Tay-Sachs disease, *PS1* mutations leading to early onset Alzheimer disease, and, eventually, various cosmetic traits. If Easy PGD did not produce the promised results for relatively straightforward genetic traits, a malpractice suit would be a plausible response (although it is unclear when the parents would be entitled to any damages).

The harder questions of disappointed parental expectations fall into two categories: unexpected surprises that were not the subject of Easy PGD predictions and possibly unreasonable surprises about Easy PGD predictions for less penetrant traits, probably particularly behavioral traits. Many bad things that afflict babies are, at least sometimes, not of any known genetic cause. Although some cases of intellectual disability have known genetic sources, such as Down syndrome, fragile X syndrome, or untreated phenylketonuria, most intellectual disability has nongenetic (and much has completely unknown) causes.[3] If parents carefully select their embryo to avoid any known genetic risks, but still end up with a child having severe developmental problems, will their relationship with that child be different because they went through Easy PGD? (If they end up with a fetus with such problems and choose to terminate the pregnancy, this may cause unhappiness but not the kind of unhappiness that is the subject of this chapter, which is about family relationships.) Maybe, but if so, how many parents would be affected, how seriously, and for how long? How badly would it affect their happiness and would their reactions have negative consequences for the child?

Those are cases where Easy PGD, at least if competently presented, has said nothing about the actual result. The counselors presumably would have said, honestly, that genetic selection cannot guarantee a healthy or normal baby, just one that does not have this set of known genetic conditions. But what about common situations where Easy PGD will say something, but something probabilistic?

Consider the parents who pick embryo #12 because it is predicted to grow into a tall, strong male, with an 80 percent likelihood of better-than-average athletic abilities and a 70 percent chance of higher-than-average intelligence. "That's Andrew Luck," the prospective father says. But maybe he turns out to be a decent but not great athlete, or a very good athlete with serious flaws, or a great athlete who would rather be an architect than play football? Will this kind of disappointment sour the relationships between parents and children? The health professionals involved in their Easy PGD presumably will not have promised the parents that they would get Andrew Luck, just that his embryo would have a better than usual chance of that kind of outcome. But how will the parents take the reality? And how will they take it out on the child?

We do not know. We do know, of course, of parents whose hopes, dreams, and aspirations for their children end up in small broken pieces. It happens all the time. Even lucky parents will rarely (probably never, guesses this parent) have children who turn out exactly how they had planned. Genetic selection is unlikely to increase the mismatch between hopes and reality—that has always been great—but it may increase the parents' confidence in their hopes. To some extent that will be justified—their genetically selected daughter will be a daughter—but to some extent it will not be—she will be highly unlikely to be the next Jane Austen.

How much extra suffering will that cause, netted against the extra happiness from those genetic traits that can be successfully predicted? It seems impossible to know. Can the unreasonable expectations of parents be limited through appropriate counseling? Probably, to some extent and in some cases, but predicting the balance is perilous.

But the parent-child relationship is not just about the parent. How will genetically selected children feel? One of the first times I talked about these issues, I was taken aback by the vehement reaction of one student. He was appalled that his parents would know all about his genome before he was even born. This struck him as an enormous invasion of his privacy, like searching his room, he said. Of course, postnatal genetic screening is likely to produce the same degree of screening, but apart from that, will, or should, children feel that Easy PGD has invaded their privacy?

Perhaps more importantly, what will be the consequences for their happiness of knowing that their parents selected them with more than the usual hope of getting the next Andrew Luck or Jane Austen? Would they feel less valued as inherently uncertain individuals—or more valued because they knew that, out of a hundred possible children, their parents chose them? Will they feel more anxious about living up to (or not) their parents' now genetically based expectations? And how important would any of those feelings be? These are tangible and potentially serious uncertainties arising from Easy PGD or any form of genetic selection.[4]

These questions are not entirely unprecedented—children selected by adoptive parents and children born from carefully selected gamete donors might have some similar reactions. The degree of harmful psychological consequences in such children does not seem great. It certainly has not halted either practice. Still, the enormously greater genetic specificity of Easy PGD compared with sperm or egg donor selection leads me to think the risks cannot be usefully assessed before the procedure is used. Perhaps the best that can be done in the beginning is to warn prospective parents of these potential perils.

The Right to an Open Future

Some have argued that we need not try to add and subtract the hard-to-predict psychic costs and benefits of genetic selection for both the parents and the children. Instead, we can rely on a principle, the principle that each child has a "right to an open future." As noted below, legal philosopher Joel Feinberg first published the idea of a right to an open future, and law professor Dena Davis developed it further in the context of genetics. It makes sense that the idea came, at least in part, from law, because it resonates with certain legal ideas.

In the common law world at least, children can "own" property but they cannot (usually) manage it until they come of age. Typically, that property is held and managed for their benefit by their parents, who have a fiduciary obligation to put the child's interest first with respect to the child's property. In effect the parents hold the child's property in trust for the child.

What about the child's interest in its own future? The child can only slowly begin to exercise, or even know, its own wishes for its future. To some extent, the parents are duty bound to ignore some of a child's early desires for its future—"cake today, cake tomorrow, cake forever" would be a very bad idea. But if the child's future is really to be its own to decide, the parents should try to maximize the child's options, not to foreclose them. Genetic selection could strongly foreclose some of those options, substituting the parents' view of a good future for the child's chance to choose—and make—his or her own future.

Feinberg first enunciated the principle of a right to an open future in his 1980 book chapter "The Child's Right to an Open Future."[5] He laid out a categorization of rights that both adults and children share (such as a right against aggression), rights that adults have (such as the right to vote), and rights possessed only by children (or adults who are as dependent as children). The last set of rights, he argued, included both rights to be supported because dependent and "rights in trust"—adult rights that the child is not yet able to exercise but that should be preserved for him or her. He uses the example that a two-month-old child, who cannot yet move on his own, has a right to walk that would be violated by the needless amputation of his or her legs.

Feinberg's seminal chapter actually focused on a U.S. Supreme Court decision, *Wisconsin v. Yoder,* where the court held that Amish parents had the right, based in their religious freedom under the First Amendment, to avoid mandatory education of their children past eighth grade.[6] Feinberg attacks the decision vigorously, setting up a truly uneven contest between a talented philosopher and the late Chief Justice Warren Burger. Feinberg argues that the decision impermissibly forecloses children's right to choose their futures, based solely on the religious desires of their parents. Feinberg recognizes the dilemmas inherent in his argument, fundamentally the inevitability that a child's future will, and must, be narrowed by parents, the state, chance, and other things. Nonetheless, he argues that the child's options should, as far as possible, be kept open.

Feinberg was a prolific and highly respected American philosopher, active from about 1960 to his death in 2004. His work on a child's right to an open future, though occasionally cited in the philosophical literature, is not mentioned in the entry about Feinberg in the *Encyclopedia of Philosophy.*[7]

Dena Davis seems to have brought Feinberg's argument to prominence in a more applied field in her 1997 article "Genetic Dilemmas and the Child's Right to an Open Future."[8] She examined the implications of genetics for three particular issues of parental selection: the choice of deaf parents to have a genetically deaf child, predictive testing of already born children for adult onset diseases, and sex selection. Davis argues that each of these practices is wrong. Deaf parents harm their (deaf-selected) children by limiting their options, predictive testing for adult onset disease takes away a child's right not to know about his or her genetic risks for no good reason, and sex selection allows sex stereotyping to begin even in utero, thus exacerbating restrictions on the child's future.

Davis's article made "right to an open future" a mainstay in discussion of reproductive genetics. It is a powerful and important contribution to the discussion, but, to me, it is unconvincing.

Feinberg, in his initial chapter, recognizes the major problem with a right to an open future—it's impossible. No one can have a fully open future. Everything about one's life guides and influences that future from the very beginning: being born male or female; rich or poor; American,

Chinese, or Nigerian. Sickness, parental health, education, accidents—they all change, irrevocably, our lives. Feinberg says the right should be maintained "as far as possible," but what can that mean? A parent choosing between violin lessons, a soccer team, or video games for a young child's afternoon pastime is closing, to some extent, that future—and not choosing is, of course, itself a choice. If the answer is merely "not too much," how can one say whether Easy PGD crosses that line? And which uses of Easy PGD would cross that line and why: avoiding early onset diseases, avoiding later diseases, choosing cosmetic traits, picking predispositions for behavioral traits, or sex selection?

But in the genetic context there is, I think, a second objection. In terms of the genetic variations we are born with, all of us have that part of our future closed roughly nine months before birth. Whether a baby inherits two Tay-Sachs alleles, a pathogenic *BRCA1* allele, or a Y chromosome is inevitably set. (Note that, as discussed in Chapter 12, truly safe and effective genome editing could reopen that closed future.) The only difference is that with Easy PGD, parents are making the choice, not so much of which trait to give their child, but which child, with those traits, they will seek to bring into existence. Does the intervention of parental choice, as opposed to the equally option-foreclosing effects of chance, make a moral difference? It may make it possible for the parents to act immorally, but the mere fact of parental choice does not seem immoral, any more than the parental choices of violin versus soccer (and maybe, or maybe not, video games).

Davis points to certain examples and says that they are bad but that does not mean that the process of choosing is necessarily bad. Easy PGD deprives a child of the "right" to suffer and die from Tay-Sachs disease or to be condemned to early onset Alzheimer disease (or, more accurately, deprives the world and the parents of such a child in favor of another child with, happily, narrower options). If it is the specific kinds of choices, and their consequences, that matter, then each kind of choice needs to be analyzed separately, and with an eye to ways of mitigating its harms.

I will revisit the issue of choosing an embryo in order to have a child with a disability later in the book, but surely the nature and extent of the disability, as well as the ability to compensate for it, would be relevant in judging parental actions. The disclosure of genetic risks to children from early testing could be avoided by not disclosing the whole genome sequencing results to the children until adulthood. And for sex selection, one would want to weigh the benefits to the parents (and the resulting child) from having the parental preferences come true against the various risks and costs, including whatever costs come from thinking about the future child as a little boy or girl from five days after conception instead of from the first determinative ultrasound.

Unusual Families Made Possible by the Techniques of Easy PGD

Thus far I have stayed within the realm of "ordinary" families with children. These will most often be a family based around one male and one female parent, though increasingly single women (voluntarily or not) have formed or sustained families with children, as have gay or lesbian couples, using either adoption or egg and sperm donation (and sometimes gestational surrogacy). Easy PGD opens, or greatly expands, seven much less conventional—or entirely novel—possibilities. These include gay and lesbian couples who are genetic parents, genetic parents of previously impossible age, (different kinds of) posthumous genetic parents, "incestuous" parents, unsuspecting genetic parents, "commercial" parents, and "uniparents."

Currently, gay or lesbian couples are like straight couples who are infertile because they lack the necessary one egg and one sperm. All such couples can use donated eggs or sperm (plus, in at least the case of a gay male couple, a borrowed or rented womb), but those eggs and sperm will not carry the genetic variations of the partner who lacks a gamete. By using gametes from a sibling, these infertile couples could have children with half of their genetic variations from one parent, one quarter from the other parent, and the other quarter from the future aunt or uncle, but this three-quarter genetic child may not satisfy the desire for a "child of one's own"—especially when iPSC-derived gametes make that possible. It is hard to see how adding some more genetic similarity of children to the parents would make same-sex parenting different in any negative way.[9]

Easy PGD might also lead to genetic parents of ages that were previously impossible. The biological clock may well stop ticking for women, at least in terms of the quality of their eggs. Instead of being forced to rely on the dwindling stock of oocytes laid down before birth, women might be able make brand new—and possibly fertile—eggs from skin cells at age forty-five, fifty, sixty, or even ninety.

We have seen geriatric genetic parents before, albeit only on the male side, so that idea is not entirely new, although elderly mothers do seem to bother some people more than elderly fathers.[10] But Easy PGD could be used to make gametes, and hence children, not only from the sexually postmature but from the sexually immature—children, toddlers, infants, or even, presumably, fetuses. Why someone would want to do that is not at all clear but the technology opens the possibility. (Making gametes from a child undergoing chemotherapy might make sense, but those could be saved for use only when the child becomes an adult.) Note that these conceptions would, at least when they occurred below the legal age for marriage or sex, be without legal consent. What are the legal rights and responsibilities of unconsenting, and possibly unknowing, genetic parents who happen to be children?

Making gametes from people of "unusual" ages also opens the door to making gametes from the dead as long as one had viable cells from them. This does not mean Jane Austen or George Washington; the cells will almost certainly need to have been taken from a person while alive or newly dead and either used quickly or subject to careful medical freezing (not to be tried at home). Either the fresh cells or carefully thawed frozen cells would then be used to make iPSCs, from which gametes would in turn be derived.

We have dealt with posthumous parents before in at least three situations—living pregnancy but posthumous birth, living conception with posthumous pregnancy, and posthumous conception and pregnancy. Thanks to nine months of gestation, babies with dead fathers have always existed. The development of life support equipment has also led, on occasion, to dead mothers—cases when a brain dead pregnant woman's body has been kept functioning long enough to permit the birth of a viable baby. The use of IVF to produce frozen embryos makes it possible for parents to be living at the time of conception but dead at the time a pregnancy is established. And now, with harvested sperm and eggs and gamete freezing (long established with sperm, just coming into clinical use with eggs), conception itself can take place after the death of one or both of the genetic parents. Deriving gametes from iPSCs just means that the creation of the gametes could take place after death—as long as good cell samples existed.

Lawsuits have already arisen from these scenarios. In one famous California case, the use of previously frozen sperm was in question when the children of a dead father objected to his mistress's plans, consistent with the wishes the father expressed before dying, to use them to make new half-siblings.[11] The use of stored sperm from dead men to create new embryos has also been litigated. Similarly, some controversies have arisen over gamete retrieval from the dead. In Israel the question whether to remove sperm from a dead soldier—a not very invasive procedure with the advantage of causing no discomfort to the corpse—for use to make children led to guidelines from the attorney general that focused on the dead man's likely consent.[12] In these contexts, litigation thus far has focused on the consent, or the intent, of the dead man. (As far as I know, the cases have not yet dealt with using eggs from a dead woman—egg freezing has been available for a much shorter time than sperm freezing, and the retrieval issues for eggs, for the dead as for the living, are much more complicated than they are for sperm.)

A fourth problem Easy PGD could raise, albeit in a somewhat novel way, is incest. What if two people otherwise within the scope of a jurisdiction's incest prohibitions decided they wanted to have children through Easy PGD? Would that count as incest? Should it?

Incest laws have regulated who can get married and who can have sex together; they have not specifically regulated who can have children with each other. Historically, that was not an issue. Until recently one could not have children together without having sex together. Incestuous marriages are invalid and often, in themselves, criminal; incestuous sexual intercourse is criminal. In spite of the fact that artificial insemination has been available for about a century, there seems to be no law on whether people who are too closely related can use it to have children together. Widespread availability of Easy PGD (or its predecessor, Easy IVF) could make this possibility more common. Should "incestuous" Easy PGD be banned, even if it involves neither marriage nor sexual intercourse?

Historically, incest prohibitions had a broad scope, including people who were not genetically related, such as the sister of a man's dead wife or a stepparent and unrelated stepchild, based largely on religion.[13] Today the restrictions are primarily about sex (or marriage) between closely genetically related people—parents and children, siblings, uncle/aunts and nieces/nephews. Many American states still forbid first cousin (third-degree relative) marriage.[14] These statutes are justified in part on grounds of likely unequal power relationships leading to abusive sex, as in parent-child incest. Genetics added a medical or biological justification to the bans, which have been thought to help limit unhealthy inbreeding.

The use of Easy PGD to create children would not, in itself, violate prohibitions on sexual intercourse between two people as incest or the concerns about disparate power relationships. The latter might still be a worry, but less so than when sexual gratification is the driving emotion. Of course, that is true today with regard to making children using artificial insemination or IVF. But today's novel methods of making children do not remove the medical concerns about incest; Easy PGD largely could. If embryos created by two close genetic relatives were screened using PGD, the known problems of inbreeding, mainly involving the higher risk of autosomal recessive diseases, would be eliminated. In that case, would there be any good reason to ban or limit Easy PGD using gametes from two people who would otherwise be within the incest prohibition?

A fifth new category of parenthood is the unsuspecting parent. If only one good cell is necessary to make gametes, cells taken from clinically removed tissue (a biopsy or even a blood sample), from discarded tissues, or from the mouth of a discarded water bottle might be used to make gametes. Presumably one would have greater success in making iPSCs and then deriving gametes if one started with a lot of cells, collected with that purpose in mind, but it is certainly not impossible that surreptitiously collected cells could be used to make gametes. What duties or responsibilities would such an unwilling or unsuspecting genetic parent have?

The sixth possibility flows from the fifth—as making eggs and sperm become easier, one can imagine someone selling their iPSC-derived gametes to make babies. Of course, that happens now to some extent, with both egg and sperm "donors." Prospective parents choose the donors they use based, in part, on those donors' characteristics. This "shopping" is more prominent with sperm than with eggs. Big sperm firms have samples from thousands of men, whose characteristics are described in varying detail in the firms' catalogs. Egg brokers typically have much smaller numbers of women available for a new egg harvest procedure at any time. (Greater use of frozen eggs could make egg brokers more like sperm banks, though the initial difficulties of donation remain much greater.)

I wonder why truly commercial gamete "donation" has not caught on yet, at least for men. (One early sperm bank specialized in Nobel Prize winners, but it flopped.)[15] Male celebrities, like famous racing stallions, could sell their sperm for vast amounts. Sperm is not among the "organs" whose sale is prohibited in the United States by the National Organ Transplantation Act—why have we not seen auctions of the sperm of male celebrities? Is some form of modesty at work? That's hard to believe in our celebrity culture. Or perhaps the answer is some concern about the legal obligations of the donor, though choosing the proper state, having the appropriate documents, and using the legally required processes should avoid parental obligations. Maybe even celebrities feel a bit odd about having too many of their genetic children, unknown to them, in the world.

Easy PGD might change at least part of this calculation. It would clearly increase the number of choices for egg donors; its effect on sperm donation is not as clear but could also be significant. With Easy PGD, you can make 100 or more embryos and so choose the best of those 100 combinations of your own egg and donor sperm. The vastly increased availability of eggs leads to a vast increase in the plausible number of created embryos, which in turn can increase the realized value of the sperm donation—as a result of embryo selection, a customer would have a greater chance of having a baby that resembles the celebrity. Should we care?

The last category is definitely novel—the "uniparent." As discussed in Chapter 8, if we can make sperm from a woman's cells and eggs from a man's cells, people could create and combine eggs and sperm both derived from themselves. Wherever the parent had the same alleles for a given gene on each chromosome, the unibaby would be identical to its parent. When the parent had two different alleles, the child would have a 50 percent chance of also having mixed alleles, but 25 percent chances each of having two copies of one allele or two copies of the other. Genetically, a uniparent would certainly be substantially different from other kinds of single parents; apart from likely ego issues, would a uniparent be any different socially from other single parents? Arguably, the uniparent could be engaging in a kind of incest never dreamt of when incest prohibitions and statutes were created, but, given the use of PGD to screen the resulting embryos for the unhealthy effects of this ultimate form of inbreeding, are there good reasons to restrict the practice?

Easy PGD would probably change, at least to some extent, the psychological and social relationships between parents and children within existing, fairly common kinds of parent-child families. It would also make possible, or more possible, unconventional or entirely new kinds of parent-child families. Would these families be happy or unhappy? And how can we possibly assess that without any relevant data? One could ask for clinical trials to watch for family stresses, but how apparent they will be, how quickly they will appear, and to what control group they should be compared would all be difficult problems. The issues around novel families may be the least predictable questions raised by Easy PGD—but not the least important.

15

FAIRNESS, JUSTICE, AND EQUALITY

Fairness, justice, equality—these terms, and the relationships between them, have been contested for at least 2,500 years. No one has yet produced a unified theory of these concepts that has gained general approval. This book will not try. Yet some of the hardest and most important issues raised by Easy PGD are issues of fairness, justice, and equality. This chapter sets out some of those concerns. To set a foundation, it looks first at just how much "better" children conceived through Easy PGD would likely be. It next considers issues raised by differential access to the technology before moving on to problems that could stem from different decisions whether or not to use the technology by parents who had equal access to it. It ends by examining how the ways in which Easy PGD might be used could exacerbate existing problems of justice or equality, with special attention to questions of disability and of sex.

The Likely Importance (or Not) of Easy PGD

Just how much difference would Easy PGD make in those born with its use compared to those born without it? Not as big as the difference between the Morlocks and Eloi, the brutish but effective and the artistic but helpless human successor species from 1895's *The Time Machine* by H. G. Wells.[1] Not as large as in the stratified world of Andrew Niccol's 1997 movie, *Gattaca*.[2] And not the "genobility" ruling over a caste society as envisioned in law professor Max Mehlman's 2003 nonfiction book, *Wondergenes*.[3] Neither the method of Easy PGD nor the underlying genetic variations are powerful enough to make that much difference.

Remember the limitations of Easy PGD. It can only select among the genetic variants that are present in the prospective parents. Even if the parents have the desired variants, if there is a 50 percent chance of getting the "right" combination, selecting one embryo that has the "preferred" set of variations in just ten of the more than 20,000 genes would require making, and testing, over a thousand embryos. Getting the right variations in twenty genes—less than one gene in a thousand—would require over one million embryos. Besides, we do not know any genetic variants that cause, or even contribute significantly, to superpowers, or even substantially enhanced powers.

If we ever do discover powerful enhancing genetic variants, or combinations of variants, germline genomic engineering, through CRISPR/Cas9 or otherwise, might provide a path to creating embryos with those combinations, but, as noted in Chapter 12 above, that technique faces many technical and safety issues. (And may be a topic for a different book.) Easy PGD will not lead, at least for many generations, to supermen and superwomen, but to people who do not have certain genetic diseases, who have a lower risk of getting other diseases, who have cosmetic features preferred by their parents, and who may have some marginal improvements in their behavioral traits. It seems (to me) realistic to say that Easy PGD might produce humans that are about 20 percent healthier and, say, 10 percent both better looking and more talented. What would follow from that?

I think not much. With a 20 percent difference, the two populations would form bell curves with substantial overlap. While most Easy PGD children may be healthier, handsomer, and smarter than their naturally conceived counterparts, many will not be. A society like *Gattaca*'s that made a prenatal gene screen a prerequisite to success would be foolishly wasting talent.

Human populations that are, on average, 10 to 20 percent apart will not be different species. Compared with our ancestors from only a century ago, we are more than 20 percent healthier, certainly as measured by life expectancy, and probably by morbidity—1900's fifty really is today's sixty, or even seventy. We are, on average, probably somewhat better looking, with fewer people carrying now avoidable or reparable deformities. We are even smarter, at least on average and as measured. The so-called Flynn effect has been observed with intelligence tests of different types and in many different societies—with test results that have improved by more than one standard deviation over the course of the last sixty or seventy years. Americans who in 1930 had an average test score of one hundred would score about eighty on today's tests.[4]

I am not sure how fully I believe in this evidence of our superiority even on IQ tests—and I'm certainly not confident we would, in a real setting, outperform or outcompete our ancestors. But, objectively, we are roughly as much "better" than they were as the children of Easy PGD would likely be "better" than the naturally conceived.

But of course, we do not even have to look into the past. Health, tested intelligence and education levels, and less certainly but possibly even appearance correlate positively in our own societies with income. The better-off live longer and healthier, get higher test scores, and receive better (or at least more) education; they also have more dentistry, cosmetic surgery, and other beauty enhancements than the worse-off. In some countries the gap is larger, in some it is smaller, but everywhere I know of, it exists. Those gaps, too, are likely to be at least of roughly the same magnitude as the gap between Easy PGD children and naturally conceived children.

Even if the change is "only" the size of existing income-based differences, adding a gap this size between people based on how they were conceived is certainly not a positive thing for equality. But note that, at least if there were genuinely equal access to Easy PGD across the economic spectrum, this gap would not correlate perfectly with the existing gaps. Some poor parents will choose Easy PGD; some rich parents will not. It is even plausible that the benefits of Easy PGD might be greater at the lower economic levels, where, for example, "healthier genes" might make up for worse health care in situations where, with good health care, the genes would not matter.

So, yes, Easy PGD is likely to cause some changes, and probably increases, in inequality. Depending on who uses Easy PGD, as well as its actual effects on outcomes, those changes might exacerbate, mitigate, or leave unchanged today's patterns of inequality. The effects of Easy PGD on inequality will bear watching and perhaps some kinds of counteracting interventions (special programs for naturally conceived children, for example). But they are not likely to bring about human speciation events, genetic or political apocalypses, or other results outside the range of our current experience.

Access to Easy PGD

Throughout the world, countries treat some kinds of goods and services differently from others. Every country considers some things so important that everyone should have access to them—for instance, clean air and water, education, food, shelter, a postal service, electricity, and, in almost all rich countries, health care. Others are luxuries, available to those who have both the ability and the desire to buy them, but not provided to all. And some things are, sometimes, viewed as so important that everyone should have the same access to them—like organs for transplant (or votes). What falls into which category varies by country, culture, and time, and some things fit uneasily between necessity and luxury, but the basic division is sound. This section looks at both the fairness questions of expanding access and of limiting it.

Universal Access

Given its real, but limited, consequences, how should we consider Easy PGD? I argued in Chapter 9 that health plans, private or public, would find it to their financial advantage to provide broad access to Easy PGD because preventing the births of sick children would likely pay for itself many times over. But my guesses about the costs of Easy PGD might be too low; my guesses about the future costs of treating DNA-linked illnesses in children might be too high; or cultural, political, or religious objections may prevent health care systems from covering Easy PGD.

What then? Presumably, some parents will not be able to afford Easy PGD. Others, who might be "able" to afford to pay for it, will choose not to. As a result, compared with a system of universal access, a higher percentage of children will be born with a "natural" random selection of their parents' genetic variations. Is that fair?

The United States allows innocent children to grow up poor, in vermin-infested homes, eating junk food, with limited health care. On the other hand, the United States makes some benefits universally available to children. Primary and (most of) secondary education are not just free but compulsory. The vaccinations most important for children's health are generally required and are available for free to poor children. Children and pregnant women are the people to whom Medicaid, the program for the poor, has been most generous. Even in the area of genetics, neonatal genetic screening is mandatory but free to those who cannot afford it.

In almost all rich countries it is hard for me to believe Easy PGD, if allowed at all, would not be made universally available. And I think it is highly like that, when Easy PGD becomes common, even the United

States will ensure its universal availability, from some combination of a sense of fairness to the children (and their parents) and its perceived public health benefits.

In part that is because of the "easy." Intervening to change social conditions is hard. It is hard to do in any sustainable way anywhere in the world; it seems particularly difficult in the United States, at least for those not lifted by a "rising tide" of economic growth. But paying for, and encouraging the use of, Easy PGD is simple. It could not only relieve the scruples of some about the justice of differential access but could also hold out at least the promise of broad public benefits, from better health to, possibly, "better" behaviors.

I think the discussion above applies, with some cultural variations, to other rich, or even middle income countries. Countries that can afford to provide Easy PGD to their prospective parents are likely to do so to all interested parents. But not all countries will be able to afford it. It is easy to see universal free access in the United States, France, Qatar, Singapore, and Japan, among many others. It is plausible to see it being provided broadly in Russia, Turkey, Mexico, Brazil, China, and perhaps even India. But it is very hard to envision it becoming widespread anytime soon in Chad, Laos, Haiti, Somalia, Bolivia, or Papua New Guinea. Even if the fairness issues around access to Easy PGD are solvable within rich or middle income countries, gaps are still likely to arise internationally. How much should we care about them is one interesting question; how much will we care about them is another. Neither question should be ignored.

Thus far I have been talking about financial access to Easy PGD, but "free" is not always the same as "accessible." People in remote areas, people with limited access to health care, people with inadequate education, teenagers—all of these people may, in reality, have less effective access to Easy PGD than the well-educated and well-off.

If one thought that, as a general matter, Easy PGD were a good thing, these differences would be another unfortunate health disparity and steps might be taken to expand effective access. One important part of expanding access would clearly be education, so that people would understand the potential benefits of Easy PGD and how it could fit into their lives. Note though that the lines between improving access through education (or, from another perspective, propaganda), "nudging" through encouragement, and, as will be discussed in Chapter 16, frank coercion may well prove difficult to draw.

Restricting Access

But saying "Easy PGD will be free" is not saying "how much" Easy PGD or "what kind" of Easy PGD will be free. Consider all the various constraints on health care coverage for IVF in Europe.[5] One might imagine governments paying only for the directly health-related portions but not, say, the sex-selection or cosmetic trait aspects. (Of course, given that all the genetic information will be the result of whole genome sequencing, the extra costs for providing more information would be tiny.) More importantly, one can see governments (or insurers) paying for only "so much" Easy PGD—which raises the question of what should be done if (rich) parents want to buy "more" Easy PGD.

Remember how quickly the number of needed embryos increases with the number of "preferred" alleles. Some things prospective parents may want, particularly some behavioral traits (intelligence) or cosmetic traits (height), may involve 100 different genes. Parents determined to have not just a "satisfactory" child but an optimal one might be willing to spend extra money—a lot of extra money—to make, and test, many more embryos.[6] (Of course, if genome editing at the embryonic stage, discussed in Chapter 12, becomes safe and effective, those parents are likely to use that rather than the clunky, for their purposes, process of genetic selection.)

If, say, creating and testing 100 embryos were to cost about $10,000 and creating and testing 1,000 embryos were to cost $100,000, would anyone pay the higher amount for some marginal improvement in their expected results? Yes, just as certainly as some people buy Teslas and Lamborghinis instead of Toyotas and Fords, or spend extraordinary money and efforts on placing their children in "the best" college, prep school, primary school, or day care. What will we do about people who want the luxury version—and does it matter just how much genetically "better" we think their children could be as a result? Will they be allowed to buy "deluxe" Easy PGD? The issue of not only providing a floor but possibly mandating a ceiling may be important and will likely be decided differently in different cultures.

The Effects of Not Using Easy PGD

Without the most severe coercion, it will never be the case that every child will be born after Easy PGD. Even if everyone within the relevant area—a nation, a region, or the whole world—had the same effective access, financial and otherwise, to Easy PGD, sometimes it would not be used. At times that would be the result of a conscious decision (or a half-conscious reluctance) based on religion, ideology, culture, or personality. Sometimes it will be just an accident. In the early days of Easy PGD, even in rich countries friendly to the procedure, only a comparatively few children will be born that way; after a decade or two, it could be more than half or even three-quarters of births. But the reality that some and not all children will be born after Easy PGD has implications for fairness that must be addressed. One is possible discrimination against people born with or without using Easy PGD; another is the risk of exacerbating existing social tensions based on what groups choose not to use Easy PGD.

Discrimination

It is possible that even without large actual differences in health, beauty, or abilities, society might consider one group or the other unjustifiably superior. Long experience teaches that human groups do not need any good reasons, let alone solid evidence, to view "the other" as inferior. This is the kind of stigmatization that the movie *Gattaca* portrays, with the social chasm between the prenatally screened "valids" and the naturally conceived "in-valids." It is also possible that the social feelings present, at least in some settings, could make Easy PGD children into the stigmatized class, the "would be" supermen or the "rich kid" picked on in the poor school.

I think this should be a concern, but not a huge one. For one thing, it would be a harsh culture that blamed the children for their parents' choices (or nonchoices). That kind of stigma is certainly not unknown—"bastards" never had any choice in the marital status of their parents at the time of their births—but, one hopes, it is less common today.

More importantly, who falls into which category will not be obvious. Presumably, children will not, in the absence of coercive legislation, carry identifying tattoos on their foreheads or badges with full or broken helices on their clothes. More importantly, as individuals they will not be phenotypically distinguishable from Easy PGD children. In both populations, some will be pretty, some won't be; some will be smart, some less so. Both populations will include people who are intellectually disabled, who get cancer, or who have early heart attacks. Although no Easy PGD children would likely be born with Down syndrome, many would be born with other forms of intellectual disability that could not be tested. Similarly, few Easy PGD children would be born with a known cancer-predisposing mutation in *BRCA1* or *BRCA2*, but about 95 percent of people who get breast or ovarian cancer have no *BRCA1* or *BRCA2* mutations. On average, the Easy PGD population should be "better," but that is only on average, not in individual cases.

There are exceptions. A few diseases, like cystic fibrosis, Huntington disease, Tay-Sachs disease, or sickle cell disease, are only caused by pathogenic genetic variants. People with those diseases will normally be identifiable as having been naturally conceived (or as a result of an Easy PGD mistake), but for most diseases, although the rates will be higher for the naturally conceived, any individual patient might be the result of either Easy PGD or sex.

Finally, for the first few generations of Easy PGD, both populations will grow up in societies where the parents and grandparents of both the naturally conceived and the Easy PGD children were naturally conceived. That should make social stigma trickier. How would the playground dialogue go? "You're a dirty natural." "Yeah, well so's your mother."

Reinforcing Group Biases

All of these factors make me think that social stigma against the naturally conceived is unlikely—unless it is coupled with a strong association between natural conception and some other possibly stigmatized group. What might be the implications if the decision whether or not to use Easy PGD correlated with some other preexisting social division?

Consider, for example, the possible role of religion. What if, in the future, the majority of the public used Easy PGD but use was much smaller in particular religious groups? This is plausible. Easy PGD would be deeply disapproved of by some fundamentalist Protestant churches and would be considered deeply sinful in Catholic doctrine (although Catholic doctrine often does not predict American Catholics' reproductive behavior).[7] A popular belief that, for example, Down syndrome cases were found almost exclusively in particular religious groups might cause social tensions. One can imagine the question. "Why should *we* pay for the disabled children *they* chose to have?"

Or the lines might not be religious, but cultural. In California, it is generally believed that women of Hispanic (mainly Mexican or, increasingly, Central American) ancestry are much less likely to undergo prenatal screening than non-Hispanic women.[8] If the same held true in use of Easy PGD, tensions over public funding for disabled people thought, accurately or not, to have "preventable" genetic diseases could build ethnic tensions.

The division could even be economic. If the poor made much less use of Easy PGD than others—whether through financial constraints, nonfinancial limitations on access, or other causes—antagonism toward the "shiftless, improvident, unthinking" poor could increase (or sympathy decline). And, of course, it is possible that religion, ethnicity, and poverty could all combine to stir up a witch's brew of antagonism. It would still be very likely that most disabled children would not come from any of these minorities. But the possible connections to more "traditional" forms of group hatred make this a substantial worry.

Worrisome Collective Effects of Parents' Easy PGD Choices

The potential consequences of Easy PGD for fairness, justice, and equality are not limited to effects caused by who chooses to use Easy PGD but also include the results of what kinds of traits parents choose, or avoid, when using Easy PGD. Those choices can reinforce differences, loading them with new and stronger power that affects not only the people born in the future with the disfavored traits but also those who have already been born with them.

This applies potentially to any trait. If preferences are fairly evenly mixed, or seen to be so, then no harm is done. If, on the other hand, parents disproportionately and visibly choose offspring with light eyes, those with dark eyes are, to some extent, being marked as inferior. If parents choose taller children, then, to paraphrase Randy Newman's satire, "short people got [less] reason to live."[9] If it turned out that there were some strong genetic associations with sexual orientation, parents choosing only straight (or gay) children could have broad ramifications. I will focus on two areas—disability and sex—where the effects seem to me most plausible and where they could well extend far beyond stigmatization to more concrete harms.

Disability

At a conference I helped organize on noninvasive prenatal testing (NIPT), one of the panelists was a young woman with spinal muscular atrophy (SMA), an autosomal recessive genetic disease that causes systemwide muscle weakening. The disease exists in several different forms and each form has a range of severity, from infant death to nearly normal life. This panelist had spent most of her life in a wheelchair but, like me, had graduated from Stanford. She looked at me during a break and said, "If you had your way, I would not have been born."

That rocked me. I am not sure whether I personally would want to abort a fetus (the relevant "treatment" in the context of NIPT, though not in PGD) with her intermediate type of SMA, but I did, and do, think parents should have that choice. And, if I were able to use Easy PGD, I would be very unlikely to choose for a baby an embryo that predictably would develop SMA. But, of course, I didn't wish that she had never been born. Did I?

Widespread use of Easy PGD (as well as the much nearer widespread use of NIPT) will have major effects on people with genetic disabilities. Fewer of them, perhaps far fewer of them, will be born.

This may be a disadvantage to society all by itself. It is possible that people with serious genetic diseases bring something to society by their existence. One rarely sees that argument pursued for some genetic conditions—Tay-Sachs disease, for example, where life expectancy is four years or less and cognitive ability quickly and progressively disappears. Although knowing and caring for a baby with Tay-Sachs disease might make someone a better person, few would argue that such an effect justifies the baby's short and painful life.

On the other hand, many parents and siblings of people with Down syndrome say that their affected family member is a wonderful, loving person who has taught them much about humanity. These people, it is argued, are not people with disabilities but a different and valuable type of person. Similarly, some of those with profound deafness (a condition that is sometimes genetic and sometimes not) argue that deafness is not a disability but a different way of being human. If humanity is diminished by the disappearance of an indigenous culture, would it not be diminished, on this view, by the disappearance of "deaf culture" through genocide that combines effective treatment of some and genetic prevention of others?

Having fewer affected people will also have more tangible, albeit collateral, effects, both on those born after Easy PGD becomes common and those born before. With fewer patients—and the expectation of still fewer patients in the future—there will be less pressure (and less funding) for research on treatments. There will be less demand for provision of social support to improve the lives of those with the condition. And it is possible there may be somewhat less sympathy and support for people with a condition or disease who have it because of their parents' choice.

And then there is my panelist's challenge. What do we mean by equality if, in fact, we say, "Her life is not worth living"? I could have replied that I did not think her life was not worth living, only that parents should be able to choose to have her but without her disability. But as she would have correctly insisted, she would not be herself except for her disability. The disability shaped her experiences and life in ways that a version of her, genetically identical in every way except for those disease-causing alleles, would never have felt.

Yet surely this proves too much. My father spent his last fifteen years with paraplegia as a result of radiation treatment for an inoperable and otherwise fatal cancer. I think in some ways his disability made him a better person, including launching him on a late life career as a legal aid lawyer. But he would have preferred to be able to use his legs and I surely was not glad that he had been paralyzed. If being disabled is equally as good as being fully abled, why do we take measures to prevent disabilities at all, through, for example, seat belts, airbags, and speed limits, let alone vaccines and medical treatments?

In a world of universal Easy PGD, my panelist may not have existed.[10] Someone else, with a slightly different mix of her parents' genetic variants but without SMA, would have existed. Would that person have been "better" than my panelist? That is, of course, impossible to say, but it is also impossible for me to object to a parental decision to have a child without a serious disability instead of a child with one.

But it is also impossible not to empathize with the panelist's feeling of personal attack. The people not born as a result of Easy PGD will not be able to complain, but the people who have been born with a condition that Easy PGD is widely used to avoid will have an understandable grievance. As may people with nongenetic disabilities, who may feel that in popular culture Easy PGD will lead to devaluing—or even dehumanizing—all those with disabilities. Although people working in bioethics have become reluctant to "play the Nazi card," we should never forget where the Nazi devaluation of "worthless eaters" led.

Sex

The other big issue raised by the ends for which Easy PGD is used is sex selection. The easiest thing to predict by Easy PGD will be "boy or girl"[11] and we know that is something many parents care deeply about. How much—and how—should we worry about parents using Easy PGD for sex selection?

The answer to that question is complicated. Sex selection has been going on for millennia, from efforts at setting the sex (some of which are now somewhat effective, through sperm sorting) to abortion based on ultrasound or prenatal genetic testing to infanticide—and now to PGD. Its meanings and consequences also vary, from its use in cultures that value males much more highly than females to its use by a family that, after four children of one sex, want one of the other sex as a contrast (or a relief).

The issue has been widely discussed, most notably in Mara Hvistendahl's book, *Unnatural Selection*—which I found in equal measures fascinating and frustrating.[12] It has also been the subject of much legislation, regulation, and guidance. Disclosing the sex of a fetus to parents is illegal in many countries, including India, China, and South Korea, although not well enforced. Aborting a fetus based on its sex is illegal in even more places, including several American states. (The constitutionality of those statutes is both unclear and untested.) Using PGD for sex selection violates the regulations of the British Human Fertilisation and Embryology Authority, unless it is done for purposes of avoiding an X-linked disease (a kind of disease that affects males much more frequently than females). The ASRM, the professional organization for fertility medicine in the United States we discussed in Chapter 11, has gone back and forth on sex selection through assisted reproduction techniques.

Opponents have raised at least four different kinds of arguments specifically against prenatal sex selection (as opposed to more general arguments against destroying embryos or interfering with "nature"): reinforcement of sex discrimination and gender stereotyping, the familial effects on the children born as a result of such selection, the diversion of needed medical resources to this unimportant use, and the broad social effects of the possible resulting imbalance. The second issue was discussed in Chapter 14. The third issue is effectively unimportant in the context of Easy PGD that is being used for any reasons in addition to sex selection—the power of Easy PGD to select against various diseases or disease risks presumably would justify that cost. I will discuss the first and fourth issues here.

Would allowing parents to select their children's sex increase sex discrimination? In some cultural contexts, it would demonstrate social acceptance for a preexisting disparity. It would, in any case, reduce if not eliminate cases where parents come to see the good in the undervalued sex as a result of unintentionally having a baby of that sex. And allowing parents to choose their babies' sexes could serve to reinforce their images of appropriate gender roles. By being allowed to choose their own little boy or little girl, whom they are choosing to do "little boy" and "little girl" kinds of things, it could reinforce the idea that there are male and female roles, in childhood and beyond, to which people should adhere.

What about the broad social consequences of a sex imbalance? It is clearly the case that sex ratios at birth have in some countries and regions deviated wildly from the natural roughly 105 boys for every 100 girls. Ratios of up to 120:100 have been documented in parts of East and South Asia, leading to hundreds of millions of "missing girls." But that might not happen in other cultures. The few American surveys do not show a pronounced male bias[13] and, at least anecdotally, it is thought that Americans other than some recent immigrants actually prefer girls.

More subtle biases, though, could have big long-term effects. If, for example, couples wanted their first child to be a boy but the second to be a girl, sometimes they might end up stopping with only one child. The result could be a substantial overall imbalance. If families with two girls will be more eager to have a boy than families with two boys were eager to have a girl, then a similar imbalance could result. It is not clear that, at least in most of the Western world, parental sex selection would lead to a powerful surplus of boys, but it certainly might produce some change from the normal birth ratio.

Even if there is a birth ratio imbalance, is that a problem? The discussions worry about too many men, not too few women, based on the existing examples, mainly in Asia, of such imbalances. Some China hawks argue that tens of millions of sexually frustrated Chinese young men will prove a destabilizing factor, both domestically and internationally. (Oddly or perhaps not, few people if any make the same argument about the equally too numerous Indian or Korean young men.) In fact it is impossible to say whether having too many young men for the available number of young women would or would not be socially disruptive. There are examples of unbalanced societies (particularly on frontiers) that were unusually violent as well as some that were not. Even today, Alaska is 52 percent male whereas the District of Columbia is 53 percent female—the percentage of its population that is female is more than 10 percent higher than in Alaska.

Some people have taken an economic approach and argued that as women become relatively rarer, their value will go up, leading to the production of more of them, and possibly their better treatment. Hvistendahl decries this argument, and yet her own book gives examples of villages that started selecting for female births in order to provide brides for other regions. She contends, probably accurately, that the women are still being exploited, but it could be better to be exploited when valued more highly than when valued less highly. Ending exploitation altogether would surely be better, but perhaps unrealistic, at least quickly, in some traditional cultures. And greater value might even contribute to the ending of exploitation.

Finally, even if male overpopulation is a problem, will it last? Again, Hvistendahl shows examples of a reversion to a lower, though still not natural, sex ratio at birth in some parts of Asia. In India and China, for example, the sex ratios in big cities and among higher-income people are retreating, though they continue to rise as the technology reaches into more rural and lower-income populations. Perhaps the most interesting example is South Korea. At its peak, in 1990, the sex ratio at birth hit 1.165 boys for every girl. By 2006, the ratio was down to 1.074, nearly the natural level. It remains around 1.07. The laws on the books did not change during that time in South Korea: sex selection was illegal at both the start and the end of the period. But something changed.

In summary, it is not clear that parental sex selection through Easy PGD would lead to a substantial sex disparity. It is not clear such a disparity would have negative social effects. And it is not clear that it would last. The issue deserves attention but probably not panic.

This part of the book is devoted to risks and costs, but it may be worthwhile to ask whether something like sex selection has any possible benefits. One benefit is just rooted in liberty—people, including parents, should be able to do what they damn well please unless there is a good reason to stop them. When it comes to raising children, we give parents very broad discretion—why not in selecting the sex of their children?

Another argument is that if parents are happier with children of a particular sex, those children may themselves have happier and better lives, either directly from better treatment or from the psychological effects of being more wanted. These arguments seem more plausible in the context of family balancing—where a couple has had, say, four children of one sex and wants to experience having one of the other. They might, however, exist even for a first child.

The implications of widespread Easy PGD for fairness, justice, and equality are complex and uncertain. Much depends on access to the technology. If access is unfair and unequal, the results are likely to be unfair, unequal, and unjust. Even with equal access, there is the possibility of unequal treatment of those conceived naturally. Some of the more vexing questions involve not just the effects of Easy PGD on those conceived either with or without its use, but on the social meanings and effects of the choices made by those using Easy PGD, particularly around disability and sex. Happily, the next chapter, on coercion, is less complicated, though not less important.

16

COERCION

"Three generations of imbeciles are enough."[1] The great "liberal" Supreme Court justice Oliver Wendell Holmes Jr. penned that memorable line in his majority opinion in *Buck v. Bell,* finding constitutional the forced sterilization of the feebleminded. Today, it is a chilling reminder of the eugenics movement. "Eugenics" is a term that has come to mean many things, all of them negative, and some of them only a synonym for "bad."

But what was the evil of eugenics—the mere fact of genetic selection, the inaccurate early genetic science that guided it, the sterilization that negative eugenics used to avoid "bad" genes, or the compulsion? I believe it was not the fact of genetic selection but the latter three issues—its inaccuracy, the use of sterilization, and, above all, the compulsion with which it was often applied.

The history of the eugenics movement makes the possibility of coercion to use Easy PGD to produce "better babies," or to avoid "worse babies," a major concern. This chapter discusses the issues of coercion. It starts by reviewing the history of eugenics and, in particular, its history in U.S. constitutional law. It then analyzes governmental coercion, both in using Easy PGD and in total or partial bans on using Easy PGD.

Eugenics

Francis Galton, Charles Darwin's first cousin, created the word "eugenics" from the Greek roots *eu* (good) and *gen* (birth)—eugenics was "good birth." The eugenics movement was popular from the late nineteenth century until nearly the middle of the twentieth century.[2] In the United States its supporters were an oddly mixed lot of economic, social, and nativist conservatives, eager to prevent the "excessive" breeding of the poor or immigrant masses, and progressives, excited at a chance to strike medical and social diseases at their perceived source.

Eugenicists divided into roughly two camps—positive and negative. Positive eugenicists, like Galton, wanted to encourage "better" potential parents, from the "better" classes, to breed early and often. Negative eugenicists wanted to work from the other direction, discouraging (usually preventing) births among those they considered criminal, diseased, or defective. Both sides were aided and abetted by the early glimmerings of genetic science, which, after the rediscovery of Mendel's laws at the very beginning of the twentieth century, jumped to the conclusion that all plausibly inherited traits were produced by simple Mendelian genetics.

At the peak of the eugenics movement, thirty-one of the then forty-eight American states had adopted some form of mandatory sterilization laws as eugenic measures. So had many Northern European countries, not only Germany but also such democratic bastions as Denmark, Sweden, and Norway, along with two provinces of Canada, Alberta and British Columbia. In the United Kingdom, the birthplace of eugenics, no compulsory eugenic sterilization laws were ever adopted, although positive eugenics advocates remained active. The Catholic Church and its supporters consistently opposed eugenics, particularly mandatory sterilization laws. In the United States and the rest of the Western world, jurisdictions with large Catholic populations generally did not adopt eugenic laws.[3]

Eugenics Reaches the U.S. Supreme Court: Buck v. Bell

The United States Supreme Court has ruled twice on statutes authorizing coerced eugenic sterilization. The first time, in 1928, was the (in)famous case of *Buck v. Bell.* The case has been widely discussed, in both legal and historical literature; I will only summarize it.[4]

Indiana passed the first U.S. statute authorizing mandatory sterilization in 1907. Other states slowly followed, but it remained unclear whether these statutes were constitutional. Although a few state courts ruled on mandatory sterilization laws in the 1910s and 1920s, their holdings were based on state law and were not consistent. No federal appellate court had ruled on the issue, including, most notably, the United States Supreme Court.

In 1924, Virginia became the twenty-first state to pass a eugenic sterilization law, which applied to "mental defectives" who were inmates of state institutions. The Virginia law followed a model statute drafted by the Eugenics Records Office in Long Island (which eventually became the renowned Cold Spring Harbor Laboratory). The model statute's proponents wanted a test case to demonstrate its constitutionality. Carrie Buck, at the time of trial an eighteen-year-old unmarried mother, confined, with her baby daughter, to Virginia's "State Colony for Epileptics and Feeble Minded," became that case.

Buck v. Bell was an awful case. Carrie Buck's lawyer was not working to help her, but trying instead to help the state get a favorable result in its test case. (He had, in fact, already voted for her sterilization as a member of the institution's board.) Buck herself was almost certainly not intellectually disabled (or feebleminded); her infant daughter was diagnosed as feebleminded in a remarkably cursory manner and the girl's later school records showed normal intelligence. Some argue that Buck probably became pregnant (possibly through rape) by the teenaged son of the upper-class foster family where she had been placed when, at three years old, she was taken from her mother. These unearthed facts are important reminders of how power may abuse any statutory scheme, but I will focus on the Supreme Court's decision on the law, pretending, as the court did, that the lower court had accurately found the "facts."

The Virginia statute allowed the superintendent of a state asylum to order a patient "afflicted with hereditary forms of . . . imbecility" sterilized if he found it was "in the best interests of the patients and of society." The superintendent made those findings about Carrie Buck and, as the statute required, the question went to the institution's board, which agreed. The state then had Buck's court-appointed guardian appeal to the local trial court, which also agreed. On further appeal Virginia's highest court affirmed unanimously in November 1925.

Buck's lawyer then took the case to the United States Supreme Court, repeating the arguments that the decision denied Buck her rights to due process and equal protection under the U.S. Constitution's Fourteenth Amendment. The court heard oral argument in the case on April 22, 1927, and issued its decision ten days later.

Justice Holmes's opinion for the court is short—five paragraphs and just over 1,000 words (shorter than this discussion). Holmes starts by considering whether the law's substance violates due process in the fourth paragraph:

> We have seen more than once that the public welfare may call upon the best citizens for their lives. It would be strange if it could not call upon those who already sap the strength of the State for these lesser sacrifices, often not felt to be such by those concerned, in order to prevent our being swamped with incompetence. It is better for all the world if, instead of waiting to execute degenerate offspring for crime or to let them starve for their imbecility, society can prevent those who are manifestly unfit from continuing their kind. The principle that sustains compulsory vaccination is broad enough to cover cutting the Fallopian tubes. Jacobson v. Massachusetts, 197 U.S. 11. Three generations of imbeciles are enough.

The fifth paragraph disposes of Buck's equal protection argument scornfully, saying it is "the usual last resort of constitutional arguments to point out shortcomings of this sort." No other justice wrote an opinion in this case. Justice Pierce Butler, the only Catholic member of the court at that time, dissented without opinion.

The opinion's tone remains chilling. Although *Buck v. Bell* has never been formally overruled, it is universally rejected and widely reviled, called by one scholar a tragedy with "the highest ratio of injustice per word ever," "a quiet evil," and an example of Hannah Arendt's "banality of evil."[5] The constitutional world it came out of is now dead; due process and (especially) equal protection mean far different things today than in 1927.

But was Holmes wrong?

On the due process question, he argues the state needs to protect itself from having to support incompetents and also to secure the safety of the population from them. He uses two accepted precedents: the draft and mandatory vaccination.

Holmes lived through two drafts, in the Civil War and in World War I. In the Civil War he was wounded three times, once nearly to the death, although as a volunteer, not a conscript. If the state could force its citizens to fight and die for it, why could it not force some of them to endure the lesser cost of sterilization? On the broad public health argument, he cites the Supreme Court's decision in *Jacobson v. Massachusetts,*[6] which upheld the criminal conviction of a man for refusing a smallpox vaccination during an epidemic. If the state can protect itself and its citizens from smallpox by a forcible but safe medical intervention, what is wrong with the forcible but safe sterilization of someone who, according to the widely believed science of the day, would inevitably produce expensive, useless, and even dangerous children?

On the equal protection claim, Holmes is right that governments often make somewhat arbitrary distinctions between people. The state has control over, and is paying room and board for, those confined to mental institutions and not to those outside them. If the state action involved mandatory vaccinations for those confined to such institutions but not for outsiders, no one would think twice.

Logically, Holmes is only wrong if having children is importantly different from conscription or mandatory vaccination. And the slow social agreement that childbearing is importantly different from other human activities is why *Buck* is now "bad law."

Cracks in the Foundation of Buck v. Bell

In American constitutional history, the 1920s are seen as a period of judicial conservatism, yet no Supreme Court era is uniform. A few years before *Buck,* the Court issued two decisions that were particularly innovative in holding statutes unconstitutional for violating the rights of families.[7]

In *Meyer v. Nebraska,* the Supreme Court confronted one of many laws passed around the time of America's participation in World War I, limiting the teaching of German to children.[8] In 1923 a seven-justice majority of the court, in an opinion written by Justice McReynolds (remembered, if at all, as a reactionary), held that Nebraska's law violated the Fourteenth Amendment, because it deprived Nebraskans of their "life, liberty, or property without due process of law." Specifically, the court noted the historically crucial role of family decisions in the upbringing and education of children. (The "liberal" Holmes dissented in the simultaneous companion case, *Bartels v. Iowa.*)[9]

Two years later, in 1925, the court returned to the question of parental control over education in *Pierce v. Society of Sisters.*[10] In *Pierce,* the court held that, in banning primary education by private schools, Oregon had violated parents' Fourteenth Amendment due process rights. The majority opinion, again by Justice McReynolds, stated, "The child is not the mere creature of the State; those who nurture him and direct his destiny have the right, coupled with the high duty, to recognize and prepare him for additional obligations." This time there were no dissents.

The Holmes opinion in *Buck v. Bell* mentions neither *Meyer* nor *Pierce.* No one seems to have argued that rights to determine how to raise children were connected with any right to have children. The court did not cite those cases the next time it faced a eugenics statute, but they were seeds worth noting.

Skinner v. Oklahoma: *Marooning* Buck

Skinner v. Oklahoma,[11] the second U.S. Supreme Court case to consider a forced sterilization law, came fifteen years after *Buck.*[12] In 1935 Oklahoma passed a "Habitual Criminal Sterilization Act," expanding its earlier law for sterilizing the intellectually disabled. The new law applied to anyone who had been previously convicted of two felonies involving "moral turpitude" and was then convicted in Oklahoma of a third such felony and sentenced to imprisonment. The state attorney general could then bring a lawsuit for a sterilization order. Convicts would get jury trials, but the juries were limited to determining whether they were "habitual criminals" as defined by the statute and could be sterilized "without detriment to [their] general health." The statute exempted some crimes: "offenses arising out of the violation of the [alcohol] prohibitory laws, revenue acts, embezzlement, or political offenses."

Skinner was convicted of the requisite third crime, and the attorney general sought and received a sterilization order. Skinner thereupon appealed to the Oklahoma Supreme Court, which upheld the order by a five-to-four vote. On June 1, 1942, the United States Supreme Court unanimously voted to overturn the order, issuing three different opinions.

Justice Douglas wrote the majority opinion. His opinion first mentions several possible objections to the statute, including "the state of scientific authorities respecting inheritability of criminal traits," but did not rely on them, instead striking the law based on the Fourteenth Amendment's Equal Protection Clause, the same "usual last resort of constitutional arguments" derided by Holmes fifteen years earlier. Justice Douglas notes, among other problems, the apparent incongruity that a clerk who takes $20 from an employer's cash register is an embezzler and exempt from sterilization but the stranger who takes the same $20 is a thief and subject to it. Douglas then writes:

> We are dealing here with legislation which involves one of the basic civil rights of man. Marriage and procreation are fundamental to the very existence and survival of the race. The power to sterilize, if exercised, may have subtle, far-reaching and devastating effects. In evil or reckless hands, it can cause races or types which are inimical to the dominant group to wither and

> disappear. There is no redemption for the individual whom the law touches. Any experiment which the State conducts is to his irreparable injury. He is forever deprived of a basic liberty. We mention these matters not to reexamine the scope of the police power of the States. We advert to them merely in emphasis of our view that strict scrutiny of the classification which a State makes in a sterilization law is essential, lest unwittingly, or otherwise, invidious discriminations are made against groups or types of individuals in violation of the constitutional guaranty of just and equal laws.

Chief Justice Stone wrote separately, concurring only in the result but not in the majority opinion. Instead, he thought, in this kind of case, the convict had to be given the right to show that his own condition was not likely to be inherited.

Justice Jackson agreed with both the majority opinion and with Justice Stone. He added, though, a broader concern, while saying that the issue did not have to be decided now because of the statute's other flaws. "There are limits to the extent to which a legislatively represented majority may conduct biological experiments at the expense of the dignity and personality and natural powers of a minority—even those who have been guilty of what the majority define as crimes."

An eight-to-one decision for a eugenics law in *Buck* became a unanimous decision against a eugenics law fifteen years later. The language of the Constitution didn't change—what did?

For one thing, the court changed. Not one of the justices who decided *Buck* was on the *Skinner* court; eight of the nine justices who heard that case had been appointed by President Franklin Roosevelt.

For another, the statute was different. Unlike the Virginia law, the Oklahoma statute did not allow the person whose sterilization was sought to avoid the procedure by showing he or she would not transmit the condition. Plus, the Oklahoma statutes' exceptions opened the possibility for absurd comparisons between crimes that would lead to sterilization and those that would not, with the unstated point that many of the latter were crimes that might be found (sometimes disproportionately) among the middle and upper classes—even legislators.

More importantly, the world changed. The science behind eugenics had become, and been perceived as, weaker, but World War II was the bigger change. The Nazi regime, using California's eugenics statute as its model, had adopted and strongly implemented a forced sterilization scheme. And although the extent of the Holocaust was still unknown, it was quite clear that the Nazis imposed a wide variety of "race"-based measures against Jews. When the majority opinion says, "In evil or reckless hands, [sterilization] can cause races or types which are inimical to the dominant group to wither and disappear," it must have had in mind evil and reckless German hands. In the middle of a war for democracy and against Nazism, this surely was a factor.

Since Skinner

The Supreme Court decided *Buck* in 1928 and *Skinner* in 1943. It has not decided another forced sterilization case since then. *Skinner* distinguished *Buck,* saying the cases were different; it did not overrule *Buck.* Yet no one doubts that, were the court to confront the Virginia statute today, it would strike it down. Because, again, the world has changed.

As the full horrors of the Nazi regime became apparent, "eugenics" turned into a dirty word, tainted as an integral piece of those horrors. Also, genetic science has made it clear that even something as "straightforward" as "feeblemindedness" is genetically complex and rarely easily predictable based on parentage. But more fundamentally, in the last seventy years the Supreme Court has taken the rights of individuals against state intervention more seriously, including in reproduction and families.

Some of this is a general increase in protection of individual rights by the postwar court—in free speech, civil rights, criminal procedure, and elsewhere. But some of it is the blossoming of the seeds planted in *Meyer* and *Pierce.* The court has, in several cases, held that some parental decisions about raising children are free from state interference, including a state law that gave visitation rights, over the parents' objection, to grandparents.[13] The court has also found constitutional significance in the rights of at least some kinds of genetic parents over adoption.[14] But, most saliently, the court has found constitutional protections for decisions about bearing children, striking down state laws banning contraceptives and then, in 1973, abortion.[15] It has reiterated, with interest, Justice Douglas's view in *Skinner* that "We are dealing here with legislation which involves one of the basic civil rights of man."

Outside the United States

Eugenics was neither an American invention nor an American monopoly. But outside the United States, as inside it, enthusiasm for eugenic sterilization did not survive the end of World War II. The laws were not immediately repealed, but fell into disuse. The world had changed.

Or had it? The first president of independent Singapore, the late, redoubtable Lee Kwan Yew, long supported "positive" eugenics. Fearing that Singapore's more intelligent citizens were not sufficiently fecund, his government supported programs to encourage college-educated women to marry and have children, from free dating services to "Love Boat" cruises.[16]

Elsewhere in Asia, negative eugenics staged at least a short-term comeback. In 1998, the People's Congress of the People's Republic of China adopted what the official translation called "The Chinese Eugenics Law."[17] This wide-ranging statute governed many aspects of prenatal and infant care. One section provided that marriage licenses would only be issued based on a physician's certification that neither member of the couple was suffering from a serious infectious disease, a serious mental illness, or a serious genetic disease. Another required all pregnant women to take any prenatal tests ordered by their physicians and to follow the physicians' advice based on the test results.

The Chinese statute became controversial before the planned meeting of the world's eighteenth International Congress of Genetics, held in Beijing in 1998,[18] with calls to boycott the meeting in protest. Although the first reaction to world pressure was simply to change the official translation of the statute's title (to "The Chinese Children's and Maternal Health Law"), the act was never implemented and was quietly repealed in 2005.

Since then, news of eugenics has been dominated by belated apologies from some jurisdictions that had engaged in eugenic sterilizations, such as Sweden, the Netherlands, North Carolina, and California, and, occasionally, compensation for the few living victims of the statutes. Negative eugenics looks dead—at least for now.

Or is it? Compulsory sterilization under official court orders still takes place. Today, though, it is undertaken on the basis that it is in the best interests of the person sterilized—usually a heavily intellectually impaired woman—who is thus spared the discomforts and risks of pregnancy. It is "voluntary," usually sought by the parents or other guardians of the incompetent patients, in the expectation that patients would want the procedure if only they were competent to understand the issues. The motive is to improve their lives, not the human gene pool—and yet the result is still the sterilization of the "feebleminded" without their consent.

Government Coercion in the Context of Easy PGD

Governments could pass coercive laws of at least four different types about Easy PGD:

- Forcing prospective parents to go through the Easy PGD process, but not forcing them to make any particular use of the information disclosed
- Forcing particular choices of embryos to transfer after Easy PGD
- Forbidding prospective parents who go through Easy PGD from making particular choices based on the Easy PGD results
- Forbidding Easy PGD

Each is a form of coercion, though each form raises different issues.

Requiring the Use of Easy PGD

Let's start with a somewhat implausible law that requires all babies to be conceived through (free and generally available) Easy PGD and some or all of the resulting genetic information to be shared with the prospective parents, but does not require the parents to use the information. We could think of this as an encouragement of better genetic health, part of the vogue for government by "nudge."[19] It is like a nutritional labeling law. Nothing requires you to read the nutritional information posted in Wendy's restaurant about "Dave's Hot 'N Juicy ¾ lb Triple" (1070 calories).[20] But some people will read the information and a few will instead choose to eat kale.

Would this be constitutional? The two Supreme Court eugenics cases are not directly helpful in providing a legal answer. Easy PGD just involves the manner of conception, not the ability to conceive (or procreate). Still, the answer today is probably no. Such a mandatory Easy PGD would most likely run afoul of a combination of the parental rights cases, old and new, and the reproductive rights cases. If the state cannot constitutionally prevent people from using contraception or having abortions, how could it prevent them from choosing to conceive children the old-fashioned way?

If posed as a public health question, like mandatory vaccination or required neonatal genetic testing, perhaps it would pass muster, but we know that most children conceived without Easy PGD—a category that includes all readers of this book—are relatively healthy. It is hard to see the Supreme Court holding that relatively small health advantages would justify that much intrusion in what it has held is a fundamental and intimate part of the human experience.

And there is another issue—how would such a law be enforced? Would women who had become pregnant the old-fashioned way be forced to terminate those pregnancies? It would take a real revolution in American constitutional law (as well as American public opinion) for that to happen. Perhaps women who became pregnant without using Easy PGD would be fined or even imprisoned afterward, but the children would still be born.

At least, they would in the United States. But the world is big. Are there countries that might pass laws requiring the use of Easy PGD and forcibly terminating offending pregnancies? Perhaps. We have already seen an interest in eugenics in several Asian cultures. Combine that with cultures—not just in Asia but also in, for example, the former Soviet Union—with little concern about abortion. Now add a history of intrusive government control over individuals' decisions and little tradition of enforced individual human rights. After all, the one child policy in China, now well into its fourth decade, has faced enforcement problems, but was only abandoned—or, more accurately, modified to allow two, but not more, children—in late October 2015, effective January 1, 2016. In that context, laws requiring pregnancies to start through Easy PGD might be plausible.

Such laws would be very intrusive, banning the time-honored and (usually) easy method of conception in return for health benefits. Those benefits, though, would be undercut by the halfway nature of the legislation—forcing parents to get information by using Easy PGD but not forcing them to take advantage of that information. They seem, to me, impossible to justify.

But if a country did adopt such a law, how should other countries and citizens of other countries react? Should they view it as a violation of international and fundamental human rights, possibly meriting sanctions? Disapprove of it but not condemn it? View it neutrally as the choice of the country involved? Agree with it, but not necessarily think all countries should do the same? These are *not* rhetorical questions—and from them we move on to another, more plausible kind of coercion.

Forcing Particular Choices Using Information from Easy PGD

Now imagine a different coercive legislative regime concerning Easy PGD, one in which information from Easy PGD is used to force particular choices—by parents, doctors, or the state—of which embryos to transfer. Those could be choices requiring selection of a particular trait—a higher-than-average chance of high intelligence—or choices banning the transfer of embryos with a particular trait—embryos that would have Down syndrome.

These are different sides of the same coin. Being forced to transfer an embryo with a projected higher-than-average intelligence is effectively the same as being forced to reject embryos with projected average or below-average intelligence. Being forced to reject an embryo that would become a child with Down syndrome is the same as being forced to transfer an embryo that could not produce such a child.

So here's the question—should the state (or anyone except the prospective parents) be able to make genomic choices for the prospective parents about "their" child? Does it matter whether we are thinking of state decisions to avoid a genetic disease (negative selection) or to get "enhanced" babies (positive selection)? The distinction between "treatment" (or prevention) of a disease, on the one hand, and "enhancement," on the other, has long been deeply problematic. Still, one might feel differently about the state selecting against recognized diseases or pathologies (though recognized by whom, and how?) or the state selecting from its idea of the top part of a normal range.

Let's assume, for now, that this distinction makes sense and start with the disease side. If the state says you cannot select an embryo that would become a child with a serious disease, is that different from the state saying you must take reasonable steps (like vaccination, decent food, protection from the elements) to protect children already born? If the state refuses to allow a couple to choose to try to have a baby with a defect that prevents it from being able to breathe, how is that different from the state prohibiting parents from suffocating their born children? Either way the state is forbidding parents from choosing dead babies. To me, this is the most attractive kind of Easy PGD ban, though I wonder how often parents would make such a choice.

But what about lesser disabilities? What if prospective parents wanted to have deaf children, perhaps because they were themselves deaf? Parents who intentionally deafened a hearing child would lose custody of the child and probably go to prison. Does it matter if the action is taken before the child is born—or even implanted? Why can't these parents seek a "child of their own," a child like themselves?

What about less than fully penetrant conditions? Let's say parents want to choose an embryo that has a higher-than-average chance of having some bad disease, but also has a very high chance of having traits the parents like, traits that others would agree are generally beneficial. Consider an embryo with a 10 percent chance—ten times higher than normal—of becoming a person with schizophrenia, but no chance of getting Alzheimer disease and a ten times higher than normal chance of exceptional musical ability, something the musician parents want? Should the state be able to forbid parents from taking that set of risks on behalf of their (very) unborn child?

What about state action requiring parents to choose embryos that rate "higher" within the normal, nonpathological human range? Consider, for example, intelligence, math ability, height, or longevity, among other possible "nondisease" traits.[21] Here the state is not acting to prevent a case of serious disease; instead, it is trying to produce a population that is "better" in ways that do not directly involve disease. Is that different?

Of course, the state often intervenes in nonmedical ways to "improve" the population, sometimes quite coercively. Education is a good thing, the state thinks, so it demands that children go to school or, at least, receive equivalent home schooling. That eats into the liberty of actual children (and their parents) to avoid state approved education in the interest of producing a "better" population. Isn't intervening in the choice among embryos less intrusive? Of course, an educated population is not only better for the state but for those children who get the education, but having a higher projected intelligence is also, presumably, not just better for the state but better for the future child. Does it matter whether the state's decision is in the "best interests of the (future) child"?

Much here depends on one's view on the roles of parents and the state in children's development. The answers are clearer if you agree with Justice McReynolds (in general, something modern constitutional lawyers and scholars rarely do) that "The child is not the mere creature of the State; those who nurture him and direct his destiny have the right, coupled with the high duty, to recognize and prepare him for additional obligations." But we don't fully believe, or act, on that statement, at least not where the parents are guilty of abuse or neglect.

Most Americans would likely see a spectrum. State intervention is least justified for enhancing purposes and most justified for preventing parental decisions that would clearly severely harm the resulting child. That may help us assess the acceptability of the state requiring the transfer of embryos with higher projected intelligence versus the possible acceptability of the state forbidding the transfer of embryos that would have Tay-Sachs disease. But what does it tell us about a future America, or about other countries in the present or future, let alone about cases that fall between probabilistic higher IQ scores and Tay-Sachs disease—cases like deafness or achondroplasia or trading off a higher schizophrenia risk against a higher projected musical ability? And does the view of the majority of people within a culture really make a difference or are these questions of fundamental human rights, rights that exist whatever a majority believes?

Now note one other consequence of this discussion. If a state allows prospective parents to choose to transfer an embryo that will develop achondroplasia but not an embryo that has three copies of chromosome 13 and thus would have Edwards syndrome, isn't the government saying that a life with achondroplasia is worth living but a life with Edwards syndrome is not? Choose any other conditions you like—a government *diktat* forbidding the transfer of embryos with one genetic condition but allowing it for embryos with another genetic condition is, in effect, a government statement about the value of those two lives. Is that something the government should say—that future person A may be born but future person B may not be? (And what would it say to anyone already alive with the disfavored condition?)

If it tried, how would the government decide which diseases or traits fall into which category? One tempting line, discussed above, distinguishes diseases from nondiseases. Neither that nor any other line would always be clear. This is not a slam-dunk argument against drawing lines, but it is a caution. The law often deals with fuzzy categories—just not usually very easily.

I am tempted to focus on the harms to be avoided by restricting parental choice. How serious are the harms to a child if parents can choose to allow that child to be born with a serious genetic condition—assuming one can weigh the balance to an embryo of being born with that condition compared with never being born? How serious are the consequences for society if parental choices lead to serious imbalances in, say, the ratio of men to women? On the other side, how serious are the harms of worsening the lot of people with a particular genetic condition by allowing parents to choose to avoid having a child with that condition? But if you go down this path, you are still left with the question of whether those consequences justify interfering with parents' ability to choose the child (or the embryonic genome) they want to raise?

These are genuinely hard questions. Think back on the examples in the last several paragraphs and ask yourself what choices can the state legitimately require or forbid. And why?

Three other aspects of Easy PGD regulation could make these restrictions more or less severe. If the state did not require that any parents wishing to conceive must use Easy PGD, as posited in the earlier section of this chapter, it would allow a loophole for parents who did not want their choices constrained—the option of making babies the blind, old-fashioned way. To what extent is the state's exercise of coercive power alleviated by the option of still having babies, but without recourse to Easy PGD (or presumably to "hard" PGD or to various methods of prenatal genetic diagnosis followed by abortion)? If the right that should be protected is a right to have children whose genetic traits you want, that's small consolation. If it is only the right to have children without state interference in their genetic makeup, it is stronger.

Second, the state might not require prospective parents actually to transfer any of the embryos they had created. If transfer were required and, presumably, subsequent abortion banned (both of unlikely constitutionality in the United States), people could be forced to have a child knowing it has genetic traits they did not choose. An option not to transfer would at least allow parents to choose not to have any child, at least from that round of Easy PGD. If the improper infringement on liberty is being forced to have a child with unwanted genes, this distinction makes some difference. If it is being prevented from having one's choices respected, it does not.

Third, the state may provide an exception for religious and possibly philosophical objections. In the United States, a religious exception might (or might not) be required under the First Amendment's guarantee of the right to "free exercise" of religion. That constitutional provision is notoriously difficult to apply. Even apart from constitutional issues, though, it may be a good idea as a matter of policy to allow objectors to opt out, for religious or philosophical reasons, as the United States did with military conscription in its most recent versions of the draft (and as most states allow for vaccination and for neonatal genetic testing).

Forbidding Prospective Parents from Using Information from Easy PGD to Make Particular Choices

The previous section dealt with regulation where the government told prospective parents they could not choose embryos with particular traits or that any embryos they chose must have particular traits. In those cases the governments were regulating the parents' choices, not their motivations. Now consider a system where the government would not ban the choice of an embryo with or without a particular trait, but would ban the choice (or the nonchoice) of that embryo because of that trait. There are three possible justifications for such a rule.

First, particularly where the trait is generally perceived as a negative one, regulation would be an effort to prevent choices that were, somehow, unfair discrimination against people with that characteristic. Second, where the trait is not stigmatized but there is reason to think parents would disproportionately choose to have, or avoid, one version of the trait (such as baby girls), such a scheme could reflect a desire to avoid imbalances caused by parental choices. And third, it might embody a sense that the particular trait in question, such as eye color, was just "too trivial" to justify such a life/nonlife decision.

Each of these rationales is plausible, though each may require parents to raise children they would rather not have had. Of course, that happens today, but in a context where parents almost never have the chance to choose traits. Allowing unhappy parents to abandon children because of undesired traits seems quite different from allowing parents to decide some embryos will be transferred and others will not. Allowing parents to abort fetuses that would otherwise be born probably falls, for most people, between those two extremes.

Assume for now that a law forbidding parents from choosing embryos on certain grounds is justifiable. It would still run into two big problems.

The first is enforceability. Barring stunning advances in mindreading through neuroscience, how could the government know why parents made a particular choice? Perhaps they did not choose the embryo at high risk for intellectual disability in order not to have a child with such a disability but for some other projected trait—or as the result of coin flips. Regulating motives is difficult.

Still, several American states currently ban women from deciding to have an abortion because of the sex or the race of the fetus. That raises the same issues of unknowable motives, but it may not matter to those legislatures. They may be satisfied with making a normative statement—we condemn this behavior even if we cannot stop it.[22]

A second path makes enforcement easier, at least in theory. One might eliminate the possibility of improper parental motive by limiting the prospective parents' knowledge of the embryo's genetic traits. This would move the site of the regulation from the parent to the clinics or physicians. They could be forbidden to disclose to the prospective parents the embryo's Down syndrome status, or sex, or eye color. Clinics and physicians, after all, have a great deal to lose if they were found to have violated such laws—not only possible criminal sanctions but business or professional licenses.

Health care professionals are likely to resent, lobby against, and possibly resist those limitations on their communication with patients. But at least in one case where the health services were paid for by the government, the federal government passed, and the Supreme Court upheld, a similar restriction. In *Rust v. Sullivan,* the Court upheld a statute banning doctors and clinics being paid by Medicaid from discussing abortion options with their pregnant patients.[23] If the government paid for Easy PGD, this decision seems on point.

Whether such a restraint on speech when the government is not paying for the service could be upheld under the First Amendment is murky, as, indeed, is the whole area of commercial speech. False or misleading commercial or professional speech can be regulated; whether truthful and accurate speech about an embryo's genetic predispositions could be regulated might well depend on how strong a court considers the government's interest in restricting the speech. This is an active area of litigation, discussed further in Chapter 18, with no clear rules yet emerging.

If those regulations were held constitutional in the United States, or permissible in other countries, one would still worry about circumvention. Some laboratories or doctors might just give the prospective parents the raw genetic results, allowing them to discern or discover the meaning of the results from someone else—like the "very hard to regulate" Internet. Or prospective parents might engage in reproductive tourism, doing Easy PGD in states or countries without such limiting laws. Or they might bribe the doctors.

But even apart from the difficulties of regulating information rather than motivation, this regulatory strategy shares a deep problem with the strategy of forbidding, or requiring, certain outcomes. It puts the state in the position of saying that certain traits are serious enough to allow parents to base a decision on them—or to receive information about them—but other traits are not.

Banning Easy PGD Entirely

One could avoid many of these questions. If PGD (easy or hard) were outlawed, only outlaws would use PGD. This would just return the situation for more than 99 percent of the world's population to today's status quo—and for everyone to the situation before 1990. But that cannot be a sufficient answer. Censoring the Internet cannot be justified just because, within living memory, there was no Internet. Once Easy PGD exists, prohibiting its use requires justifications. What might those be?

Safety would be a clear justification. If research reveals that Easy PGD has very serious safety risks for those children born as a result of it, governments could well be justified in protecting these as yet unconceived children, who have no choice in the matter, from a grave risk of harm. Would the family risks discussed in Chapter 14 justify a complete ban on Easy PGD? What about the fairness concerns from Chapter 15?

And remember that the question is not whether you want to use Easy PGD in making your own children or even whether your neighbors should want to use Easy PGD. Instead, it is whether the government is justified in preventing anyone from using Easy PGD. Barring a very serious safety problem, I suspect the only plausible justifications for that kind of ban need to be sought in deeper moral or principled arguments against Easy PGD—which are discussed in the next chapter.

To say something is "coercive" is not to say it is wrong. The FDA's prohibition of the use of unapproved drugs is coercive, but is an excellent idea. Any government action, or inaction, about Easy PGD can be viewed as "coercive"—the absence of a prohibitory statute "coerces" people to live in a world where other people are using Easy PGD. And governments are by no means the only possible sources of coercion. We can imagine employers, health payors, parents-in-law, or even spouses trying to force, or at least influence, a choice involving PGD. But coercion is always a cost to liberty, something that needs to be justified, whatever direction it pushes.

17

JUST PLAIN WRONG

I'm not sure how to characterize the last major category of objections. These objections are from people who think the whole idea—Easy PGD, PGD, or any genetic selection—is "just plain wrong." Sometimes they say, in a triumphant voice, "well, that's just eugenics," as if a label is a trump card, or even an argument. For some it is against God's will; for others, it is unnatural. Still others will ask, "Are we wise enough to make these decisions about our children and the future of our species," or will just say this is "playing God." Often people will use bits and pieces of all of these concerns in trying to find a way to express their deep and visceral nervousness with the idea of humans making genetic selections. And still others have argued that this unease is a repugnance that is itself a normative argument against genetic selection. This chapter will survey that unease in four of the categories mentioned above: God's will, unnaturalness, ignorance (and its complement, humility), and repugnance. I conclude (you may not) that there is not a convincing argument among them—at least, not convincing enough to justify banning someone else's use of Easy PGD even if fully sufficient to justify someone's personal decision not to use it.

Earlier chapters have talked about issues of safety, family relationships, fairness, and coercion. I believe each holds some serious problems that a system of Easy PGD should worry about. I do not think any of the concerns in this chapter raises serious problems, except for the reality that their popular appeal could have political consequences that lead to harmful or ineffective regulation of Easy PGD.

Let me be clear about two things I am not saying. First, if a person chooses not to use Easy PGD based on these arguments—or any arguments—I have no objection. Individuals are certainly entitled, particularly with respect to reproduction, to base their decisions on religious beliefs, philosophical perspectives, or emotional reactions I do not share. My question is not about individual decisions but about arguments that would justify imposing one's decision on others who disagree.

Second, even if a state or a country were to decide to ban Easy PGD entirely based on the concerns discussed below, I would not condemn that decision. I would not applaud it, as I do not think it would be justified on the basis of the consequences of Easy PGD. But countries need not always act rationally. They may and, to the extent they reflect their different cultures, should have different laws. When my fellow Californians voted for an initiative to ban the sale or consumption of horsemeat from humanely slaughtered animals, I chalked it up to the democratic process reflecting popular tastes.[1] If other countries allow their people to eat meat from dogs and cats, once I choke down my rising gorge, I will not condemn them. The French like eating snails, Americans like watching baseball, and the Japanese like some very odd television shows. That's fine.

I am not a complete relativist. Local customs and mores do at some point have to take into account some universal human rights. Genocide is wrong, murder is wrong, slavery is wrong, racism is wrong, napalming babies is wrong. But do not press me on where that list comes from or what other categories fall into it.[2]

One could argue that parental choice over reproduction is one such inalienable human right or, conversely, that the freedom from other people's use of Easy PGD is such a right. I am more drawn to the former than to the latter, but neither crosses whatever indistinct line separates my preferences from universal rights. So if Vatican City, or Germany, or South Dakota decide to ban Easy PGD for what I consider illogical and unjustified reasons, so be it—though, as discussed in the next chapter, they may have trouble enforcing those laws. But they should be recognized as the products of local cultural preferences and tastes, not as the expression of self-evident truths.

Against God's Will

Some religious arguments against Easy PGD will overlap with or even totally adopt the concerns discussed below, particularly about naturalness and humility. Some religions, such as Buddhism, do not even have a "God" whose will is to be obeyed. The same is true of some cultural traditions, such as Confucianism, which may or may not be considered "religion" but that fill many of its roles. This section will lay out some religious objections that are specific to Easy PGD or its component parts and then discuss how much weight such objections should be given in setting regulations and policies.

I will say from the start that I do not find the religious arguments attractive. I am, largely, a consequentialist—whether an action is or is not ethical is mainly (but not entirely) determined for me by its likely consequences. Most religions are "deontological"—whether an action is ethical or not depends on whether it fulfills or violates a duty of some sort. But I have always found it hard to argue with, or even discuss, these duties. If a fundamentalist says that something is wrong because of his interpretation of a (usually ambiguous) verse in the Bible, beyond perhaps pointing out some unexpected (il)logical consequenes of his interpretation, I cannot engage. He may be right and some of these proscriptive ethical positions might be divinely required, but, barring divine revelation to me, I have little to say about them.

Even religious arguments based on arguments rather than on dicta are often hard to dispute. For example, the basic Catholic position on reproduction, as I understand it, is that the unitive and procreative significance of marital acts (sex) may not permissibly be separated. Why not? Because by "safeguarding both these essential aspects, the unitive and the procreative, the conjugal act preserves in its fullness the sense of true mutual love and its orientation toward man's exalted vocation to parenthood."[3] How can one debate that—with empirical evidence about "the sense of true mutual love"? I suspect it is not a position where empirical findings would be considered relevant.

More surprisingly, I would note that finding "religious" positions about these issues turns out to be surprisingly difficult. Few religions have central authorities that pronounce on these questions. No one person or group "speaks for" Islam, Judaism, Hinduism, or most branches of Christianity.

Perhaps this should not be startling—the Torah, the Bible, the Quran, and other holy books from the past really could say nothing directly about technologies to be developed thousands of years later. Still, the absence of positions on various aspects of assisted reproduction or genetic selection by contemporary American denominations is disconcerting. There are no centrally expounded or generally adopted statements about the morality of these technologies from the American Methodists, the Southern Baptists, the Missouri Synod Lutherans, the Church of Jesus Christ of the Latter Day Saints, the Union for Reform Judaism, or most other such groups.

Happily for people working in bioethics, the Catholic Church is not only hierarchically organized, but has presented Catholic views on various issues involved in human reproduction. It is an additional blessing that those arguments tend not to rely on citations to scriptural passages, but to stress logical arguments. This chapter does not focus on analyses from the Vatican because of a bias for or against Catholicism, but because those analyses exist, are intellectually serious, and are easy to find.

Catholic dogma, as it currently exists, would clearly oppose Easy PGD—and does oppose most of its predecessor technologies—for at least two reasons: interference with the natural act of reproduction through sexual intercourse within a marriage and the creation and destruction of "unused" embryos.

The Catholic position on the crucial role of marital sexual intercourse was set out by Pope Paul VI in the encyclical *In Humanae Vitae* in 1966.

> This particular doctrine, often expounded by the magisterium of the Church, is based on the inseparable connection, established by God, which man on his own initiative may not break, between the unitive significance and the procreative significance which are both inherent to the marriage act.
>
> The reason is that the fundamental nature of the marriage act, while uniting husband and wife in the closest intimacy, also renders them capable of generating new life—and this as a result of laws written into the actual nature of man and of woman. And if each of these essential qualities, the unitive and the procreative, is preserved, the use of marriage fully retains its sense of true mutual love and its ordination to the supreme responsibility of parenthood to which man is called.[4]

This doctrine has been applied to forbid not just contraception but also almost any form of assisted reproduction, from the simplest "turkey baster" versions of artificial insemination to IVF, even when done within a heterosexual marriage. Only limited kinds of interference with natural reproduction are allowed. Moral methods of dealing with infertility include "natural" family planning, which involves timing sexual intercourse around the most fertile periods of a woman's menstrual cycle; surgeries to remove blockages in either the male or female reproductive systems that prevent conception after sexual intercourse; fertility drugs that encourage egg ripening; and "lower tubal ovum transfer," where a woman's egg is moved from the ovaries to below a blockage in the fallopian tubes.[5]

The Catholic ban on destruction of embryos poses an independent barrier to Easy PGD. As mentioned in Chapter 11, Catholic dogma is often misunderstood to hold that embryos, whether the result of sexual intercourse (marital or otherwise) or through extracorporeal fertilization, are ensouled at the moment of conception. In fact, the Church does not now have a definitive conclusion about when ensoulment takes place. (For hundreds of years the most common Catholic position was that ensoulment took place several weeks after conception.) The official position is more complicated. Given the uncertainty of the timing of ensoulment, the Church believes that caution requires acting as though an early, extracorporeal embryo has a soul. Some also argue that even if it does not have a soul, cutting short a process that would, or at least might, lead to its receiving a soul—and hence immortality—is also profoundly wrong.[6]

IVF can never be acceptable under current Catholic doctrine. It would, however, be possible to use IVF without violating the specific Catholic ban on destroying embryos. One would merely have to transfer for possible pregnancy every embryo that is created. In Italy, as noted in Chapter 11, that was a legal requirement. That approach, though, cannot work with genetic selection. The "selection" part of genetic selection inevitably means that some embryos will be selected and used to try to make babies, and others will not. This is particularly true if Easy PGD involves the creation of not a handful of embryos for each couple but of scores, or of hundreds.

A few other religious traditions have established guidelines for reproductive technologies, although without the absolute nature of the Catholic positions. For example, in the United States the Rabbinical Assembly's Committee on Jewish Law and Standards has provided recommendations to Conservative Jews about the propriety of some kinds of assisted reproduction. It has approved artificial insemination, with sperm from the husband or a donor, as well as IVF, egg donation, and surrogacy.[7] Up to three embryos may be transferred at any time for possible implantation and embryos may be frozen. Guidance issued in 1996 even approved the use of preimplantation genetic diagnosis, at least for serious genetic diseases.

Islam provides another example of religious views on some reproductive technologies. There is no central authority for Islam overall, or even for most of its branches. But some leaders or institutions are considered particularly persuasive.[8] In 1980 Sheik Gad El-Haq Ali Gad El-Haq of Al-Azhar University in Cairo issued a fatwa, a judgment or interpretation of Islamic law, on reproductive technologies. The fatwa approved of artificial insemination and IVF when the sperm and egg came from a husband and wife. It banned third-party gamete donation and, in particular, sperm banks, but allowed embryo freezing although only for the use of the original husband and wife (embryo donation was forbidden). This fatwa did not have authoritative force, but was widely adopted by other groups in other countries that issued similar fatwas. In the late 1990s, however, a split opened in the Islamic response when Ayatollah Khameini of Iran ruled that third-party gamete donation was permissible.

These three different examples of religious positions provide an important lesson—the religious positions, even in those few cases where authoritative or broadly accepted positions can be found, are different. There is no one religious position on these questions. And, indeed, even within religions with a clear position on the questions, members of the religion may not agree with that position. Catholics in the United States widely ignore the official Vatican position on contraception and infertile Catholic couples often use officially sinful IVF.

Individuals making decisions on what they want to do surely may use religious positions on reproductive technologies. So may health care providers in deciding what kinds of care they want to provide (at least to some extent). Whether and to what extent they should be used in creating legislation, regulations, or public policy is more complicated.

It depends in part on the country. In the United States, adoption of a religious position by the government because it is a religious position would probably violate the First Amendment, as being the "establishment" of a religion. For example, if Massachusetts, the state with the highest percentage of Catholic residents, were to ban Easy PGD on the stated ground that it violated Catholic doctrine and was therefore sinful, that ban would almost certainly violate the Establishment Clause. If, on the other hand, it were to ban Easy PGD because it interfered inappropriately with the role of natural marital relations in human reproduction, that facially secular reason would probably not violate the federal Constitution, even though it was shared by a religion.

But not every country has legally enshrined the idea of religious freedom and included in that freedom the equivalent of an Establishment

Clause. Indeed, even today many countries have official or established religions, which are given different levels of preference and respect. Examples range from Argentina, England, and Denmark to Saudi Arabia, Iran, and Cambodia. Unless the existence of countries with established religions violates a universal human right or some more specifically adopted international or regional standard (such as a European Union provision), it seems legitimate for such states to legislate based on the positions of their official religions.

Unnatural

The religious argument against Easy PGD has a secular equivalent: whether or not it is against God's will, it is against the will of Nature, or of natural selection, or of Charles Darwin. This seems to me another religious argument, but with Nature or natural selection in the role of the deity. But, if so, at least it is a deity whose commands, or desires, are to be deduced from a source other than the deity's revelations, directly or through a prophet or holy book. Its laws are to be read in the "book of nature" that is itself.

George Bernard Shaw encapsulated my thoughts on this point in *Caesar and Cleopatra* when he has Caesar say, "Pardon him, Theodotus: he is a barbarian, and thinks that the customs of his tribe and island are the laws of nature."[9] (Caesar was referring to a British legionnaire's revulsion at the fact that Cleopatra, following the Egyptian custom for royalty, was married to her brother.) Confusion of the familiar and the parochial with the laws of nature is all too common.

Philosophers have written for centuries on this impulse as the "naturalistic fallacy," the argument that just because something "is," it therefore "ought" to be.[10] The problem in the fallacy seems to me to be so self-evident that I will not discuss it in any further, but will instead add two arguments related to it: that if "nature" is taken more broadly than just its human aspect, Easy PGD is not particularly unnatural and that, if confined to humans, the argument is almost always deeply hypocritical.

Looking beyond humans, as set out in Chapter 2, "nature" (at least as we know it on earth) provides many forms of reproduction. Most life on this planet, including almost all of the trillions of our bodies' cells, reproduces clonally. Then there are species, microscopic or not, that switch back and forth between reproducing clonally and reproducing sexually. Among sexually reproducing species, some are made up almost entirely of asexual individuals with only a tiny fraction of the individuals able to be either genetic mother or father (think of bees and ants). In others, including some vertebrates, an individual changes from one sex to another during its lifetime, sometimes several times. Some species, like malaria and toxoplasmosis, can only reproduce with the help of other species, by going through different parts of their life cycles—parts in which they look like entirely distinct species—while inside different host species.

Some species make thousands or even millions of offspring and pay no attention to them once launched, the so-called "k" strategy. Other species, notably but by no means only humans, make only a few offspring and lavish parental attention on them (the "p" strategy). Some species regularly practice infanticide or fratricide, "sorocide," or "siblicide." Some species avoid incest, whereas others reproduce regularly, or only, through what we would consider incest.

In some species a few dominant males hoard all reproductive possibilities by acquiring "harems." In others, the males are tightly controlled by the females, existing only to mate and die. In still others, the females mate indiscriminately with whoever happens to be in the right place at the right time. (Often, no two of the many kittens in one litter will share the same father.) Many species broadcast their gametes far and wide, to be fertilized by any complementary gamete they happen to meet. Sometimes, though, mating takes place only (or nearly only) between couples that have pair bonded for life. Looking at our own species, we see examples of equivalents of many of these various nonhuman reproductive behaviors. Maybe one approach to human reproduction is most common, but in the real world, the variation across, and within, human cultures is vast.

We presume too quickly that what we are familiar with—what we see in humans and, to a certain extent, in our pets and livestock (a significant number of which, in the United States, are now conceived through artificial insemination and a few by cloning)—is "Nature's law." But the biosphere is stranger than we dream. It is true that no species in nature has reproduced through Easy PGD (although tens of thousands of humans have been born as a result of "hard" PGD). But given the vast range of reproductive behaviors, something that combines a couple's egg and sperm but then intentionally selects which embryos should become offspring is nowhere near the far edge of "naturalness."

Ah, but some will say, these nonhuman examples are beside the point. It is the nature of humans, the crown of creation (or the tiptop of the evolutionary tree), which should concern us. And occasional pathological deviations from the natural human path indulged in by some perverted individuals (or cultures) should be dismissed. The existence of humans with one arm does not mean that humans are not naturally "supposed" to have two arms. At this point I generally lose all patience and shout (or want to shout) "Hypocrite!" This is particularly true if the argument comes from someone who has flown in an airplane to present a talk on PowerPoint, with a sound system amplifying his voice, in a building with artificial lighting and ventilation.

Unless one takes a very broad view of "nature," civilization is not natural. Our earliest hunter-gathering *Homo sapiens* ancestors did not wear clothes, raise crops or livestock, or go to school to learn reading, writing, and arithmetic, let alone fly, drive, or use computers. Neither did our "natural" ancestors practice much in the way of effective medicine—antibiotics, modern childbirth, and the thousands of other ways in which we prevent human suffering and protect human life, all of which are deeply unnatural.

If you accept the convenient unnatural parts of our civilization but not Easy PGD, you have to provide a line that distinguishes between the "unnatural" you accept and the "unnatural" you abhor—and provide a convincing justification for the line. I might accept the argument from, say, the Old Order Amish, who believe that God has commanded them to follow the ways of Palestinian farmers of the early Christian era, even though actually their way of life is more like that of German peasants of the early Reformation. (Interestingly, the Old Order Amish are quite excited about modern medicine, and particularly genetic medicine, as they have a high incidence of several genetic diseases for which they would like prevention, treatment, or cure.) It is hard for me to accept it from anyone who broadly accepts modern civilization.

Again, the response might be "well, that's fine for all those other parts of human life, but reproduction is so central, so crucial, that any change in it is wrong." Again, I might consider taking that argument seriously from someone who, like the Catholic hierarchy in the Vatican, abjures modern contraception, artificial insemination, fertility treatments, or IVF (albeit, in the Catholic hierarchy's case, they abjure these for other, noncelibate people). But for others, I would like to see that person reject prenatal care, prenatal ultrasound, cesarean section, aseptic delivery, and even forceps. None of those is natural and millions of woman and babies have lived longer, healthier, and better lives because of their adoption.

In sum, I do not find the "naturalness" argument convincing on almost any level, when leveled by almost any person. If people want to avoid Easy PGD for their own use, based on some incoherent feeling of naturalness, I would not deny them that right. But I would demand a much more coherent and convincing argument before endorsing their power to restrict anyone else's use of the technology.

Humility in the Face of Ignorance

Another argument in this same vein may seem a bit more reasonable, though I find it ultimately unconvincing. "We do not know enough to make such a big change and, in light of our ignorance, we should act humbly and not proceed."

To the extent that this is an argument for caution in making important changes, I embrace it. Thinking about important changes before flying into them makes sense. Even more sensible is monitoring the outcomes and adjusting policies as a result. I view this book as part of that process, particularly as it calls, loudly, for close FDA regulation over the safety of Easy PGD for the children it is used to produce. Weigh the risks and benefits, then watch as they unfold and modify as needed.

But that is not the same as saying, ignore the present assessment of risks and benefits, and the uncertainties attached to them, by refusing to go forward at all. This is the strong form of the so-called "precautionary" principle. That argument has been subject to much discussion and criticism, which I will not repeat in detail here. It is clearly true that we cannot predict all future consequences of our actions with certainty. It is also clearly true that we cannot let that uncertainty keep us from acting.

I will make only two points. First, this is not what we do in the rest of our lives. We invent new medicines, create new gadgets, put out new hardware and software, and generally change the world every day without stopping to say, "Here's a speculative way in which *Angry Birds* could lead to untold human suffering and so it should not be distributed." The main difference between the rest of the world and Easy PGD is that, as a medical procedure, it would be the subject of regulation and of the need for proof of safety and efficacy before it was adopted. That is more of a precautionary principle than exists almost anywhere else in our law and culture.

Second, strong application of the precautionary principle is impossible. It may be that doing nothing will itself lead to terrible results. A sufficiently clever (and motivated) person could, no doubt, construct a scenario where slowing the increase in carbon dioxide levels led to disaster. Prove, beyond a reasonable doubt, that sustainable energy or energy conservation will not be disastrous. I do not think one can.

Caution and humility are important, as far as reasonable. It might well make sense for some couples, or some countries, to "wait and see" how Easy PGD works for other people or societies before trying it themselves. But caution and humility, like anything else, turn bad when carried to unreasonable extremes. They are not powerful arguments against Easy PGD.

Repugnance

I have saved this argument for last, probably because I so strongly despise it. I often disagree with Professor Leon Kass, the bioethicist who for six years chaired President George W. Bush's Commission on Bioethics. I do not despise him, but I do despise this argument he made, most notably in a much-discussed 1997 article in *The New Republic* entitled "The Wisdom of Repugnance."[11] Kass's argument, motivated by the announcement of the birth of Dolly the cloned sheep and aimed at human reproductive cloning, is basically that feelings of repugnance should be a warning to us of the "wrongness" of particular actions. Just as our repugnance leads us to avoid unhealthy contact with vomit, puss, or crap, we should listen to our untutored, not thought out, visceral reactions against new technologies. He stated:

> Revulsion is not an argument; and some of yesterday's repugnances are today calmly accepted—though, one must add, not always for the better. In crucial cases, however, repugnance is the emotional expression of deep wisdom, beyond reason's power fully to articulate it. Can anyone really give an argument fully adequate to the horror which is father-daughter incest (even with consent), or having sex with animals, or mutilating a corpse, or eating human flesh, or even just (just!) raping or murdering another human being?[12]

But in what cases is repugnance an expression of such a deep wisdom, "beyond reason's power fully to articulate it"? It is not clear—except that the repugnance at human cloning belongs in the same category with father-daughter incest and bestiality. And, as a result, it is not clear from his article whether Kass views (or wants us to view) repugnance as an independent moral ground for rejecting things, or just as a warning to look more closely. Either way, the argument, in the context of Easy PGD (and other issues) has serious problems.

First, we know that feelings (or expressed feelings) of repugnance vary both over time within a culture and between cultures (as well as within individuals in a culture). Some things we now hold of great importance, even to the extent of constitutional protection, would have been viewed by our not-too-distant ancestors as repugnant. Interracial marriage and unmarried sex, whether heterosexual, gay, or lesbian, was a crime in many American states during my lifetime; the very idea was repugnant. Gay marriage, recently unthinkable, is now a reality in much of the West and a constitutional right in the United States. Equal, or even close to equal, rights for women was repugnant to many before the twentieth century (and in some cultures remains repugnant today). Even the idea of equal rights for non-Christians, such as Kass, was repugnant to many Europeans well into the twentieth century.

But the Kass argument has a further problem with Easy PGD. I do not believe most people, let alone an overwhelming majority of people, at least in most contemporary Western cultures, would view Easy PGD with repugnance. Hard PGD has been used for over twenty-five years, legally in most countries. Even if we were to accept (as I do not) Kass's argument against human reproductive cloning, Easy PGD is a much, much less radical transformation of human reproduction, adding just one step to IVF. Kass opposed IVF in the 1970s.[13] He has not publicly done so in recent years and if he did, he would find few allies outside the Vatican. Even if we were to take "The Wisdom of Repugnance" seriously, in a strong form, it is, at least in most Western societies, unlikely to be a telling argument against Easy PGD.

As you can tell, I am not moved by the arguments in this chapter. I recognize that reasonable people, including you, might be so moved. If these arguments, even after my discussion of them, affect your decision about whether to use Easy PGD, or whether to recommend that family, friends, or others use Easy PGD, I think that is and should be your right. I only ask that you consider whether those arguments are strong enough to impose decisions on those who do not share your views.

18

ENFORCEMENT AND IMPLEMENTATION

Any utopia might be possible—except for the mundane problems of enforcement and implementation. Those problems must always be considered seriously before making any major change and Easy PGD is no exception. This chapter discusses the many problems of enforcement as well as the single most serious implementation problem of Easy PGD.

Enforcement

Laws are easier to pass than to enforce. This is not a definitive argument against laws. It is clearly wrong to require perfect enforcement of any law. I am willing to say that laws against murder are a good idea, even though, in spite of their ubiquity, murders still happen. Even laws that cannot be enforced at all may not be useless. They may express a society's moral sentiments in a way that influences people even if they cannot be enforced. They also may, of course, provide good publicity and campaign contributions for lawmakers seeking reelection. At the same time, passing unenforceable laws can make the law (or the government) look incompetent, unfair, or heartless.

It is prudent, therefore, to consider the likely success, costs, and consequences of efforts to enforce new laws. Laws banning, or limiting, Easy PGD may well prove particularly difficult to enforce for at least three reasons: constitutional limits, practical enforcement limits, and reproductive tourism. In addition, some ancillary prohibitions will prove necessary. All are discussed below.

Possible Constitutional Limits on Enforcement

This section will talk only about constitutions in the United States, for the good reason that they are the only ones I come close to understanding. In other countries both the substantive arguments and the procedures for resolving them will be different. Even when the inquiry is narrowed to whether American constitutions might make laws banning or limiting Easy PGD, the bottom line is several definite "maybes."

Federal or State Law?

Notice that I talked about "constitutions." Each state has a constitution, which, to the extent it is consistent with federal law, controls its laws. A state law that violates the federal constitution, a federal statute, or even a regulation by a federal agency cannot stand, as a result of the Constitution's "Supremacy Clause" (reinforced by the Civil War).[1] But because it violates no federal constitutional provision, statute, or regulation, Oregon can ban customer self-service at gasoline stations, New Hampshire can allow helmetless motorcycle riding, and California can ban the sale of horsemeat for human consumption. Nevertheless, each of those statutes can still face challenges as violating its state's constitution.

This distinction leads to an important question—would laws restricting Easy PGD be federal laws or would they have to be state laws? In American constitutional theory, the federal government has only the powers delegated to it in the Constitution. States have the underlying broad general powers that belong to any sovereign government, sometimes called "the police powers," except to the extent the Constitution has given them exclusively to the federal government or forbidden them to the states.

For the first two-thirds of the Constitution's existence, practice roughly tracked the constitutional theory, but the last third has seen the growth of federal power beyond the borders contemplated by the Constitution's framers. This has come largely through the congressional powers to regulate interstate commerce (the "commerce power") but also to tax and to spend.

It would be hard to claim the federal government has the constitutional power flatly to ban Easy PGD, but it is not hard to argue that it has the power to pass a law that forbids Easy PGD as it affects interstate commerce, or that puts a prohibitive tax on its use, or that bans receipt of some federal spending by people participating in Easy PGD. (Such statutes might also be upheld under Congress's power to enforce the Fourteenth Amendment, particularly if the statute only banned the use of Easy PGD to select fetuses based on race or sex, although that argument has not been very successful recently.)

One federal law, discussed at length earlier, already governs some aspects of Easy PGD—the Federal Food, Drug, and Cosmetics Act. It uses the commerce power to prohibit the introduction into interstate commerce of certain drugs, devices, biological products, and other goods unless permitted by the FDA. And it is clearly constitutional.

But what about a federal law that just wanted to ban all Easy PGD when used to select among embryos for nondisease traits. Would that be within the federal government's powers? Probably not unless it relied on the federal government's commerce, taxing, or spending powers, but would even those powers be enough?

In law school I was taught that Supreme Court decisions during and after the New Deal had made any federal action possible under the commerce power. Since then the Supreme Court has twice disproven my teachers, holding in 1995 that the commerce power did not justify a ban on possession of some guns within a certain distance of schools[2] and in 2000 that it could not support certain provisions of the Violence against Women Act.[3] (These were the first cases since the 1930s where the Supreme Court had found against Congress's commerce power.) And in the main Obamacare cases, it held that the commerce power could not be used to force people to participate in interstate commerce (though the taxing power could).[4]

In the medical world, the federal Partial Birth Abortion Ban Act is the closest we have come to such a statute. The act made it illegal for "Any physician who, in or affecting interstate or foreign commerce, knowingly performs a partial-birth abortion and thereby kills a human fetus."[5] In 2007 in *Gonzales v. Carhart,* the Supreme Court, by a five-to-four vote, upheld that act against claims that it infringed the federal constitutional right to an abortion.[6]

Neither the majority opinion nor the dissent discussed the Commerce Clause "hook." In a brief concurring opinion, however, Justice Thomas, joined by Justice Scalia, wrote, "I also note that whether the Act constitutes a permissible exercise of Congress' power under the Commerce Clause is not before the Court. The parties did not raise or brief that issue; it is outside the question presented; and the lower courts did not address it."[7] And so neither did Justices Thomas and Scalia.

So the federal government might try to regulate Easy PGD on nonsafety grounds through the Commerce Clause (or taxing or spending powers). Whether it would be upheld or not is uncertain even today, let alone after twenty to forty more years of constitutional law (and Supreme Court justices).

In any event I suspect states will lead the way. We already see much more state than federal action about abortion and reproductive rights. The breadth of views encompassed in the whole country and the relative ease of blocking legislation at the federal level have, as a political matter, made the federal government less active, leaving the field to the states. Some states encourage prenatal genetic testing; several other states have already passed bans on some forms of selective abortion, prohibiting abortion based on the fetus's sex, race, or, in North Dakota, its likely disability status. State laws would not face the same questions as congressional action would about the power to enact such legislation, but they would face identical issues about whether they violated individuals' rights.

Under today's American constitutional law, state or federal limits or bans on Easy PGD would face two and possibly three kinds of claimed violations of the federal Constitution: that they violate substantive due process guaranteed by the Fifth and Fourteenth Amendments, the Equal Protection Clause of the Fourteenth Amendment, or, in some cases, the freedom of speech guaranteed in the First Amendment.[8] Of course, today's constitutional law is not controlling; the constitutional law of twenty to forty years in the future will be. But we can only work with what we have. The rest of this section will examine each of those three federal constitutional objections.

Substantive Due Process

Over the last century and a half the U.S. Supreme Court has created the seemingly oxymoronic "substantive due process" doctrine. It stems from the Constitution's guarantees, in the Fifth and, after 1868, the Fourteenth Amendments, that neither the federal government (the Fifth Amendment) nor any state (the Fourteenth) may deprive any person "of life, liberty, or property, without due process of law." On its face this seems like a guarantee that appropriate procedures will be used ("due process"), but the court has held it to include certain substantive rights that are not contained in the constitutional text.

In the late nineteenth and early twentieth centuries, the Supreme Court used substantive due process most notoriously to strike down reformist labor legislation, such as special protections for working women and children, as well as general limits on working hours. This period is sometimes called the *Lochner* era, after *Lochner v. New York,* a 1905 decision that held a state limiting the working hours of bakers violated those bakers' constitutionally protected right to contract.[9]

During the New Deal, the court ultimately rejected the expansive use of substantive due process to invalidate economic legislation; "Lochner" eventually became a pejorative. But in the 1960s the Supreme Court again started finding constitutional protection for rights not specifically mentioned in the document. In *Griswold v. Connecticut* in 1965 the court struck down a Connecticut statute that prohibited any use of any form of contraception.[10] The majority opinion, written by Justice Douglas, based the decision on a right to privacy, found in the "penumbras" and "emanations" of specific constitutional rights. Justice Harlan concurred in the result but not in the court's opinion, arguing that the protection came not from an inferred privacy right, but from the Due Process Clause itself, which protects rights "implicit in the concept of ordered liberty." Other decisions involving contraception, abortion, and the rights of the family followed, but ultimately the Harlan position, that these are part of general substantive due process, was widely, though not universally, accepted.

Two strands, and possibly a third, of substantive due process decisions might be applied to invalidate laws banning or limiting Easy PGD. One is directly about control over reproductive decisions. *Skinner v. Oklahoma,* discussed in Chapter 16, held that the right to procreate was a fundamental right (albeit in a decision under the Equal Protection Clause).[11] The contraception cases have left almost no room for governmental regulation of contraceptives for reasons other than safety and efficacy. The abortion cases have followed a more convoluted path, but, since *Roe v. Wade* in 1973, the majority of the court has consistently recognized that women have a constitutional right to terminate a pregnancy up to the point where the fetus is viable.

Easy PGD is a method of reproduction that involves choices by people about childbearing. One might take a broad view of the contraception and abortion cases and say that they encompass a "right to reproductive liberty" that could include other reproductive choices. A broad reading of a constitutional right to reproductive liberty might make some or all government restrictions on Easy PGD unenforceable.[12]

On the other hand, other, narrower interpretations of these rights are possible (including the position, taken by a substantial minority of the Supreme Court since *Roe,* that the Constitution does not protect abortion). One could focus on the physical intrusiveness of pregnancy to women or the privacy violations involved in monitoring the use of contraceptives. How the Constitution will be read on this point in twenty to forty years depends on, among other things, doctrinal developments in constitutional law between now and then and, as a practical matter, to some extent on public views then about Easy PGD.

The second approach draws from cases that focus on family and protect some aspects of family or parental decisions about child rearing from state intervention. *Meyer v. Nebraska* and *Pierce v. Society of Sisters,* discussed in Chapter 16, find that parents have some rights over their children's education that the state cannot take away. These were joined in the post-*Lochner* era by two more Supreme Court decisions. In *Moore v. City of East Cleveland,* in 1977, the court invalidated a zoning ordinance that prohibited nonfamily members from occupying the same dwelling unit but defined family so narrowly that a grandmother could not raise her grandchild.[13] Twenty-three years later, in *Troxel v. Granville* the court invalidated a Washington statute that gave visitation rights to grandparents over the objections of a child's parents, once again without a majority opinion.[14] Four justices, the odd combination of two "conservatives," Chief Justice Rehnquist and Justice O'Connor, and two "liberals," Justices Ginsburg and Breyer, held that the statute violated substantive due process by not giving any special weight to the parents' concerns. (Justices Souter and Thomas agreed with the result, but without joining the plurality opinion.)

Other cases also provide a special status for some parental decisions about children, sometimes alone or sometimes in conjunction with other constitutional rights, such as the right to free exercise of religion, as in 1972's *Wisconsin v. Yoder,* involving Amish parents' rights to control their children's education.[15] How far these cases, and the doctrine behind them, reach remains deeply unclear. Could substantive due process protections for parental decision-making or for the family unit limit the power of states to prevent prospective parents from selecting their future babies based on genomic tests? Maybe.

A third possible argument could come from a broad reading of cases about sexual behavior and marriage. The Supreme Court has held that governments cannot constitutionally criminalize same-sex sexual activity between consenting adults. It has held, in a different line of cases, that governments cannot prohibit people of different races and, very recently, people of the same sex from getting married. Easy PGD is neither a form of sexual behavior (witness the title of this book) nor (necessarily) about marriage, yet, depending on how these doctrines evolve, they might provide some support for a right of intimate or family decisions that could protect Easy PGD.

The Equal Protection Clause

The Fourteenth Amendment forbids any state to "deny to any person within its jurisdiction the equal protection of the laws." In 1954 the Supreme Court held this constraint also bound the federal government, through the Fifth Amendment's Due Process Clause. Like substantive due process, equal protection has had a twisting path through the case law. It was initially eviscerated, then revived in cases that benefited corporations, and finally, after World War II, it rose to be the main constitutional provision supporting civil rights movements. Though clearly initially inspired by the problems facing African Americans, the clause has long been used more expansively.

But just how expansively has been a problem. Many laws treat two people unequally. Federal income tax rates are not the same for every person; social security benefits vary enormously from person to person; even Medicare coverage varies to some extent from region to region. During the 1960s and 1970s the Supreme Court evolved a structure for thinking about the Equal Protection Clause. Most distinctions would be upheld as long as the court could find they had some "rational relationship" to the problem they were aimed at. Other distinctions, though, would be looked at more closely, particularly those based on "suspect classifications," like race and religion, or "fundamental interests" like marriage, the right to vote, or access to courts. When these kinds of distinctions are involved, the law undergoes "strict scrutiny" or "heightened scrutiny," and the government must show that it had a compelling interest at stake and that it chose a narrowly tailored method to achieve that interest.

Over the years, this framework for Equal Protection Clause analysis has become more complicated and less clear, but the approach is still largely followed. Who wins depends on the level of scrutiny: in cases with "rational relationship" scrutiny, the government almost always wins; in cases with "strict scrutiny," it almost always loses.

A straight ban on Easy PGD would probably not evoke a plausible Equal Protection Clause argument. Everyone would be covered. Some prospective parents who needed Easy PGD (or its equivalent) to avoid the risk of transmitting a serious genetic disease might argue that they were unequally affected by this facially equal ban and should be able to invoke the clause. (This is strikingly similar to judicial decisions affecting Italian and German laws against PGD, which struck them down only in the cases of possible transmission of serious genetic disease.)[16] The argument is plausible, but probably would be unsuccessful.

A more nuanced limitation, one that allowed Easy PGD to avoid some conditions but not others, would be more likely to bring an equal protection challenge. If prospective parents were allowed to use Easy PGD to avoid Tay-Sachs disease but not to avoid Down syndrome, parents who wanted to use it for Down syndrome might complain they were denied the equal protection of the law. So might parents denied the right to use it to avoid a late onset condition, like Alzheimer disease, or the chance to choose their child's hair color or sex. Or someone denied the chance to be a uniparent.

If the court viewed the right to make decisions about reproduction as a fundamental interest or right, it would subject such statutes to strict scrutiny and likely invalidate them. But what is the right? To have a healthy child? And, if so, what counts as healthy? Or to make any possible decision about your child's genetic makeup? The possible judicial approaches are many and complex. At this point, it is only safe to say that the Equal Protection Clause might, or might not, make some state statutes unenforceable, but that it would more likely apply when statutes were nuanced, allowing some uses and denying others.

The First Amendment

States might like statutes that ban clinicians from giving prospective parents genomic information, because, as discussed in Chapter 16, they are easier to enforce. But by banning health care providers from informing their patients of facts (often medically relevant), they could run into the First Amendment: "Congress shall make no law . . . abridging *the freedom of speech*." You may not be surprised to learn that its application is also unclear.

Governments cannot abridge the freedom of speech, but they can regulate the practice of medicine, as well as exercise control over what doctors being paid government money say. The Supreme Court has not spoken on the direct point of whether a government can generally forbid a health care provider from speaking freely to a patient. Yet. But such cases may be getting closer.

Recently, the state of Florida decided to prohibit physicians from asking their patients about gun ownership. This statute, apparently inspired by the National Rifle Association, responded to the views of some physicians and physician groups that guns are a public health issue. The lower court upheld the statute, finding it, in a case entitled *Wollschlaeger v. Governor* (but widely known as "Docs v. Glocks"), a constitutional regulation of the practice of medicine. The appellate courts have, thus far, agreed.[17]

At nearly the same time, the Ninth Circuit upheld a California law that prohibits licensed health care workers, upon penalty of professional discipline (including possibly loss of their license), from engaging in "Sexual Orientation Change Efforts" on minors. The doctors again claimed a First Amendment right; the court disagreed, though basing its decision in large part on a finding that no medically safe and effective "sexual orientation change efforts" exist.[18]

And on the third hand, the federal appellate courts have split over whether states may force physicians to convey specific information to patients seeking abortions. In December 2014 the Fourth Circuit struck down a North Carolina statute requiring that such a patient be shown a current ultrasound of the fetus and, if she averts her eyes, have the fetus described to her.[19] The court based its decision on the doctor's First Amendment right not to be forced to speak. Other federal circuits had recently reached opposite decisions on related statutes by which the states tried to compel physicians to deliver anti-abortion messages to their patients.[20]

This may all be relatively unimportant, though, depending on how Easy PGD is financed. As mentioned in Chapter 16, in 1991 in *Rust v. Sullivan* the U.S. Supreme Court upheld federal regulations banning physicians from even discussing the possibility of abortion with their patients—if the patients were covered by a partially federally financed family planning program. The court reasoned that the doctors could say anything they wanted, but not on the government's dime.[21]

Practical Limits to Enforcement

Let's assume a statute limiting or forbidding Easy PGD is passed and upheld as valid under the relevant constitution. A naïve view might be that law forbids it and so it ceases. But, of course, in reality, laws are often disobeyed.

Some kinds of laws are relatively easy to enforce because their violation will provoke complainants and witnesses eager to see action and the violators will be easy to find. Consider a raucous late-night party violating noise restriction ordinances in a quiet residential neighborhood. Others will be more difficult, especially those where the "victims" want the crime to be committed. "Victimless" crimes, like illegal drug sales and prostitution, are notoriously hard to enforce. Practical enforcement problems will vary in predictable ways depending on which aspects of Easy PGD are allowed and which are restricted, but also on whose actions are prohibited.

Consider how, and against whom, enforcement would take place. Surely we would not kill a child born as a result of unlawful Easy PGD and it is quite unlikely we would force such a pregnancy to be terminated. Any embryos that had not yet been transferred to a womb might be destroyed, although this would be controversial.

More likely is criminal action against the parents who sought the PGD. This would likely have bad effects on any children they have, but we already prosecute parents of young children. Between a reluctance to break the law, stringent penalties for getting caught, and a reasonable chance of getting caught, such sanctions might dry up parental demand. (It may be worth noting, though, that criminal penalties in India and China for sex-selective abortion have coexisted with very high rates of such abortions.)

Criminal or civil action against the various licensed professionals whose help is needed to perform Easy PGD seems most likely. Those will generally be legitimate businesses, with real addresses, customers, and interests in maintaining their legal business. (Although back-alley PGD facilities might be possible, this seems unlikely.) This situation does not guarantee perfect enforcement, but it makes it better.

Now let's consider not on whom enforcement will fall, but what actions will be prohibited. A government might take aim at Easy PGD by restricting gamete creation, PGD in general, PGD for particular purposes, or selection based on particular PGD results.

Limiting the "Easy" part of Easy PGD—the creation of gametes from iPSCs—could be fairly easy to enforce if the ban were broad. Even in a world where iPSCs were being regularly created and transformed into heart, liver, or brain cells, the specific derivation to gametes (and their subsequent maturation) would likely require specialized expertise and "tools"—in this case, probably specific fluids and biochemical factors for culturing the cells and turning them into mature gametes. That should make violators easier to find.

On the other hand, if gamete production were allowed for some purposes, such as infertility, but not for others, such as Easy PGD, enforcement becomes harder. A couple wanting gametes for Easy PGD only has to convince a physician that they are infertile. The diagnosis of infertility may well change in the next few decades, but, now at least, it is failure to conceive after twelve months of unprotected sex—something easy to claim and hard to disprove.

A ban, complete or partial, on PGD seems more likely. (It is genetic selection, after all, not derived gametes, that worries people.) A complete ban on PGD could also be reasonably enforceable. The specialized expertise and equipment required would be, primarily, the methods of doing blastomere biopsy and, possibly (depending on how the technology develops), the methods of doing single-cell whole genome sequencing.

Once again, this only works well if the procedure is totally banned. If we want to allow PGD for any purposes, such as avoiding fully penetrant, early onset, invariably fatal diseases, the procedures will exist. A couple could then seek PGD for one of the permissible reasons, but how could their use of the resulting information be controlled?

One option would be to ban whole genome sequencing, at least in the context of PGD. Instead, the only DNA sequenced would be DNA related to the conditions for which testing was allowed. This is plausible. It forfeits the economics of whole genome sequencing but in twenty to forty years, testing a few specific genomic regions should also be cheap and easy. Presumably, whole genome sequencing would be widely used in the rest of the health care system, but enforcement might be able to do a good job of keeping biopsied embryonic cells, which are likely to be removed and stored in specialized clinics, from being sequenced.

A second option would be to allow whole genome sequencing as part of PGD, but only permit clinics to give certain kinds of results to the prospective parents for use in selecting among embryos. The information about all the genetic traits would exist as a result of the whole genome sequencing, but if the parents did not know it, they could not use it.

This method relies on the clinics which have this knowledge not sharing it with the parents. Because of principle or of bribes, some of those with the information might choose to let it slip.

The last general method would be to prohibit selection among embryos based on all or, more likely, some genomic variations. This might range from bans on sex selection, through bans on behavioral or cosmetic selection, to bans on selection based on specific health traits. In the absence of an information ban, though, these prohibitions, as discussed in Chapter 16, face a strong enforcement problem—the prospective parents could claim they had other reasons for selecting the specific embryo.

Reproductive Tourism

Another great constraint on enforcing laws against Easy PGD is the existence of other jurisdictions with different laws. This is not a new problem. If you want to bet legally on sporting events, visit Nevada (or the UK). If you want to smoke marijuana legally (at least under state law), visit Colorado or Washington. And, historically, if you wanted an easy divorce, you moved (briefly) to Nevada.

Medicine has not been exempt from this jurisdiction shopping. Harvard law professor Glenn Cohen recently published a comprehensive analysis of medical tourism.[22] He considers patients going to other countries to get services that are legal in their home countries, but are available elsewhere for lower cost or at higher quality. But he, as have others before him, also considers patients crossing borders for medical services illegal or otherwise unavailable in their home jurisdictions, including end-of-life care, unapproved procedures, organ transplants, abortions, and fertility treatments. Easy PGD tourism poses a huge problem for jurisdictions trying to ban or limit the procedure.

One solution is for every jurisdiction to adopt the same limits on Easy PGD. If all jurisdictions have the same laws, jurisdiction shopping becomes irrelevant. Given the likely very different views in different countries about Easy PGD, the idea that they would unanimously ratify a treaty with specific limits on Easy PGD seems nearly impossible. Although countries might adopt a treaty with some vague limiting language, making it more likely to be widely acceptable, their interpretations of that language might well differ dramatically. For example, the Council of Europe adopted a protocol banning "human cloning," but expressly let member states define "human clone." Some decided to ban all human somatic cell nuclear transfer; others only banned human reproductive cloning.[23]

In a world where some countries allow (or encourage) Easy PGD, what can a country that wants to restrict it do? It might make it a crime for its citizens or nationals to obtain services illegal at home from overseas. For the most part, countries regulate the behavior of people in their geographical jurisdictions; their citizens who want to misbehave overseas generally may do so. But there are exceptions. Australia, Canada, the United Kingdom, and the United States, among other countries, make it a domestic crime for their nationals to engage in paid sex with a child anywhere in the world.[24]

But if a couple goes on a foreign vacation and the woman comes back a week or two later pregnant, how will the home jurisdiction ever know? Or even if it does find out about the pregnancy, how will it know whether Easy PGD was used overseas? The enforcement problems seem overwhelming.

Of course, the cooperation of the country where Easy PGD is legal could make the restrictive country's task much easier. Other countries could make Easy PGD legally available only to their own residents, as have four U.S. states that have legalized physician-aid-in-dying—Oregon, Washington, Vermont, and, most recently, California. They might also put a time requirement on residency; for example, originally Nevada required six months residency to be eligible to use its liberal divorce laws. (The push for "divorce tourism" led Nevada to shorten its residency to three months in 1927 and to six weeks in 1931.)

Such residency requirements could prove a legal problem in some parts of the world. The European Union expressly gives citizens of its member countries the right to get medical treatment in any of its member countries, a right enforced most dramatically to forbid the Irish government from prohibiting pregnant Irish women from traveling elsewhere to obtain abortions illegal in Ireland. A similar principle in American constitutional law that guarantees a right to travel between states has been used to strike down some durational residency requirements, but not others. Whether it would apply to efforts by one state to limit a medical procedure to its own residents (with or without a durational period to the residency) is unclear.

Or a jurisdiction could agree to provide information about nonresidents who have used Easy PGD to their home countries. This would be rather like tax and banking treaties by which countries agree to provide relevant information on bank accounts to the account holder's home country. Banking havens have not, however, disappeared; secrecy-seeking clients have shifted to other countries.

The creation of Easy PGD havens seems quite possible. Some countries will seek to benefit financially from supplying a highly desired service to people from other jurisdictions. Nevada's easy divorce laws and very low residency requirements were not random but rather were successful efforts to bring "divorce tourists" and their money to the state. France bans IVF for people who are not in a stable heterosexual relationship; as a result, IVF clinics in Belgium do booming business with French lesbians.[25] If Indiana were to limit Easy PGD substantially, Illinois would likely seize the opportunity. And if major nations were to limit it, some smaller jurisdictions (the Grand Cayman Islands, perhaps) would likely fill the gap.

Enough sustained pressure from enough powerful nations might lead to some kind of global ban or regulation, though guaranteeing effective enforcement would be harder. It has not happened yet with banking secrecy, the slave trade, or the proliferation of nuclear weapons. It seems unlikely to happen with a relatively accessible technology like Easy PGD.

At an extreme, that pressure might include armed intervention. At least one author writing about the prospects for genetic enhancement argued that armed military intervention might be necessary to keep rogue countries from practicing this kind of positive eugenics.[26] That kind of action, if effective, would prevent all Easy PGD, including its use by reproductive tourists. It also seems, to me, vastly out of scale to the threats Easy PGD presents, but predictions about distant future decisions by national leaders can come with no guarantees.

I think the most likely outcome is a world where some countries encourage Easy PGD, others allow it, others limit the practice, and still others ban it, but Easy PGD tourism will not be effectively deterred. In the more restrictive countries, the rules would reduce their nationals' use of Easy PGD. The time, hassle, and expense of traveling for the procedure, plus the need to be willing to violate domestic laws, will also cut down on the number of those who use the procedure.

The price of the procedure makes this particularly true. I have argued that countries will want to subsidize Easy PGD in order to minimize their own long-term health care costs, but they will only want to subsidize it for their residents. The interjurisdictional issues return us to another version of the domestic enforcement reality.

Making Easy PGD illegal, in whole or in part, will reduce the number of people who use it and will make them more likely to be people who have strong financial resources. This raises some important fairness issues. They will also be people with a willingness to disregard the law, which may raise other issues. Would either of those developments be a good result?

Ancillary Prohibitions

In addition to perhaps inspiring bans or limitations on Easy PGD directly, widespread use of Easy PGD is quite likely to lead to ancillary prohibitions, such as those discussed in Chapter 14 on unknowing parenthood based on cell theft, sales of gametes, incest, or "uniparenting." Banning the use of "stolen" cells has the problem that, without disclosure to someone of the stolen nature of the cells to be used to make gametes, knowing that a crime has been committed may prove very difficult. One would probably want to focus enforcement of this ban on those facilities that could transform cells into iPSCs and then into gametes by requiring that they have good evidence of consent to genetic parenthood, at least for cell samples that are not medically removed skin biopsies.

Banning the sale or purchase of gametes (or cells for derivation of gametes) has the problem of all "victimless" crimes. If the seller and the buyer are happy, who will bring the transaction to the attention of law enforcement? Widely advertised gamete sales could be blocked; small-scale sales would be much harder to find. And if discovered, the parties might disclaim any sale. Payment for cells might often be traced, but probably not always.

Consensual incest would be even harder to prevent, especially if the parents were willing to claim that one of the gametes came from an unrelated party. Prohibiting uniparenting may be more feasible, especially if one clinic is collecting or producing both the egg and the sperm from the same person. In both of these last two cases, however—incest and uniparenting—some enforcement might be possible by examination of the baby's or embryos' genomes. When the embryos get the PGD part of Easy PGD, high levels of similarity between their two genomes (maternal and paternal) would be a powerful sign that a unibaby was the goal; lesser but still substantial similarity would be at least some evidence of incest, perhaps enough to prompt further investigation. This would presumably also become clear after a child was born, as its routine neonatal whole genome sequencing should pick up signs of close parental relationship.

Implementation: The Need for Guidance

One last issue demands discussion—how could Easy PGD be effectively implemented in a way that produced the best or, at least, the "least bad" results? The problems of regulation, monitoring, and so on seem solvable, but I worry greatly over another problem—the problem of sufficiently informed prospective parents.

Think back to the table in the Second Interlude, the one providing the genetic traits of five generally acceptable embryos. Do you feel competent to make a good choice among those embryos? I don't. And we have had the benefit of reading this book and thus knowing much more about these issues than most people. How can people make choices that are good—for them and for their prospective children?

They need help. The adoption of Easy PGD will demand a vast expansion in the number of professionals trained to help regular people make decisions involving genetics. Genetic counselors typically obtain master's degrees after training in both genetics and in counseling. They would try to help explain to prospective parents what the various diseases and their risks mean, as well as to point out possible advantages or disadvantages of some of their choices. The counselors would not "direct" decisions. Genetic counseling, which got its start in prenatal testing, has a strong tradition of being nondirective. But they would help.

But for genetic counseling to help, we will need genetic counselors. Currently about 3,000 genetic counselors are active in North America and they are busy working with existing and expanding genetic testing for fetuses, infants, cancer patients, adults with bad family disease histories, and research subjects. For twenty years we have been saying that we will need to expand their numbers dramatically but that has not happened, perhaps because the profession is neither high status nor well paid. For parents to make choices that seem good to them, and to understand the consequences of those choices ("No, choosing #12 does not guarantee an All Pro quarterback"), we will need more counseling.

If, though, three million couples end up using Easy PGD each year in the United States, we would need a lot more counselors. And it will make sense for prospective parents to get help well before they are called upon to pick embryos. The counseling function will need to take place both before and after the actual testing. It will almost certainly need to include Internet, video, and other methods of extending the reach of genetic counselors. Face-to-face counseling will still be crucial—video programs presented over the Internet may, even in twenty to forty years, still have difficulties in reading viewers' reactions—but it will need to be supplemented if prospective parents are to have any chance of understanding what they are doing.

CONCLUSION

CHOICES

The future is coming; there is no way to prevent it. The question is whether and how to try to shape it. I believe that future will include substantial use of Easy PGD or, if not that particular technology, some method that allows intentional selection or modification of babies' genetic traits—that allows those to be chosen. What do we want that future to be and how can, or should, we go about trying to achieve that goal?

This final chapter does not argue for any particular answers to those questions. Instead, it sets out futures I think some readers will prefer and then analyzes paths toward those futures. At the very end, I will tell you my own preferences, but mine are not especially important. I am not going to be making those decisions—instead, they will be made by a wide variety of we's in different countries, cultures, and families. And what we decide will shape, in some ways, the future of humanity. My goal has been to help those many us's decide. This chapter discusses the ends we may want to reach and the means we might want to use to reach them, as well as two crosscutting issues of regulation, before giving my own thoughts.

Ends

Where do we want to go? More accurately, where might different people want to go with this technology? Many possible endpoints could be chosen, but I think they fit into three categories: no Easy PGD, some regulated Easy PGD, or unregulated PGD.

No Easy PGD

Some people will want a future with no Easy PGD or other genetic selection (or modification) technologies. Their preferences may have many sources: religion, a desire for naturalness, a romantic idealization of randomness, a Burkean kind of conservatism, a concern about the inability of humans to handle the technologies ("playing God"), the "yuck" factor, or something else. (I am focusing on opposition to genetic selection, but some opponents might also oppose making gametes from stem cells to treat infertility caused by the absence of gametes.)

On one, very personal level, this position does not ask much. If the preference is for a world where you, or perhaps your family or loved ones, do not have to use these technologies, we just need to ban coercion, frank or implicit, to use Easy PGD. (Banning "implicit" coercion, though, might prove difficult.) If, on the other hand, the preference is to live in a world (or a country) where no one uses these technologies, the task becomes harder.

An additional factor complicates the "no" position. Related technologies already exist and are being used. Making eggs and sperm from stem cells has not yet been done in humans, but IVF has been used for over thirty-five years and has produced millions of children—many now adults. Genetic selection has also been used for decades. Fetal genetic testing followed by termination dates back to the late 1960s; the first PGD baby was born in 1990. And more broadly, less accurate "genetic selection" through mate selection goes back beyond the birth of our own species.

No one will seriously try to remove mate selection as a source of genetic selection, but should people who oppose Easy PGD permit the existing technologies? Catholic doctrine does not. It holds all these technologies sinful, both because of their displacement of natural procreation and the destruction of embryos or fetuses they entail. But removing an existing technology is more difficult than preventing the entry of a new one. Opposition to Easy PGD could include a commitment to ending the use of any genetic selection techniques or could be limited to stopping the adoption of new technologies.

Some, Regulated, Easy PGD

The second end would be to allow some uses of Easy PGD or to allow Easy PGD subject to some conditions, or both. This goal, though, is really many different policies, as different people may want to regulate

Easy PGD in many different ways, ways that could be used singly or in myriad combinations.

Some will want to regulate it to make sure it is safe for the children born using this method. Others will want to regulate the genetic traits that can be selected, for or against: "Serious diseases, yes; cosmetic traits, no." Some will want to regulate who can and cannot use the method or, perhaps more accurately, whose cells can and cannot be used to make gametes. Others will try to ensure fair access, and perhaps use, of the procedure, while still others will be concerned to prevent coercion in the use (or perhaps the nonuse) of Easy PGD. Many of the above can be, and will be, combined. And like the first goal, these various conditions or limitations on the use of Easy PGD could be chosen for an individual, family, community, or everyone.

Unconstrained Easy PGD

The third end would be completely unregulated Easy PGD, justified by either a special reproductive liberty or liberty more broadly. It could also come in variants from individual to universal with all stops (community, state, nation, region, etc.) in between. In some places, this would require only a continuation of existing nonregulation; in most countries, however, some regulatory structures would have to be abolished, or at least modified, to permit fully free Easy PGD. It isn't clear to me whether anyone would truly seek this result; it seems at the least unlikely to be a common position.

Means

Choosing ends is important in deciding how to respond to Easy PGD and similar technologies. Reasonable people can, I think, choose any of the ends set out above, depending on the person's assessment of the information about Easy PGD set out in this book, along with his or her beliefs and personality. But one more factor should be considered: the means. Ends may, at least in some cases, be modified by the ease or costs of the means of attaining them—not that ends justify the means but the means may modify the ends. And so to the means we turn.

Reaching the three kinds of goals discussed above could involve at least seven different approaches: limiting research, banning the practice of Easy PGD, requiring proof of safety, restricting the choices made using the technology, regulating whose cells are used in the technology, avoiding differential access, and addressing coercion. Only the first two speak directly to the goal of preventing Easy PGD; the rest concern regulation of the technology. And, of course, none of them is appropriate to the third end, unrestricted Easy PGD. Each is discussed below.

Preventing Easy PGD

Two plausible methods exist to prevent the use of Easy PGD: stopping the research necessary for its development and directly banning its use for all purposes. The first seems impossible at this point; the second seems unlikely.

Limiting Research Needed for Easy PGD

In theory, limiting the research necessary to create clinically available Easy PGD, either through banning the research or by forbidding its public funding, is an attractive approach. Successfully preventing a "bad" technology from coming into existence will almost certainly work better than trying to prohibit the use of an existing technology. That strategy has two big problems in this case. The first is that it's already too late. The second is that the relevant research will almost certainly continue for purposes other than Easy PGD.

Two technologies make up the core of Easy PGD: cheap genome sequencing and production of gametes, particularly eggs, from stem cells (probably iPSCs). Already whole genome sequencing is available for about $1,500 per genome, depending on whom you believe and how you add up costs. Less powerful but still useful technologies like whole exome sequencing are cheaper. Large-scale genomic analysis of multiple embryos is thus already possible, at least for people willing to spend tens of thousands of dollars. And such people exist.

Deriving gametes from iPSCs has not yet been done in humans. But the basic research into differentiating iPSCs into different cell types, including gametes, has been and is still going on. The gap between what is known and what is needed seems small; successfully preventing research to close that gap would be difficult.

No one is going to stop all research on better methods of whole genome sequencing, on the meaning of genome sequences, or on the derivation of specific cell types from iPSCs just to try to stop, or slow, Easy PGD. Each of these research fields has broad, promising, and uncontroversial applications. For each of them, Easy PGD is a "secondary use," something that tags along with the research's primary goal. The specific derivation (and subsequent care and handling) of gametes from iPSCs is a partial exception, but it does have value other than Easy PGD—providing an effective treatment for some infertile couples.

A ban on research that does not require huge investments and that can be done with Spartan facilities and commonly available equipment seems unlikely to succeed. So does prohibiting public funding in a field with so much potential profit (and so much basic research already done). Whether stopping research is ever a good idea is controversial. Some have even urged that the U.S. Constitution has, or should have, a "freedom of research" read into its First Amendment.[1] But in this case an effective ban on the research is so impracticable that this question does not have to be addressed.

Banning All Use of Easy PGD

In the right jurisdiction, a complete ban might work. In fact, plain PGD is banned already in at least seven countries: Algeria, Austria, Chile, China, Ireland, the Ivory Coast, and the Philippines.[2] In those countries, Easy PGD necessarily would already be illegal.

On the other hand, many countries are not on that list, including some that are typically quite reluctant to adopt new reproductive or genetic technologies, like Germany and Italy. Each of those countries has stringent laws regulating assisted reproduction. In 2012 the European Court of Human Rights held that Italy's ban on PGD violated the European Convention on Human Rights, at least when enforced against parents seeking to use PGD to avoid bearing a child with a genetic disease.[3] (This puts in doubt the bans in Austria and Ireland.) The German Federal Court of Justice interpreted its law against PGD in 2010 to not apply to cases where it was needed to avoid "serious genetic illness," leading to passage in 2011 of a statute permitting such uses."[4]

A complete ban on Easy PGD would require new legislation in almost all countries and, as already seen, could violate constitutional protections or regional or international human rights treaties in some countries. The existence of countries where the procedure was legal would surely inspire "reproductive tourism." In the European Union, for example, the right to travel within the EU for medical purposes appears to protect that practice.

It is also hard to see a complete ban, one that includes a prohibition on its use to avoid serious genetic diseases, being maintained even within one country if, in other countries, Easy PGD were common and perceived as successful. Although some prospective parents in restrictive countries would presumably travel for their procedures, others would lobby for liberalization of the ban. They might be joined by those concerned about public health in the restrictive countries, or even by nationalists, worried about their homeland "falling behind." The main hope for those seeking to avoid Easy PGD, or to turn back the clock to eliminate all PGD, would seem to be an international treaty banning the procedure, one that had nearly complete ratification and effective enforcement. As discussed in Chapter 18, that is a tall order.

Permitting Some Easy PGD

While preventing all Easy PGD seems difficult, regulating it seems more plausible. The kinds of regulation that Easy PGD might call forth already exist for other forms of assisted reproduction in many countries, though Easy PGD does raise some novel issues. This section discusses five different ways people might want to regulate Easy PGD: safety, the choices available in Easy PGD, whose cells can be used in Easy PGD, differential access to the technology, and coercion.

Safety

Requiring Easy PGD to be proven safe (and, less crucially, effective) seems the easiest regulation to impose—after all, it already exists in many countries. For example the U.S. FDA claims jurisdiction over all human cells that are more than minimally manipulated if used in a medical context. It would regulate the "Easy"—the safety of using stem cell–derived gametes to create babies. But regulation also could look at the "PGD." The United Kingdom has a specific agency, the Human Fertilisation and Embryology Authority, that regulates assisted reproduction, including all uses of PGD.

The hard questions, as discussed in Chapter 13, involve how to regulate the procedure's safety. How much work with nonhuman animals (and of what kinds) should be required before human trials are allowed?

How many babies should be studied for adverse health effects and for how long—months, years, decades, lifetimes? How safe would Easy PGD have to be—as safe as "normal" reproduction, as safe as IVF, as safe as current PGD? Would its health benefits, the avoidance of genetic diseases, be allowed to act as a "set off" to ways in which it increased health risks?

These questions do not have easy answers, although we may get some help from the United Kingdom. Its recent parliamentary approval of attempting mitochondrial transfer for making human babies should lead to some careful thought about, and some experience in applying, safety testing in human reproduction. However we respond, the answers will have huge effects on when, if ever, Easy PGD might be allowed. The need to regulate the safety of Easy PGD should not be controversial, but, as always, the devil haunts the details.

And safety regulation would not only have to be implemented but would have to be enforced. Eager (or desperate) early adopter parents could travel to less restrictive countries to receive Easy PGD or could look for domestic black market providers. When the object of safety regulation is a baby, the case for tough enforcement is stronger than when a competent adult chooses to try even a quack remedy.

Limiting Choices Available in Easy PGD

Many will want to limit the uses made of Easy PGD. These issues, explored mainly in Chapter 15, pose many difficult choices. Remember my earlier categorization in Chapter 7 and in the Second Interlude of the five kinds of information Easy PGD could provide—serious and highly likely early onset diseases, other diseases, cosmetic traits, behavioral traits, and sex. Each category could be the subject of regulation.

Selecting against serious and highly likely early onset diseases will probably be the least controversial use of Easy PGD. Courts have already struck down PGD bans in Italy and Germany that did not allow this use of PGD. Defining the category, though, is a problem. What diseases qualify—or, more accurately, which genomic sequences will lead to diseases—will not be obvious. How serious, how likely, how early an onset? Or even how likely or how early an onset?

But another problem in drawing the line is that having such a line effectively means the government says that "these embryos would have lives at least potentially worth living, those embryos would not." That seems a very difficult, and uncomfortable, position for a government to take about what could become its citizens. People with the genetically inherited conditions for which Easy PGD is allowed could fairly argue that such a line devalues their lives as "not worth living."[5]

Another possible line could be drawn between "diseases" and other traits. This also has its line-drawing problems. What is a "disease" and what isn't can often be controversial. Consider autism and the arguments of some people with autism that they are just examples of beneficial "neurodiversity."[6]

The "disease" category could also take in some very minor conditions, such as genetically determined color vision variants. This is a genetic "condition," detectable by PGD, but it seems a weak reason to avoid selecting an embryo. Similarly, a higher risk of having a disease late in life could count—but how much higher a risk? Should the difference between a 15 percent lifetime risk of type 2 diabetes and an 18 percent (or 12 percent) risk count? And, perhaps more importantly, any "qualifying" diseases could serve as a legal excuse to avoid selecting an embryo that is really disfavored for other reasons, like looks, behavior, or sex.

Cosmetic traits and behaviors share the same line-drawing problems. Red hair and freckles constitute a cosmetic genetic trait but also confer a higher risk of melanoma. Is a higher risk of having a neurotic personality a behavioral trait or a disease? But perhaps the more important question about these categories is the justification for totally banning their use in selection. If a couple makes 100 embryos, why not let nondisease traits be tiebreakers—though how would one know if they were truly just tiebreakers?

The last category is the easiest to detect, often one of the most important to prospective parents, and one where we already have experience with arguable problems from genetic selection: sex. As discussed in Chapter 15, using largely ultrasound and abortion, millions of people around the world have acted on a preference for baby boys rather than baby girls. The result has been serious imbalances in sex ratios at birth in many countries and regions with more than 120 boys being born for every 100 girls as opposed to the usual ratio of about 104 to 100.

This clearly reflects a bias against female infants, whatever its cultural or economics roots. Whether it reinforces or, perhaps, in the long run, undercuts that bias is also unclear. Bans of sex-selective abortion have proven popular, although (and perhaps because) their enforcement has been spotty. They have existed for decades in China and India at the same time those countries have shown unnaturally high sex ratios at birth. They have even been adopted by several American states, in spite of the lack of any evidence of an overall sex ratio imbalance.

Easy PGD allows sex selection without abortion; embryos of the "wrong" sex would just not be transferred, at least not yet. (As many parents want children of each sex, but a boy first, subsequent use of these female embryos is certainly plausible.)

It is worth noting that in the United States, these direct regulations of parental choice among embryos might be challenged on federal constitutional grounds. As discussed in Chapter 18, the claims are not frivolous, although they may not appeal to the current U.S. Supreme Court. But, of course, that court will change in the next few decades; so may its view of the Constitution.

And any of the kinds of choices discussed above could be limited not just by direct bans but by banning information sharing. Thus, doctors might be allowed to tell parents only about the traits of their embryos for which selection was allowed. Among other advantages, this would prevent prospective parents from using a mild medical condition as an excuse for a selection really based on banned grounds, like sex.

Whose Cells Can Be Used in Easy PGD

Another form of specific regulation would focus not on the choices made by those using Easy PGD but on whose cells are used to make the gametes used in Easy PGD. Today, some countries put similar restrictions on IVF, limiting it to heterosexuals, to couples (no single women), to married people, or to women under a certain age. Other restrictions are sometimes placed on who can use other kinds of assisted reproduction, from bans on anonymous sperm donors to limits on the number of children a single donor's sperm can "father." Easy PGD would raise these questions, but also some novel ones, arising from the relative ease of getting cells that could be used to make iPSCs compared with directly getting gametes. Ponder some of the following questions.

Will age limits on genetic parenthood need to be created or extended? Although safety studies would be needed, it seems quite plausible that usable eggs could be made from skin cells of ninety-year-old women—or nine-month-old baby girls.

Number limits may also become more important. Although, in theory, men can provide enough sperm to sperm banks to father thousands of children, women can provide only a few eggs. Using the iPSC method, both men and women could provide an indefinitely large number of gametes. A particularly popular sperm or egg "donor" could have a vast number of genetic children. And some particularly popular "donors" might decide to go into the business of selling their derived eggs and sperm. Should people be allowed to sell their gametes to the highest bidders?

Similarly, gametes might be derived from the cells of the dead—the recently dead or, if tissue samples had been well preserved, the long dead. (Again, safety studies would be important.) Countries are beginning to face this issue with frozen embryos and sperm "donation" from recently dead men or the use of previously frozen sperm from less recently dead men. As egg freezing becomes more common, these issues will arise with eggs as well. No consensus seems to have emerged yet, and very little regulation, about dead people conceiving children. Easy PGD will make these issues more pressing.

Perhaps the issue around the dead should revolve around consent. Did they knowingly and intentionally give their consent to become posthumously conceiving parents? And what role should we grant the parties' marriage, or long-term partnered relationship, or previous children in determining, or affecting, consent? A focus on consent may have other applications. What about people whose cells were stolen and then used, without their consent, to make gametes? "Stealing" eggs and sperm from someone is difficult; "stealing" a few living cells would be easy—perhaps as easy as looking for living cells on the arm of a chair, a discarded drink can, or a toothbrush.

Finally, if, as discussed in Chapter 8, cells from men could be made into eggs and cells from women could be made into sperm, other issues will be raised. Some may be concerned about giving gay and lesbian couples the ability to have children who are equally genetically related to each parent. Another, wilder concern, is the "uniparent"—the person who uses his (I'm guessing most likely "his") skin cells to make both egg and sperm and, with a gestational surrogate, produces a "unichild." Should that new, and indeed previously inconceivable—in both senses of the word—kind of reproduction be banned?

Differential Access to Easy PGD

As discussed in Chapter 15, some would want to regulate Easy PGD to avoid distributional inequalities, either in its availability or in its use.

Financial concerns could be one differential barrier to access. In Chapter 9, I argued that insurers and public health care financing programs will provide Easy PGD free of charge to prospective parents because of its power to lower health care costs, but that could be wrong, particularly in such a politically charged field.

If it is expensive to prospective parents and some cannot afford it, should regulation intervene, either to forbid it to the rich or to provide it to the poor? Just how useful Easy PGD proves to be may make a difference. If Easy PGD children have huge advantages over others, regulation (including required subsidies) will seem more compelling than if the advantages are fairly minor.

What is easily affordable in the rich world may not be affordable at all in the poor world, or at least in the poor parts of the poor world. The upper classes in Guatemala, Burkina Faso, or Nepal may be able to afford Easy PGD; the poorest in those and other poor countries won't. Should the world require that all prospective parents, in all countries, have access to this technology and, if so, how?

The hardest question might be what, if anything, should be done to try to minimize differential use of Easy PGD. Even if it were financially accessible to all, not everyone will use it. And probably people from some religions, ethnic groups, or cultures will be less likely to use it. Would it be appropriate for a government to encourage (or, indeed, to require) its use?

Addressing Coercion

Some might want to regulate Easy PGD with regard to coercion. As discussed in Chapter 16, defining coercion gets tricky. The first reaction is to think it means preventing people from being forced to use Easy PGD or, possibly, forcing people to make particular uses of Easy PGD. Either action smacks of government-enforced eugenics, which is about the only thing in this field on which (almost?) everyone agrees. (It's bad.) One could easily imagine forbidding a government from forcing anyone to use Easy PGD or to make particular choices based on the information it provides. But that's not the only question about coercion that needs addressing.

Why stop with governments? Should private actors—insurance companies, health plans, employers, mothers-in-law—be forbidden from coercing someone to use Easy PGD? Should the government and private actors be forbidden from encouraging people to use Easy PGD, which might be viewed as an implicit form of coercion? What about the health plan (private or public) that agrees to cover Easy PGD and pregnancy, but refuses to pay for care for problems that could have been prevented if the parents had based their selections on the Easy PGD results? All these are ways in which the use of Easy PGD would be encouraged, if not fully required.

And is coercion, even state-sponsored coercion, always wrong? In the United States, states require that newborns be screened for particular genetic diseases, that school children (usually) be vaccinated, and that parents appropriately care for their children's health, at the risk of having the children taken away. If Easy PGD really is easy, safe, and effective, why shouldn't a government require it or require that a prospective parent who uses it avoid intentionally choosing an embryo that is certain to have a very serious genetic disease? Those are also questions respecting coercion.

And, finally, coercion can be turned on its head. What if the government coercion is seen not as forcing people to use Easy PGD but as prohibiting people from using Easy PGD? Or what if it's not forcing people to avoid selecting embryos with trisomy 13 but instead forcing them to select embryos that do not have trisomy 13. Viewed in this light, "coercion" implicates all of the possible regulations discussed in this section. And what if avoiding all regulation, when redefined as coercion, becomes the object of those who might want completely unfettered Easy PGD?

Two Broader Issues

Two questions cut across all the positions: questions of jurisdiction and questions of regulatory dynamics.

For better or for worse, there is no world government. Governments exist at all levels, from hamlets to provinces to nation-states and, arguably, to regional and international organizations such as the European Union, the North Atlantic Treaty Organization, the General Agreement on Tariffs and Trade, and the United Nations. The degree of influence or power that one government or government-like organization has on another, either de jure or de facto, differs dramatically.

This complicates the regulation of Easy PGD because, as described in Chapter 18, a government with restrictive laws has to worry about its nationals traveling to less restrictive jurisdictions to use Easy PGD. This raises two different questions: what level of government should regulate Easy PGD and what should a restrictive government try to do about its nationals going to less restrictive countries for kinds of Easy PGD it prohibits.

Where should regulation be sought? The more universal the level, the more difficult it usually is to get meaningful regulation. International agreements typically require consent from all the national governments involved. Unanimity among governments is rare, particularly for regulation with enforcement teeth—purely symbolic agreements are often easier to achieve. Getting an enforceable treaty regulating Easy PGD adopted widely, let alone universally, is likely to prove very difficult, particularly as different countries have already shown vastly different regulatory preferences with related technologies.

In some countries subnational jurisdictions may have the power to regulate PGD—in the United States the fifty states have clearer power than the federal government, something even truer in the Swiss confederation. And regional organizations like the European Union play an important and effective role in some places. For the most part, though, nation-states are the likely site of effective regulation. But that raises the problem, discussed earlier, of reproductive tourism and the difficulties in coping with it.

Times change, facts change, but regulatory systems do not always change with them. That can be a curse, or a blessing. Just how fixed should an Easy PGD regulatory regime be? If you have principles you think are truly timeless and important, you may want to carve them into regulatory stone. Otherwise, you may want to provide more flexibility for future, and different, generations. In any event, the actual effects of Easy PGD may diverge—whether for good or for bad—from predictions.

These are the last specific questions about regulating Easy PGD. How easy or hard should it be to change the regulatory regime? And what mechanism, if any, should be created to collect and analyze information about the effects of Easy PGD?

Choices: Yours and Mine

So there are your options, at least as I see them. Based on what you now know about Easy PGD and its possible (and plausible) regulation, what are your choices—what do you want to do?

If your goal is to try to prevent all use of Easy PGD, good luck. In the unlikely event you were able to get a universal international treaty that could truly be enforced, you might succeed, at least for a while. Short of that, a geographical ban could work fairly well, though, given reproductive tourism, not perfectly, at least in some particularly friendly jurisdictions. And only for a while.

If your goal is to try to prevent any regulation at all of Easy PGD, you'll need even more luck. At a bare minimum, safety regulation and limits on some cells as sources for gametes (stolen cells, for example) seem both inevitable and appropriate. And in many jurisdictions, regulation of at least some categories of parental choices will be politically almost impossible to avoid.

Most of the real choices are in the middle, for people who want or are willing to have Easy PGD used, at least when safe, by some people for some purposes. If you fall in that group, you have many decisions to make.

This book is not about my choices; it is about your choices—or, more accurately, our choices and those of our near-term successors. But it seems too coy not to tell you what I currently think I would do. I do stress, though, that you should give my choices no particular weight. Many of them depend on principles and personality traits that differ from person to person. After you've read this entire book, I can honestly say none of my choices depends on knowledge I have that you don't. I've put it all out.

So what would I do?

First, I would regulate the safety of the procedure very strongly, though not so strongly as to make approval impossible. I would require substantial preclinical work, in nonhuman primates among other nonhuman animal models, and with in vitro embryos that were not transferred for implantation. And I would require several years of follow-up of several hundred children born from the procedure before any possible approval. (How many "several"? I'm not sure.) I would also require continuing surveillance of Easy PGD children.

Second, I would not regulate any choices appropriately informed prospective parents make about the genetic traits of their own future children. I assume that the parents would almost always want the best for their children, although I know the world is big, some parents are crazy, and some bad decisions would be made. For me, the issue is whether I trust parents to make good decisions about the traits they want in their children more often than I trust governments. And the answer is yes, but just barely.

I am particularly conflicted about letting parents knowingly select an embryo with a severe genetic disease, such as Tay-Sachs disease. I would intervene to stop such a decision—if I thought it could be stopped without interventions spilling over into less clear-cut cases. I am consoled by the thought that very few prospective parents would make that choice.

I am also conflicted about sex selection, especially as I suspect a disproportionate number of parents in many countries would, at least at first, choose boys and that this would have some harmful effects. Again, though, I am reluctant to allow any intervention, particularly because I suspect this preference might well be self-correcting. I might not, however, fight very hard against efforts to prohibit either selection for terrible diseases or for sex, at least in places where sex choice seemed likely to be heavily skewed.

Third, I would largely ban making someone a genetic parent without his or her consent. That would necessarily eliminate using cells from children (at least those too young to consent), the incompetent, or those among the dead who had not already consented to such uses. (I might make an exception for some people who died without warning and who could be reliably expected, from other evidence, to want posthumously conceived children.) This ban would also prohibit using stolen or otherwise unauthorized cells. The entire requiring of consent might be best implemented by demanding that a clear provenance be provided with all gametes used in Easy PGD.

I find the idea of individual sperm and egg sellers distasteful, especially in a context where gamete donation would almost never be necessary. I am not, however, sure I have a good reason to ban it. Similarly, any potential uniparent seems to me likely to be outrageously egomaniacal and just plain silly, but I am not sure that is a good reason to ban the process. (There may, however, be good health reasons to require the prospective uniparent to put his embryos through PGD.)

I would subsidize Easy PGD to make it affordable to everyone—eventually, to everyone in the world. I would not require worldwide access before allowing it to be used anywhere. I really do not think the advantages to the Easy PGD children are enormous; for a while, poor countries might help their children more through nongenetic interventions, like clean water, electricity, good schools, and basic health care.

I would prohibit governments, or others, from forcing people to use Easy PGD or, consistent with my views on parental choice, forcing people to select, or not to select, particular future traits if they use Easy PGD. Some public health benefits might be lost, but, when it comes to making decisions about their not-yet-born children, I would rather let parents choose.

For the most part, I would work at a national or subnational (state or province) level. Different countries and cultures can reasonably have different views on these issues. I would probably support an international treaty banning governments from forcing people to use Easy PGD or to select, or avoid, particular traits, but I would not be optimistic about the treaty's chances of adoption or effective enforcement. I would mildly oppose efforts to ban reproductive tourism, which could provide a useful safety valve for restrictive countries.

Finally and, to me, most importantly, I would create a structure for monitoring the effects of Easy PGD, in the United States and elsewhere and for making regular policy recommendations about its use. I feel very strongly about this because successfully predicting the future is a chancy business. My guesses about how Easy PGD will affect the world could be wrong, as could anyone else's. I might even support a "sunset" provision for Easy PGD legislation, requiring that it be readopted, revised, or eliminated every few decades. The existence of a sunset provision could make it much easier for a possibly temporary majority to undo what I might consider good legislation. Still, I do believe regulatory schemes should be dynamic, paying attention to the actual facts on the grounds and their implications for policy. Sunset provisions can encourage that responsiveness.

But there is another, deeper, reason. I have few principles I am confident should apply in all cultures, to all situations, and over all of time. I believe, quite deeply, in things that almost all of my ancestors only 250 years ago would, no doubt, have found appalling, such as racial, sexual, and sexual-preference equality and freedom of, and from, religion. Few of my great-great-great-grandparents would probably have agreed. What things do I believe today that my grandchildren, let alone my great-great-great-grandchildren, will view as bizarre? I don't know, but I suspect there are some. And the world of fifty, one hundred, or two hundred years from now will be *their* world, not mine. Or yours. They should have the right and the power to run it.

Those are my choices, and my reasons for them, at least today. I cannot guarantee they will be the same tomorrow, neither do I think you should necessarily make the same ones. But you do need to think about what choices you would make. Because choices will be made. And if informed people do not participate in making those choices, ignorant people will make them. That cannot be encouraging.

You are now among the most informed people in the world about Easy PGD and humanity's reproductive future. I charge you—I beg you—use that information. Pay attention to these issues, think about them, talk about them with others. Help us all to shape a world where these new technologies bring as much benefit, with as little harm, as humanly possible.

THE END

of the End of Sex.

And its beginning.

NOTES

Introduction

1. Aldous Huxley, *Brave New World* (London: Chatto & Windus, 1970).
2. *Gattaca,* directed by A. Niccol (Los Angeles: Columbia Pictures Corp., 1997), motion picture.
3. William Shakespeare, *The Tempest*, ed. Robert Langbaum (New York: Signet Classic, 1964), 5.1.182–184.

1. Cells, Chromosomes, DNA, Genomes, and Genes

1. Samuel Butler, *Life and Habit* (London: A. C. Fifield, 1910), 134. Apparently, despite Butler's disclaimer, it is his statement. As far as I can tell, no one has yet identified an earlier source for it.
2. Richard Dawkins, *The Selfish Gene* (New York: Oxford University Press, 1976).
3. I am not providing references for the basic science in this book; I will provide references for some more specific points. My knowledge of the science has been built over twenty-five years of reading books and articles and (often) talking with scientists (and specifically confirmed by having specialist friends look over the relevant chapters). If you are interested in getting more detail on the general biological information in this and following chapters, most of it can be found in any introductory biology text, such as Geoffrey Cooper and Robert Hausman, *The Cell, A Molecular Approach,* 6th ed. (Sunderland, MA: Sinauer Associates, 2013); David E. Sadava, David M. Hillis, H. Craig Heller, and May Berenbaum, *Life—The Science of Biology,* 10th ed. (New York: W. H. Freeman, 2012); Keith Roberts et al., *Molecular Biology of the Cell,* 6th ed. (New York: Garland Science, 2015), to name a few. These books are intended for freshman and sophomore undergraduates. If you want a somewhat more accessible

source, try René Fester Kratz, *Molecular and Cell Biology for Dummies* (Hoboken, NJ: Wiley Publishing, 2009).

4. This is actually a very long and still continuing debate where the answer, of course, depends on your definition of "alive." See Luis P. Villareal, "Are Viruses Alive?," *Scientific American* 291(6) (2004): 100. I think the closest analogy may be a naval mine, deployed in the ocean, that is magnetically attracted to passing ships. It attaches, delivers its payload, and wreaks havoc.

5. Slime molds have their own complicated relationships and variations. Plasmodial slime molds occur when individual cells come together and fuse into one enormous—as much as a meter across—cell with thousands of nuclei. See "Introduction to the 'Slime Molds,'" University of California Museum of Paleontology, available at http://www.ucmp.berkeley.edu/protista/slimemolds.html.

6. Skeletal muscles in humans, like our biceps, are made up of muscle fibers that are the result of the fusion of many individual cells. These muscle cells have multiple nuclei that are located on the edges of the cells. See "General Anatomy of Skeletal Muscle Fibers," *GetBodySmart,* available at http://www.getbodysmart.com/ap/muscletissue/fibers/generalanatomy/tutorial.html.

7. I cannot resist recommending Watson's memoir of the discovery of DNA. Although some have questioned the book's accuracy, it is an engaging look, from Watson's perspective at least, at the discovery of the structure of DNA. James D. Watson, *The Double Helix: A Personal Account of the Discovery of the Structure of DNA* (New York: Athenaeum, 1968). A fuller discussion of the discovery of DNA's structure—and much else in the history of molecular biology, can be found in Horace Freeland Judson, *The Eighth Day of Creation: Makers of the Revolution in Biology* (New York: Simon & Schuster, 1979) or its expanded edition (Plainview, NY: Cold Spring Harbor Press, 1996). A good, though also perhaps biased, source for the story of Rosalind Franklin is Anne Sayre, *Rosalind Franklin and DNA* (New York: Norton, 1975).

8. James D. Watson and Francis H. C. Crick, "A Structure for Deoxyribose Nucleic Acid," *Nature* 171 (1953): 737–738.

9. The Genome Reference Consortium's human web page, available at http://www.ncbi.nlm.nih.gov/projects/genome/assembly/grc/human, provides more information about its activities, including about the Human Reference Sequence.

10. Oswald T. Avery, Colin M. MacLeod, and Maclyn McCarty, "Studies on the Chemical Nature of the Substance Inducing Transformation of Pneumococcal Types: Induction of Transformation by a Desoxyribonucleic Acid Fraction Isolated from Pneumococcus Type III," *Journal of Experimental Medicine* 79 (1944): 137–158. This classic paper is also a nice example of how much more abbreviated scientific publications have become over the years; its twenty-one pages today would have to be contracted to four or five with a major loss of detail.

11. For a clear and useful discussion of the somewhat controversial question of how much of our DNA is used, see Tabitha M. Powledge, "How Much of Human DNA Is Doing Something?," *Genetic Literacy Project,* available at http://www.geneticliteracyproject.org/2014/08/05/how-much-of-human-dna-is-doing-something/.

12. Arthur Conan Doyle, *The Sign of the Four* (1890), Chapter 6 (emphasis in original).

2. Reproduction: In General and in Humans

1. For useful and very accessible introductions to this question, see Carl Zimmer, "On the Origin of Sexual Reproduction," *Science* 324 (2009): 1254–1256; Zimmer, "Why Is There Sex? To Fight the Parasite Army," *The Loom* (Blog) (2011), available at http://blogs.discovermagazine.com/loom/2011/07/07/why-is-there-sex-to-fight-the-parasite-army/#.Vcpi7EXviZo. Another useful source (yes, even academics use it) is *Wikipedia,* see "Evolution of Sexual Reproduction," available at https://en.wikipedia.org/wiki/Evolution_of_sexual_reproduction.

2. See Lutz Becks & Aneil F. Agrawal, "Higher Rates of Sex Evolve in Spatially Heterogeneous Environments," *Nature* 468 (2010): 89–92. The paper is discussed in Jef Akst, "Why Sex Evolved," *The Scientist* (2010) available at http://www.the-scientist.com/?articles.view/articleNo/29307/title/Why-sex-evolved/.

3. This "intersex" condition is fascinating and increasingly politically controversial, down even to the question of how rare it is. Estimates range from about 0.1 to about 1.7 percent. For those interested, I highly recommend a book by one of my Stanford colleagues, Katrina Karkazis, *Fixing Sex: Intersex, Medical Authority, and Lived Experience* (Durham, NC: Duke University Press, 2009). This book also will not discuss the fascinating and complex issues raised by transgendered people, who believe that their "true" sex is, or ought to be, other than the one implied by the reproductive organs, or the chromosomes, they were born with. Both conditions, though, are further evidence of the complexities of both sex and gender.

4. The replacement approaches have not prevented continuing controversies, including one resolved (at least for the time being) by the Court of Arbitration for Sport in July 2015. See Juliet Macur, "The Line between Male and Female Athletes Remains Blurred," *New York Times,* July 28, 2015, B12.

5. That is the general consensus. One researcher, Dr. Jonathan Tilly, has claimed that primary oocytes can occasionally be created after birth, at least in mice, and that he has isolated egg stem cells both from adult mice and from the ovaries of adult women. See Yvonne A. R. White et al., "Oocyte Formation by Mitotically Active Germ Cells Purified from Ovaries of Reproductive-Age

Women," *Nature Medicine* 18 (2012): 413–421. These findings are very controversial in the field; I think it is fair to say the jury is still out.

3. Infertility and Assisted Reproduction

1. Anjani Chandra, Casey E. Copen, and Elizabeth Hervey Stephen, "Infertility and Impaired Fecundity in the United States, 1982–2010: Data from the National Survey of Family Growth," *National Health Statistics Report* 67 (2013): 1–18.

2. The following discussion of the history of artificial insemination is largely, but not entirely, based on Alan F. Guttmacher, "Artificial Insemination," *DePaul Law Review* 18 (1969): 566–583. For a history of artificial insemination in humans, see Willem Ombelet and Johan Van Robays, "Artificial Insemination History: Hurdles and Milestones," *Facts Views Vis Obgyn* 7(2) (2015): 137–143. For its history in nonhumans, see R. H. Foote, "The History of Artificial Insemination: Selected Notes and Notables," *Journal of Animal Science* 80 (2002): 1–10.

3. The Pancoast story, however, was only reported in a letter to a medical journal twenty-five years later, after Pancoast's death, from one of his former students. This may undercut its credibility. Addison Davis Hard, letter to the editor, "Artificial Impregnation," *The Medical World* (1909), discussed in A. Gregoire and Robert Mayer, "The Impregnators," *Fertility and Sterility* 16 (1965): 130–134.

4. Oddly, its use in nonhuman animals remains controversial in at least one context. Horses conceived by artificial insemination are not recognized by the American Thoroughbred Association and hence cannot race in thoroughbred races. The justification given concerns some vital spark provided by the stallion; it is widely believed that this is an effort to affect the stud market by limiting the number of mares any one stallion can impregnate.

5. Caitrin Nicol Keiper, "Brave New World at 75," *The New Atlantis*, no. 16 (Spring 2007): 41–54.

6. Barry D. Bavister, "Early History of *In Vitro* Fertilization," *Reproduction* 124 (2002): 181–196. Bavister calls this work "more convincing" than other early claims but not conclusive. Pincus also claimed, clearly incorrectly, in a 1936 book to have produced a living rabbit through parthenogenesis. The rabbit appeared in *Look* magazine, and Pincus did not get tenure at Harvard; some have inferred a causal connection between those events. Pincus went on to greater fame as one of the parents of the oral contraceptive pill.

7. Editorial, "Conception in a Watch Glass," *New England Journal of Medicine* 217 (1937): 678. A watch glass is a circular, slightly convex-concave piece of glass used by chemists to hold liquids or solids, or to prevent solids from

contaminating beakers. They resemble the glass that forms the front of pocket watches. "Watch Glass," *Wikipedia,* available at https://en.wikipedia.org/wiki/Watch_glass.

8. Harvard professor John Rock apparently admitted years later than he had been the author. John D. Biggers, "IVF and Embryo Transfer: Historical Origin and Development," *Fertility* 16 (2013): 5–15, available at http://www.ivfonline.com/Portals/0/SiteAssets/Images/Fertility%20Magazine/FWMag_Vol_16.pdf.

9. Robert G. Edwards, Barry D. Bavister, Patrick C. Steptoe, "Early Stages of Fertilization in Vitro of Human Oocytes Matured in Vitro," *Nature* 221 (1969): 632–635.

10. Robin Marantz Henig, *Pandora's Baby* (Cold Spring Harbor, NY: Cold Spring Harbor Laboratories Press, 2006).

11. Jennifer F. Kawwass et al., "Safety of Assisted Reproductive Technology in the United States, 2000–2011," *Journal of the American Medical Association* 313 (2015): 88–90. These figures are much lower than earlier estimates of a greater than 1 percent hospitalization rate as a result of IVF. Still, they report that sixteen women have died, probably as a result of the stimulation, between 2000 and 2011 (though out of 1.1 million cycles).

12. Division of Reproductive Health, *Assisted Reproduction Technology: 2012 National Summary Report* (Atlanta, GA: National Center for Chronic Disease Prevention and Health Promotion, Centers for Disease Control and Prevention, 2014), available at http://www.cdc.gov/art/reports/2012/national-summary.html. Subsequent references to recent U.S. results for IVF are also from this source.

13. See the Progyny website, available at https://www.progyny.com/, and especially its discussion of their Eeva test.

14. There is also some evidence that maternal mortality throughout pregnancy is higher for IVF pregnancies than normal ones, as much as seven times higher according to one Dutch study. It is hard to know what to make of that result, as women using IVF for reproduction will be different from those who get pregnant naturally, from average age to health conditions that might contribute to their fertility problems.

15. Donna Rosata, "How High-Tech Babymaking Fuels the Infertility Market Boom," *Money,* July 9, 2014, available at http://time.com/money/2955345/high-tech-baby-making-is-fueling-a-market-boom/.

4. Genetics

1. See the nice discussion of Mendel's work in A. J. F. Griffiths et al., "Mendel's Experiments," *An Introduction to Genetic Analysis* (New York: W. H. Freeman, 2000). The section discussing his experiments is online; see "Mendel's

Experiments," NCBI, available at http://www.ncbi.nlm.nih.gov/books/NBK22098/.

2. Ironically, Mendel's own reported statistics are too good to be (most likely) true. He found an almost perfect 3:1 ratio of dominant to recessive traits in second-generation crosses, a perfection that one famous statistician, R. A. Fisher, concluded in 1936 was so statistically improbable that he accused the father of genetics of fraudulently "cleaning up" his results. Other, more recent authors have defended Mendel, concluding that Fisher greatly overstated Brother Mendel's sins, if any. See a nice discussion of the controversy (with a proposed solution) in Moti Nissani, "Psychological, Historical, and Ethical Reflections on the Mendelian Paradox," *Perspectives in Biology and Medicine* 37 (1994): 182–198.

3. In his great new book, Philip Reilly tells the Queen Victoria story with a twist I had never heard. Philip R. Reilly, *Orphan: The Quest to Save Children with Rare Genetic Disorders* (Cold Spring Harbor, NY: Cold Spring Harbor Laboratories Press, 2015), 68–74. Robert Massie, a parent of a child with hemophilia learned about Victoria's legacy of hemophilia, especially in Tsar Nicholas II's family. He not only became a leader in the hemophilia support movement but also the very successful biographer of Nicholas and his wife (Victoria's granddaughter), Alexandra—because of his own family interest in hemophilia. Robert Massie, *Nicholas and Alexandra: An Intimate Account of the Last of the Romanovs and the Fall of Imperial Russia* (New York: Atheneum, 1967). He ended up writing many books about the 300-year Romanov dynasty, winning a Pulitzer Prize for his biography of Peter the Great.

4. This sentiment is reflected in my favorite quotation from the very quotable Supreme Court justice Oliver Wendell Holmes Jr.: "The life of the law has been experience, not logic." Holmes was a brilliant and fascinating figure—the leading icon of American legal history and a person about whom I have strong and conflicting views.

5. Charles Davenport was the founder and longtime director of the Eugenic Records Office at Cold Spring Harbor. (He also directed two other laboratories there that became the current Cold Spring Harbor Laboratories.) Davenport believed that simple Mendelian genetics underpinned the facts that some families had many sailors ("thalassophilia") or that some populations, such as Gypsies or Comanches, were nomads. Daniel Kevles, *In the Name of Eugenics* (Cambridge, MA: Harvard University Press, 1998).

6. The National Institutes of Health currently puts the risk of breast cancer before age seventy at 55 to 65 percent for people carrying high-risk *BRCA1* mutations and 45 percent for those carrying *BRCA2* mutations. For ovarian cancer the risks of cancer by age seventy are 39 percent for *BRCA1* and 11 to 17 percent for *BRCA2*. "BRCA1 and BRCA2: Cancer Risk and Genetic Testing," National Cancer Institute, available at http://www.cancer.gov/about-cancer/causes-prevention/genetics/brca-fact-sheet.

7. A 2008 blog post from *23andMe* shows its assessment of the relative risks conferred by having a variety of SNPs associated with Crohn's disease. Of the twenty-three SNPs they show, only one has a relative risk of more than 1.33 and its risk is only 1.54. That means that if the population risk were 0.7 percent, having a SNP that gave you a 1.54 relative risk would put your risk all the way up to about 1.08 percent, which does not seem very useful information. "SNPwatch: Number of SNPs Associated with Crohn's Disease Triples," *23andMe* (blog), available at http://blog.23andme.com/23andme-research/snpwatch/snpwatch-number-of-snps-associated-with-crohn%E2%80%99s-disease-triples/. The higher numbers came from adding together risks from some of the SNPs, though no one has any idea whether their risks are simply additive.

8. Thomas D. Bird, "Early-Onset Familial Alzheimer Disease," *GeneReviews,* available at http://www.ncbi.nlm.nih.gov/books/NBK1236/. The article also discusses two rarer genes that, when mutated, are highly penetrant for early onset Alzheimer disease, *Presenilin 2* and *Amyloid Precursor Protein* gene.

9. A good popular discussion of the *APOE-4* allele, its discovery and effects, is found in Laura Spinney, "Alzheimer's Disease: The Forgetting Gene," *Nature* 510 (2014): 26–28. For more detail, see Shivani Garg, "Alzheimer Disease and APOE-4," *MedScape,* available at http://emedicine.medscape.com/article/1787482-overview. More detail still can be found in Thomas D. Bird, "Alzheimer Disease Overview," *GeneReviews,* available at http://www.ncbi.nlm.nih.gov/books/NBK1161/.

10. Jim Haggerty, "Do People Inherit Schizophrenia," *PsychCentral,* available at http://psychcentral.com/lib/do-people-inherit-schizophrenia/. This reference gives not only risks for identical twins but also for many other family relationships.

11. "GIANT Study Reveals Giant Number of Genes Linked to Height," Broad Institute, available at https://www.broadinstitute.org/news/6119. This is a press release about a scientific paper from researchers at the Broad Institute, not usually the best source for unbiased information. This article is striking, though, as it says the study findings roughly double the known number of gene regions that influence height—to about 400. The striking part is that even that large number is only expected to increase the explainable percentage of height from 12 to about 20 percent. The underlying paper is Andrew R. Wood et al., "Defining the Role of Common Variation in the Genomic and Biological Architecture of Adult Human Height," *Nature Genetics* 46 (2014): 1173–1186.

12. The genetics literature contains substantial discussion about "missing heritability" and how either to find it or explain it away. See, e.g., Or Zuk, Eliana Hechtera, Shamil R. Sunyaev, and Eric S. Lander, "The Mystery of Missing Heritability: Genetic Interactions Create Phantom Heritability," *Proceedings of the National Academy of Sciences* 109 (2011): 1193–1198.

13. "Facts about Down Syndrome," CDC, available at http://www.cdc.gov/ncbddd/birthdefects/downsyndrome.html.

14. "Patau Syndrome," National Health Service, National Genetics and Genomics Education Center, available at http://www.geneticseducation.nhs.uk/genetic-conditions-54/691-patau-syndrome-new; "Edwards Syndrome," National Health Service, National Genetics and Genomics Education Center, available at http://www.geneticseducation.nhs.uk/genetic-conditions-54/651-edwards-syndrome-new.

15. Jeannie Visootsak and John M. Graham, Jr., "Klinefelter Syndrome and Other Sex Chromosomal Aneuploidies," *Orphanet Journal of Rare Diseases* 1, no. 8 (2006): 1.

16. For an example of the complexity, and the promise, of studying the relationship of CNVs to disease (in this case, autism), see Jason L. Stein, "Copy Number Variation and Brain Structure: Lessons Learned from Chromosome 16p11.2," *Genome Medicine* 7 (2015), available at http://www.genomemedicine.com/content/7/1/13.

17. Jill U. Adams, "Imprinting and Genetic Disease: Angelman, Prader-Willi and Beckwith-Weidemann Syndromes," *Scitable,* available at http://www.nature.com/scitable/topicpage/imprinting-and-genetic-disease-angelman-prader-willi-923.

5. Genetic Testing

1. "Timeline: History of Genetic Genealogy," International Society of Genetic Genealogists, available at http://www.isogg.org/wiki/Timeline:History_of_genetic_genealogy. For a discussion and critique of early genetic genealogy, see Henry T. Greely, "Genetic Genealogy: Genetics Meets the Marketplace," in Barbara A. Koenig, Sandra Soo-Jin Lee, and Sarah Richardson, eds., *Revisiting Race in a Genomic Age* (New Brunswick, NJ: Rutgers University Press, 2008), 271–299. For some background on 23andMe, see "23andMe," *Wikipedia,* available at https://en.wikipedia.org/wiki/23andMe.

2. "Warning Letter," U.S. Food and Drug Administration, November 22, 2013, available at www.fda.gov/ICECI/EnforcementActions/WarningLetters/2013/ucm376296.htm. For contrasting views of the FDA action, compare Patricia J. Zettler, Jacob S. Sherkow, and Henry T. Greely, "23andMe, the Food and Drug Administration, and the Future of Genetic Testing," *JAMA Internal Medicine* 174 (2014): 403–404, with Robert C. Green and Nita A. Farahany, "Regulation: The FDA Is Overcautious on Consumer Genomics," *Nature* 505 (2014): 286–287.

3. Clyde A. Hutchinson III, "DNA Sequencing: Bench to Bedside and Beyond," *Nucleic Acids Research* 35, no. 18 (September 2007): 6227–6237.

4. Mayo Clinic, "Down Syndrome: Risk Syndromes," available at http://www.mayoclinic.org/diseases-conditions/down-syndrome/basics/risk-factors/

con-20020948. For a more detailed breakdown of risk of down syndrome by parental age, see J. K. Morris, D. E. Mutton, and E. Alberman, "Revised Estimates of the Maternal Age Prevalence of Down Syndrome," *Journal of Medical Screening* 9, no. 1 (March 2002): 2–6. There are also indications that risk of trisomy rises with paternal age as well. See Harry Fisch, Grace Hyun, Robert Golden, Terry W. Hensle, Carl A. Olsson, and Gary L. Liberson, "The Influence of Paternal Age on Down Syndrome," *Journal of Urology* 169, no. 6, (June 2003): 2275–2278. A recent Washington Post article discussed these health risks. Ana Swanson, "Why Men Should Also Worry about Waiting Too Long to Have Kids," *Washington Post*, October 27, 2015, available at https://www.washingtonpost.com/news/wonkblog/wp/2015/10/27/men-have-biological-clocks-too-so-why-does-no-one-talk-about-them/.

5. Neural tube defects occur when the fetus does not form skin over parts of its central nervous system, the brain, or the spinal cord. This allows the amniotic fluid to be in contact with the uncovered part of the central nervous system, which in turn causes damage. If the open area is along the spinal cord, it can lead to spina bifida, which causes spinal cord damage leading to loss of sensation and motor control. If the open area is in the head, it can lead to anencephaly, an invariably fatal condition where a child is born with the large and crucial part of the brain called the cerebrum either missing or greatly shrunken. "Neural Tube Defects," March of Dimes, available at http://www.marchofdimes.org/baby/neural-tube-defects.aspx.

6. The current maternal serum screening test is known as the "quad screen." It looks to the levels of four chemicals in the mother's blood to diagnose neural tube defects, Down syndrome, and Trisomy 18. "Quad Screen," *Mayo Clinic Medical Laboratories*, available at http://www.mayomedicallaboratories.com/test-catalog/Clinical+and+Interpretive/81149. The quad screen began to replace the "triple test," which had been used since the 1980s and only looked to three chemicals, in the early 2000s, since it offered greater accuracy than the triple test. Michael R. Lao, Byron C. Calhoun, Luis A. Bracero, Ying Wang, Dara J. Seybold, Mike Broce, and Christos G. Hatjis, "The Ability of the Quadruple Test to Predict Adverse Perinatal Outcomes in a High-Risk Obstetric Population," *Journal of Medical Screening* 16, no. 2 (June 2009): 55–59.

7. The remarkably successful Tay-Sachs screening story is told in Philip R. Reilly, *Orphan: The Quest to Save Children with Rare Genetic Disorders* (Cold Spring Harbor, NY: Cold Spring Harbor Laboratories Press, 2015), 89–91.

8. The story behind Dor Yeshorim, the Jewish organization that provides genetic matchmaking services, can be found in Kara Stiles, "Tinder for Tay-Sachs," *Tablet,* available at http://www.tabletmag.com/jewish-life-and-religion/187177/tinder-for-tay-sachs.

9. See the Counsyl website, https://www.counsyl.com/.

10. Some people have tried to find and test the actual fetal cells in the pregnant woman's blood before they die. Such cells are known to exist and were the subject of great excitement in the mid-1990s when they were first discovered. There are true problems, though, with fetal cell testing. First, finding the fetal cells is like finding a needle in a mammoth haystack; they make up less than one cell in a million in the pregnant woman's blood. Second, some fetal cells actually seem to engraft in the pregnant woman's body and produce daughter cells for years or even decades to come. A mother may be carrying around inside her cells from all of her children. That is a sweet thought, but it means that if you find a whole fetal cell in a pregnant woman's bloodstream, how do you know it is from this pregnancy and not an earlier one?

11. This study, published in April 2015, used the parents' genomes to help infer the embryos' genomes. Akash Kumar et al., "Whole Genome Prediction for Preimplantation Genetic Diagnosis," *Genome Medicine* 7(35) (2015), doi 10.1186/s13073-015-0160-4. In February 2015, another study showed that de novo mutations could be detected as well, ones that could not be anticipated from knowledge of the parents' genomes. Brock A. Peters et al., "Detection and Phasing of Single Base de Novo Mutations in Biopsies from Human In Vitro Fertilized Embryos by Advanced Whole-Genome Sequencing," *Genome Research,* available at http://genome.cshlp.org/content/early/2015/02/05/gr.181255.114.abstract. Which method will work best in twenty to forty years—if either—is unknowable, but the progress to date makes such embryonic whole genome sequencing seem very plausible.

12. These safety issues are discussed in more detail in Chapter 13.

13. Division of Reproductive Health, *Assisted Reproduction Technology: 2012 National Summary Report* (Atlanta, GA: National Center for Chronic Disease Prevention and Health Promotion, Centers for Disease Control and Prevention, 2014), available at http://www.cdc.gov/art/reports/2012/national-summary.html.

6. Stem Cells

1. The information in the text about Carrel's cell culture work, and more, comes from Jan A. Witkowski, "Alexis Carrel and the Mysticism of Cell Culture," *Medical History* 23 (1979): 279–296. His *Wikipedia* entry adds some further interesting aspects to his life. "Alexis Carrel," *Wikipedia,* available at https://en.wikipedia.org/wiki/Alexis_Carrel.

2. Lijing Jiang, "Alexis Carrel's Immortal Chicken Heart Cultures (1912–1946)," *The Embryo Project Encyclopedia,* available at https://embryo.asu.edu/pages/alexis-carrels-immortal-chick-heart-tissue-cultures-1912-1946.

3. Bill Cosby, "Chicken Heart—aka Bill's Fainting Ma," *Wonderfulness* (Burbank, CA: Warner Brothers, 1966), 33⅓ rpm.

4. Zane Bartlett, "Leonard Hayflick (1928–)," *Embryo Project Encyclopedia,* available at https://embryo.asu.edu/pages/leonard-hayflick-1928.

5. Craig R. Whitney, "Jeanne Calment, World's Elder, Dies at 122," *New York Times,* August 5, 1997, available at http://www.nytimes.com/1997/08/05/world/jeanne-calment-world-s-elder-dies-at-122.html; "Jeanne Calment," *The Economist,* August 14, 1997. Calment died nearly twenty years ago, but no one has yet come within three years of her age record. See "List of the Verified Oldest People," *Wikipedia,* available at https://en.wikipedia.org/wiki/List_of_the_verified_oldest_people.

6. Rebecca Skloot, *The Immortal Life of Henrietta Lacks* (New York: Crown Publishing, 2010).

7. See Samuel E. Senyo, Richard T. Lee, and Bernhard Kühn, "Cardiac Regeneration Based on Mechanisms of Cardiomyocyte Proliferation and Differentiation," *Stem Cell Research* 13 (2014): 532–541.

8. See M. J. Evans and M. H. Kaufman, "Establishment in Culture of Pluripotential Cells from Mouse Embryos," *Nature* 292 (1981): 154–156; G. R. Martin, "Isolation of a Pluripotent Cell Line from Early Mouse Embryos Cultured in Medium Conditioned by Teratocarcinoma Stem Cells," *Proceedings of the National Academy of Sciences* 78 (1981): 7634–7638. For more on the early history of stem cell research, see Davor Solter, "From Teratocarcinomas to Embryonic Stem Cells and Beyond: A History of Embryonic Stem Cell Research," *Nature Reviews Genetics* 7 (2006): 319–327.

9. For a nice biography of Thomson, see Ke Wu, "James Alexander Thomson (1958–)," *The Embryo Project Encyclopedia,* available at https://embryo.asu.edu/pages/james-alexander-thomson-1958. See also Gina Kolata, "The Man Who Started the Stem Cell War May Help End It," *New York Times,* November 22, 2007, A1.

10. James A. Thomson et al., "Isolation of a Primate Embryonic Stem Cell Line," *Proceedings of the National Academy of Sciences* 92 (1995): 7844–7848.

11. James A. Thomson et al., "Embryonic Stem Cell Lines Derived from Human Blastocysts," *Science* 282 (1998): 1145–1147.

12. At about the same time, John Gearhart, then at Johns Hopkins University, developed cells with similar properties based not on cells from the inner cell mass of a blastocyst, but on primordial germline cells, the very cells we discussed in Chapter 4 that give rise, ultimately, to eggs or sperm. Michael J. Shamblott et al., "Derivation of Pluripotent Stem Cells from Cultured Human Primordial Germ Cells," *Proceedings of the National Academy of Sciences* 95 (1998): 13726–13731. Although these cells had already begun to differentiate, Gearhart showed that they were broadly pluripotent, able to form many different cell types. (Gearhart's work was also funded by Geron.) It was initially

unclear whether Thomson's cells or Gearhart's cells would prove more useful (and hence be more commonly used), but the hESCs won out.

13. President Bush announced his policy at a press conference at his ranch in Texas, almost exactly before September 11, 2001, made stem cell research a much lower priority in his administration. For most of his two terms, the policy existed only as his statement in the press conference. "President Discusses Stem Cell Research," The White House, August 9, 2001, available at http://georgewbush-whitehouse.archives.gov/news/releases/2001/08/20010809-2.html. See the discussion by Samuel Philbrick, "President George W. Bush's Announcement on Stem Cells, 9 August 2001," *The Embryo Project Encyclopedia,* available at https://embryo.asu.edu/pages/president-george-w-bushs-announcement-stem-cells-9-august-2001.

14. The full text of the proposition can be found here: Proposition 71: The California Stem Cell Research and Cures Initiative, available at https://www.cirm.ca.gov/about-cirm/history. For more background, see Ceara O'Brien, "California Proposition 71 (2004)," *The Embryo Project Encyclopedia,* available at https://embryo.asu.edu/pages/california-proposition-71-2004.

15. South Dakota Code, 34-14-16, makes the knowing conduct of nontherapeutic research that destroys human embryos a Class A misdemeanor.

16. Stem Cell Treatment for Eye Diseases Shows Promise," *New York Times*, January 23, 2010, available at http://www.nytimes.com/2012/01/24/business/stem-cell-study-may-show-advance.html; Andrew Pollack, "Stem Cell Trial Wins Approval of FDA," *New York Times*, July 30, 2010, available at http://www.nytimes.com/2010/07/31/health/research/31stem.html. For a general overview of approved trials, see Wikipedia, "Human embryonic stem cells clinical trials," available at https://en.wikipedia.org/wiki/Human_embryonic_stem_cells_clinical_trials.

17. "Human Leukocyte Antigen," *Wikipedia,* available at https://en.wikipedia.org/wiki/Human_leukocyte_antigen.

18. "Compatible Living Donor Kidney Transplant," Renal and Pancreatic Transplant Center, Columbia University Medical Center, available at http://columbiasurgery.org/conditions-and-treatments/compatible-living-donor-kidney-transplant.

19. I. Wilmut et al., "Viable Offspring Derived from Fetal and Adult Mammalian Cells," *Nature* 385 (1997): 810–813.

20. The use of mammary tissue is why Dolly was named Dolly—after American country singer Dolly Parton, whose truly substantial musical (and business) talents were sometimes overshadowed by her substantially enhanced mammary organs.

21. Keith Campbell et al., "Sheep Cloned by Nuclear Transfer from a Cultured Cell Line," *Nature* 380 (1996): 64–66.

22. Dolly the sheep was euthanized in February 2003 as a result of a progressive lung disease. She is now on display on a rotating platform, protected within a glass class, in the National Museum of Scotland in Edinburgh. If you are in Edinburgh, she's worth a visit.

23. For an overview of state laws banning human reproductive cloning, see Nactional Conference of State Legislatures, "Human Cloning Laws," available at http://www.ncsl.org/research/health/human-cloning-laws.aspx.

24. "List of Animals That Have Been Cloned," *Wikipedia*, available at https://en.wikipedia.org/wiki/List_of_animals_that_have_been_cloned.

25. J. A. Byrnet et al., "Producing Primate Embryonic Stem Cells by Somatic Cell Nuclear Transfer," *Nature* 450 (2007): 497–502.

26. The religious sect known as the Raelians still claim that they started cloning humans in 2002 but, out of respect for the clones' privacy, they won't let anyone see or test them.

27. There is probably a good book in the whole sordid Woo-Suk Hwang scandal, but I have not found it yet. In the meantime, for some good background, see Rhodri Saunders and Julian Savulescu, "Research Ethics and Lessons from Hwanggate: What Can We Learn from the Korean Cloning Fraud," *Journal of Medical Ethics* 34 (2008): 214–221.

28. Masahito Tachibana et al., "Human Embryonic Stem Cells Derived by Somatic Cell Nuclear Transfer," *Cell* 153 (2013): 1228–1238.

29. Monya Baker, "Stem Cells Made by Cloning Adult Humans," *Nature News*, April 28, 2014, available at http://www.nature.com/news/stem-cells-made-by-cloning-adult-humans-1.15107.

30. Kazutoshi Takahashi and Shinya Yamanaka, "Induction of Pluripotent Stem Cells from Mouse Embryonic and Adult Fibroblast Cultures by Defined Factors," *Cell* 126 (2006): 663–676.

31. Kazutoshi Takahashi et al., "Induction of Pluripotent Stem Cells from Adult Human Fibroblasts by Defined Factors," *Cell* 131 (2007): 861–872. Jamie Thomson's article on deriving human iPSCs was released on the same day as Yamanaka's, but it was following Yamanaka's mouse work, so Thomson does not get equal credit. Junying Yu et al., "Induced Pluripotent Stem Cell Lines Derived from Human Somatic Cells," *Science* 318 (2007): 1917–1920. (I do think Thomson deserved some of the Nobel Prize in 2012 that went to Yamanaka and early frog cloning pioneer John Gurdon. I suspect the politics of human embryonic stem cells kept him from winning it.)

32. This process is being called "direct reprogramming." See Konrad Hochedinger, "Reprogramming: The Next Generation," *Cell Stem Cell* 11 (2012): 740–743; Eduardo Marbán and Eugenio Cingolani, "Direct Reprogramming: Bypassing Stem Cells for Therapeutics," *Journal of the American Medical Association* 314 (2015): 19–20.

33. Somewhat disconcertingly, the Japanese iPSC trials were stopped in 2015 and are to be restarted using hESC-derived cells. Although the reason is not entirely clear, coming regulatory changes in Japan are expected to encourage use of hESCs, taken from banks containing cells with many different HLA combinations, instead of iPSCs. Paul Knoepfler, "Historic turning point for IPS cell field in Japan?" The Niche (blog), available at https://www.ipscell.com/2015/08/ipscstudysuspend/.

7. Genetic Analysis

1. See Erwin L. van Dijk, Helène Auger, Yan Jaszczyszyn, and Claude Thermes, "Ten Years of Next-Generation Sequencing Technology," *Trends in Genetics* 30 (2014): 418–426; Elaine R. Mardis, "Next Generation Sequencing Platforms," *Annual Review of Analytic Chemistry* 6 (2013): 287–303.

2. "Remarks Made by the President, Prime Minister Tony Blair of England (via Satellite), Dr. Francis Collins, Director of the National Human Genome Research Institute, and Dr. Craig Venter, President and Chief Scientific Officer, Celera Genomics Corporation, on the Completion of the First Survey of the Entire Human Genome Project," The White House, Office of the Press Secretary, June 26, 2000, available at https://www.genome.gov/10001356; Nicholas Wade, "Genetic Code of Life Is Cracked by Scientists," *New York Times*, June 27, 2000, available at http://partners.nytimes.com/library/national/science/062700sci-genome.html.

3. "International Consortium Completes Human Genome Project," National Institutes of Health, April 14, 2003, available at https://www.genome.gov/11006929; Nicholas Wade, "Once Again, Scientists Say Human Genome Is Complete," *New York Times,* April 15, 2003, available at http://www.nytimes.com/2003/04/15/science/once-again-scientists-say-human-genome-is-complete.html.

4. Cristian Tomasetti and Bert Vogelstein, "Variation in Cancer Risk among Tissues Can Be Explained by the Number of Cell Divisions," *Science* 347 (2015): 78–81.

5. Hugo Y. K. Lam et al., "Performance Comparison of Whole-Genome Sequencing Platforms," *Nature Biotechnology* 30 (2011): 78–82; Frederick E. Dewey et al., "Clinical Interpretation and Implications of Whole-Genome Sequencing," *Journal of the American Medical Association* 311 (2014): 1035–1045.

6. This kind of sequencing leads to discussion of "coverage" as a marker for accuracy. The "coverage" is the number of times, on average, a particular sequence was read as part of the whole genome sequencing. Coverage of thirty or forty times is thought to be fairly good; sixty or seventy is better. The coverage is, however, just an average; in any given whole genome sequencing process,

some areas of the genome will be sequenced many more times than the average and some many times fewer.

7. After a long head start, "personalized" medicine as the name for individualized clinical care based on a patient's own genome (or his or her tumor's own genome) seems likely to be overtaken by "precision" medicine, thanks to President Obama's "Precision Medicine Initiative." "Fact Sheet: President Obama's Precision Medicine Initiative," The White House, Office of the Press Secretary, January 30, 2015, available at https://www.whitehouse.gov/the-press-office/2015/01/30/fact-sheet-president-obama-s-precision-medicine-initiative. I think either one is misleading without the word "genomic" in them—medicine is and should be personalized and precise in many ways having nothing to do with knowing the patient's genomic sequence.

8. Dmitry Pushkarev, Norma F. Neff, and Stephen R. Quake, "Single-Molecule Sequencing of an Individual Human Genome," *Nature Biotechnology* 27 (2009): 847–850.

9. Division of Reproductive Health, *Assisted Reproduction Technology: 2012 National Summary Report* (Atlanta, GA: National Center for Chronic Disease Prevention and Health Promotion, Centers for Disease Control and Prevention, 2014), available at http://www.cdc.gov/art/reports/2012/national-summary.html.

10. See Jemma Evans et al., "Fresh versus Frozen Embryo Transfer: Backing Clinical Decisions with Scientific and Clinical Evidence," *Human Reproduction Update* 20(6) (2014): 808–821; Laird Harrison, "Frozen Embryos Found to Be More Successful Than IVF," *Medscape Medical News*, February 18, 2013, available at http://www.medscape.com/viewarticle/779505. The latter mainly talks about the subsequent health of babies born, not the success rates in achieving live births.

11. That is my own rough estimate, made by adding the number of children born with chromosomal disorders (nearly 1 percent) to those with some of the less uncommon genetic diseases (cystic fibrosis, sickle cell disease, beta thalassemia) and the small numbers with the thousands of rare genetic diseases. Some estimates are higher. See "Inherited Disorders," *NetWellness*, available at http://www.netwellness.org/healthtopics/idbd/2.cfm. The CDC keeps track of the number of children born with birth defects. It estimates that those born with major structural birth defects or genetic disease amount to about 3 percent of U.S. births, but many major structural birth defects do not involve any known, substantial genetic component. L. Rynn, J. Cragan, and A. Correa, "Update on Overall Prevalence of Major Birth Defects—Atlanta, Georgia, 1978–2005," *Morbidity and Mortality Weekly Report* 57 (2008): 1–5.

12. *Wikipedia* has a good explanation of red hair and its genetics. "Red Hair," *Wikipedia*, available at https://en.wikipedia.org/wiki/Red_hair.

13. See Razib Khan, "Heritability of Behavioral Traits," *Discover* (Blog), June 28, 2012, available at http://blogs.discovermagazine.com/gnxp/2012/06/heritability-of-behavioral-traits/#.Vcq_lUXWkbU.

14. Jae Yeon Cheon, Jessica Mozersky, and Robert Cook-Deegan, "Variants of Uncertain Significance in BRCA: A Harbinger of Ethical and Policy Issues to Come?," *Genome Medicine* 6 (2014): 121; Jacob S. Sherkow and Henry T. Greely, "The History of Patenting Genetic Material," *Annual Review of Genetics* 49 (2015): 161–182.

8. Making Gametes

1. In December 2015, while I was reviewing the copy edits of this book, I saw for the first time an article by Professor Sonia Suter: "*In Vitro* Gametogenesis: Just Another Way to Have a Baby?", *J Law Biosci* 2 (2015), doi: 10.1093/jlb/lsv047. Her article explores several of the issues raised in this book and adds one I had not imagined – "multiplex parents" – which I will discuss briefly in Chapter 12.

2. Shirish Daftary and Sudip Chakravarti, *Manual of Obstetrics,* 3rd ed. (New Delhi: Elsevier, 2011), 1–16.

3. See the discussion in Ri-Cheng Chian and Patrick Quinn, *Fertility Cryopreservation* (New York: Cambridge University Press, 2010), 242–243.

4. Nao Suzuki et al., "Successful Fertility Preservation Following Ovarian Tissue Vitrification in Patients with Primary Ovarian Insufficiency," *Human Reproduction* 30 (2015): 608–615; Dominic Stoop, Ana Cobo, and Sherman Silber, "Fertility Preservation for Age-Related Fertility Decline," *Lancet* 384 (2014): 1311–1319.

5. "What Is In Vitro Maturation and How Does It Work?," Human Fertilisation and Embryology Authority, available at http://www.hfea.gov.uk/fertility-treatment-options-in-vitro-maturation.html. A perhaps more questionable source—a fertility clinic trumpeting its first IVM birth—says in 2015 that more than 3,000 babies have been born after in vitro maturation. "Marin Fertility Clinic Announces First West Coast IVM Baby," Marin Fertility Clinic, available at http://www.prnewswire.com/news-releases/marin-fertility-center-announces-first-west-coast-ivm-baby-300067645.html.

6. Alok Jha, "Twins Born after New Fertility Treatment," *The Guardian,* October 25, 2007, available at http://www.theguardian.com/science/2007/oct/25/1. I calculated the national total by using the HFEA's clinic search and looking for clinics that provide IVM in each of the nation's regions. "Advanced Clinic Search," Human Fertilisation and Embryology Authority, available at http://guide.hfea.gov.uk/guide/.

7. Jha, "Twins Born."

8. "Do Eggs Matured in the Laboratory Results in Babies with Large Offspring Syndrome?," *ObGyn.net,* June 30, 2010, available at http://www.obgyn.net/hpv/do-eggs-matured-laboratory-result-babies-large-offspring-syndrome. As far as I can tell, the report referred to in this article was presented at a scientific meeting but has not been published in the scientific literature.

9. It is not clear why FDA approval would not be required for the reasons discussed in Chapter 10 below, as the eggs would appear to be cells that are more than minimally manipulated, but it has not been. On the other hand, the FDA has largely stayed out of IVF procedures.

10. It would be possible to do Easy PGD without freezing ovarian slices, but this would have a serious disadvantage—it would require a new surgical procedure to retrieve immature oocytes with each effort at IVF, and the number of oocytes retrieved would be a crucial determinant of the number of embryos eventually created and genetically tested. The advantage of taking and freezing ovarian slices is that the surgical procedure would only have to be done once in order to provide more eggs than a woman would ever want to use. The disadvantage is that it would require that both ovarian freezing and in vitro maturation be perfected, singly and in combination. If our time frame is several decades away, it seems likely to be tried and to reach reasonable effectiveness, perhaps sooner than making oocytes from stem cells.

11. The only partial exception (other than cloning, discussed later) is if the two people who want to do Easy PGD make an embryo that is then destroyed to make hESCs, which are in turn transformed into eggs or sperm. Those hESCs would not have only half the genome of one parent; instead, they would have half the genome of each parent—the parents would, in effect, be making sperm or eggs from one of their genetic "children" in order to make more children.

12. Niels Geijsen et al., "Derivation of Embryonic Germ Cells and Male Gametes from Embryonic Stem Cells," *Nature* 427 (2004): 148–154 (mice); Amander T. Clark et al., "Spontaneous Differentiation of Germ Cells from Human Embryonic Stem Cells in Vitro," *Human Molecular Genetics* 13 (2004): 727–739.

13. K. Kee et al., "Human DAZL, DAZ and BOULE Genes Modulate Primordial Germ-Cell and Haploid Gamete Formation," *Nature* 462 (2009): 222–225.

14. Katsuhiko Hayashi et al., "Reconstitution of the Mouse Germ Cell Specification Pathway in Culture by Pluripotent Stem Cells," *Cell* 146 (2011): 519–532.

15. Katsuhiko Hayashi et al., "Offspring from Oocytes Derived from In Vitro Primordial Germ Cell–Like Cells in Mice," *Science* 338 (2012): 971–975. Both sets of Saitou experiments and their possible clinical relevance (to humans, not mice) are nicely discussed in David Cyranoski, "Stem Cells: Egg Engineers," *Nature* 500 (2013): 392–394.

16. Naoko Irie et al., "SOX17 Is a Critical Specifier of Human Primordial Germ Cell Fate," *Cell* 160 (2014): 253–268. See the discussion of this result, and the earlier Saitou work, in David Cyranoski, "Rudimentary Egg and Sperm Cells Made from Stem Cells," *Nature News*, December 24, 2014, available at http://www.nature.com/news/rudimentary-egg-and-sperm-cells-made-from-stem-cells-1.16636.

17. For a somewhat technical discussion of the history and mechanisms of iPSCs, see Matthias Stadtfeld and Konrad Hochedlinger, "Induced Pluripotency: History, Mechanisms, and Applications," *Genes and Development* 24 (2010): 2239–2263.

18. Hayashi et al., "Reconstitution."

19. Hayashi et al., "Offspring."

20. Interestingly, the secret was using the notorious stimulant caffeine to delay the start of cell division. Masuhito Tachibana et al., "Human Embryonic Stem Cells Derived by Somatic Cell Nuclear Transfer," *Cell* 153 (2013): 1228–1238. The work was replicated within a year by Robert Lanza's group; see Young Gie Chung et al., "Human Somatic Cell Nuclear Transfer Using Adult Cells," *Cell Stem Cell* 14 (2014): 777–780.

21. "X-Inactivation," *Wikipedia,* available at https://en.wikipedia.org/wiki/X-inactivationX mosaicism. X-inactivation is responsible for the color patterns of calico cats, which are always female. Some of their coat color genes are on their X chromosomes; the different colored patches are derived from ancestral cells that turned off one or the other of the two X chromosomes when the two chromosomes had genes for different colors.

22. It appears that eggs in the ovaries of women with Turner syndrome die early in life. There is interest in removing and freezing ovarian slices from girls with Turner syndrome to be used later in life to reestablish their fertility. See Jacqueline K. Hewitt et al., "Fertility in Turner Syndrome," *Clinical Endocrinology (Oxford)* 79 (2013): 606–614.

23. A. H. Sathananthan, "Mitosis in the Human Embryo: The Vital Role of the Sperm Centrosome (Centriole)," *Histology and Histopathology* 12 (1997): 827–856.

24. A relatively accessible discussion of imprinting in mammals can be found in Denise P. Barlow and Marisa S. Bartolomei, "Genomic Imprinting in Mammals," *Cold Spring Harbor Perspectives in Biology* (February 2014): doi:10.1101/cshperspect.a018382.

9. Research Investment, Industry, Medical Professionals, and Health Care Financing

1. Peter Gillman, "Supersonic Bust," *The Atlantic* 279(1) (1977): 72–81. Though written just as the planes were entering commercial service and long before their eventual retirement after a terrible 2000 crash, this piece lays out the many weaknesses, technical and otherwise, of the project.

2. See, for example, Cheryl Swanson, "The Best Stocks in Gene Sequencing," *The Motley Fool,* June 18, 2015, available at http://www.fool.com/investing/general/2015/06/18/the-best-stocks-in-gene-sequencing.aspx; Luke Timmerman,

"DNA Sequencing Market Will Exceed $20 Billion, Says Illumina CEO Jay Flatley," *Forbes* (blog), April 29, 2015, available at http://www.forbes.com/sites/luketimmerman/2015/04/29/qa-with-jay-flatley-ceo-of-illumina-the-genomics-company-pursuing-a-20b-market/.

3. John K. Amory, "George Washington's Infertility," *Fertility and Sterility* 81 (2004): 495–499.

4. The development of cross-sex gametes may benefit from a similar consumer-driven research approach supported by gays and lesbians who want to be parents.

5. This is possibly because the IVF industry has nothing to offer them, other than donor gametes. Absence of eggs or sperm will rarely be listed as a reason for an IVF cycle, unless donor gametes are being used. Donor eggs almost always meet an absence of a patient's own eggs, but donor sperm, in IVF or through (much cheaper and easier) artificial insemination, will often be used by single women and not because of a particular male's lack of enough or effective enough sperm.

6. "Sperm Banking: Background Fundamentals," Xytec Cryo International, available at https://www.xytex.com/sperm-donor-bank-about/about-sperm-banking.cfm. This source both mentions the 30,000 figure generally used and cites unpublished research by the American Association of Tissue Banks for a number between 4,000 and 5,000.

7. The incidence of multiple sclerosis comes from Ann Pietrangelo and Valencia Higuera, "Multiple Sclerosis by the Numbers: Fact, Statistics, and You," *Healthline,* available at http://www.healthline.com/health/multiple-sclerosis/facts-statistics-infographic. It claims about 200 cases are diagnosed each week, which leads to about 10,000 a year. Cancer statistics are from "Cancer Facts & Figures 2015," American Cancer Society, Cancer Facts and Figures 2015, available at http://www.cancer.org/research/cancerfactsstatistics/cancerfactsfigures2015/. HIV incidence is from "HIV in the United States, at a Glance," CDC, available at http://www.cdc.gov/hiv/statistics/basics/ataglance.html.

8. Just how accurate these percentages are is unclear, but presumably as same-sex attraction has become more acceptable, people have reported gay, lesbian, and bisexual identification more honestly. One recent survey found 1.6 percent of Americans identify as gay or lesbian and another 0.7 percent as bisexual; see Sandhya Somashekar, "Health Survey Gives Government Its First Large-Scale Data on Gay, Bisexual Population," *Washington Post,* July 15, 2014. Another reported that 3.4 percent of adult Americans identified themselves as gay, lesbian, or bisexual; see Gary J. Gates and Frank Newport "3.4 Percent of U.S. Adults Identify as LGBT," Gallup, October 18, 2012, available at http://www.gallup.com/poll/158066/special-report-adults-identify-lgbt.aspx. A third report found 1.7 percent of people identifying as gay or lesbian with another 1.8 percent as bisexual; see Gary J. Gates, "How Many People Are Lesbian, Gay, or Bisexual?,"

Williams Institute, Los Angeles, available at http://williamsinstitute.law.ucla.edu/research/census-lgbt-demographics-studies/how-many-people-are-lesbian-gay-bisexual-and-transgender/. Of course, whether bisexual people would want cross-sex gametes will depend on the person and his or her relationships.

9. For an excellent analysis of all parts of the "baby business," see Deborah Spar, *The Baby Business: How Money, Science, and Politics Drive the Commerce of Conception* (Cambridge, MA: Harvard Business Review Press, 2006).

10. Tara Siegel Bernard, "Insurance Coverage for Infertility Treatments Varies Widely," *New York Times,* July 26, 2014, B1. See also "Health Insurance 101," RESOLVE, available at *Insurance Coverage*, available at http://www.resolve.org/family-building-options/insurance_coverage/health-insurance-101.html. (RESOLVE offers patient support from the National Infertility Association.)

11. "Fellowships Overview," Society for Reproductive Endocrinology and Infertility, available at http://www.socrei.org/detail.aspx?id=3146. The numbers were gathered by perusing the Fellowship Directory, reachable from that page.

12. Frank Newport and Joy Wilke, "Desire for Children Still the Norm in U.S.," Gallup, September 23, 2013, available at http://www.gallup.com/poll/164618/desire-children-norm.aspx.

13. The doctrine was arguably established in *Central Hudson Gas & Electric Corp. v. Public Service Commission,* 447 U.S. 557 (1980), but continues to expand.

14. Michael Kornhauser and Roy Schneiderman, "How Plans Can Improve Outcomes and Cut Costs for Preterm Infant Care," *Managed Care Magazine,* available at http://www.managedcaremag.com/archives/1001/1001.preterm.html.

10. Legal Factors

1. For an interesting discussion of variations across the European Union, see K. Berg Brigham, B. Cadier, and K. Chevreul, "The Diversity of Regulation and Public Financing of IVF in Europe and Its Impact of Utilization," *Human Reproduction* 28 (2013): 666–675.

2. Human Fertilisation and Embryology Association. *About the Human Fertilisation and Embryology Association.* October, 2009, available at http://www.hfea.gov.uk/docs/About_the_HFEA.pdf; Francesco Paolo Busardò et al., "The Evolution of Legislation in the Field of Medically Assisted Reproduction and Embryo Stem Cell Research in European Union Members," *BioMedical Research International,* 2014, Article ID 307160, 9–10, available at http://dx.doi.org/10.1155/2014/307160.

3. Busardò et al., "Evolution," 5.

4. Busardò et al., "Evolution," 6, 7. These restrictions, adopted in 2004, were stricken down by the Italian Constitutional Court in 2009, though it is not clear

how practice has evolved. Italian Constitutional Court, No. 151/09, available at http://www.cortecostituzionale.it/actionSchedaPronuncia.do?anno=2009&numero=151. More provisions of the 2004 law, concerning gamete donation, were invalidated in 2014. Italian Constitutional Court, 2014; Giuseppe Benagiano, Valentina Filippi, Serena Sgargi, Luca Gianaroli, "Italian Constitutional Court removes the prohibition on gamete donation in Italy," *Reproductive BioMedicine* 29, no. 6 (December 2014): 662–664.

5. Busardò et al., "Evolution," 2.

6. The general situation in Australia is described by the Australian National Health and Medical Research Council, Assisted Reproduction Technology on their wesite. "Assisted Reproductive Technology," National Health and Medical Research Center, available at https://www.nhmrc.gov.au/health-ethics/ethical-issues/assisted-reproductive-technology-art South Australia and the Northern Territories are restrictive. See also Karin Hammerberg, *IVF and Beyond for Dummies* (Milton, Queensland, Australia: Wiley, 2011).

7. For a good discussion of the Octomom situation, both its facts and an ethical analysis, see Bertha Alvarez Manninen, "Parental, Medical, and Sociological Responsibilities: 'Octomom' as a Case Study in the Ethics of Fertility Treatments," *Journal of Clinical Research and Bioethics* S1:002 (2011), doi: 10.4172/2155-9627.S1-002.

8. American Society for Reproductive Medicine. *Oversight of Assisted Reproductive Technology*. Birmingham, AL: American Society for Reproductive Medicine, 2010. https://www.asrm.org/uploadedFiles/Content/About_Us/Media_and_Public_Affairs/OversiteOfART%20%282%29.pdf.

9. The ASRM report does not totally ignore the possibility that some more regulation might be useful. Specifically, it calls for more insurance coverage for IVF on the theory that insurance would alleviate the economic pressure to transfer too many embryos. (It would also, of course, expand the market for fertility clinics.)

10. Federal Food, Drug, and Cosmetic Act, Section 201(g)(1), codified at 21 U.S.C. §321(g)(1).

11. Ibid., Section 201(h)(1)(B, C), codified at 21 U.S.C. §321(h).

12. Public Health Service Act, codified at 42 U.S.C. §262(i).

13. Patricia A. Zettler, "Toward Coherent Federal Oversight of Medicine," *San Diego Law Review* 52 (2015): 427–500.

14. 21 U.S.C. §396. See Rebecca Dresser and Joel Frader, "Off-Label Prescribing: A Call for Heightened Professional and Government Oversight," *Journal of Law, Medicine and Ethics* 37(3) (2009): 476.

15. John Murray, "CDRH Regulated Software: An Introduction," FDA, available at http://www.fda.gov/downloads/Training/CDRHLearn/UCM209129.pdf. For discussion of a new frontier in FDA regulation of software, see Komal Karnik, "FDA Regulation of Clinical Decision Support Software," *Journal of Law and Biosciences* (2014), doi:10.1093/jlb/lsu004.

16. Framework for Regulatory Oversight of Laboratory Developed Tests; Draft Guidance for Industry, Food and Drug Administration Staff, and Clinical Laboratories; Availability, 79 Fed. Reg. 59776-59779 (October 3, 2014). A copy of the draft guidance can be found at http://www.fda.gov/downloads/medicaldevices/deviceregulationandguidance/guidancedocuments/ucm416685.pdf.

17. "FDA Allows Marketing of Four 'Next Generation' Gene Sequencing Devices," FDA, News Release, November 19, 2013, available at http://www.fda.gov/NewsEvents/Newsroom/PressAnnouncements/ucm375742.htm. See also Francis S. Collins and Margaret A. Hamburg, "First FDA Authorization for Next-Generation Sequencer," *New England Journal of Medicine* 369 (2013): 2369–2371.

18. Raëlians believe the intelligent aliens who founded their religion had declared in discussions with their prophet, Raël (formerly a French race driver named Claude Vorhilon), that cloning humans was a religious duty. The religion set up a corporation, called Clonaid, to bring about human reproductive cloning. Charlie Jane Anders, "Meet the Raelians: Inside the World's Strangest—and Nicest—UFO Sex Clone Religion," *io9,* November 21, 2011, available at http://io9.com/5860418/meet-the-raelians-inside-the-worlds-strangest--and-nicest--ufo-sex-clone-religion. More generally, see the Raelian website, Message from the Designers, available at http://www.rael.org/home.

19. Seed, a nuclear physicist in Illinois, announced in December 1997 that he was going to clone himself before the process could be made illegal. Seed looked perfect as a mad scientist, but his plans had a smidgen of credibility because he and his physician brother had been involved in embryo transfer work in the 1980s. See Elizabeth Price Foley, "The Constitutional Implications of Human Cloning," *Arizona Law Review* 42 (2000): 647–730, 648–649.

20. "Reported Efforts to Clone Human Beings," Center for Genetics and Society, July 29, 2004, available at http://www.geneticsandsociety.org/article.php?id=383.

21. Among others, see Richard A. Merrill and Bryan J. Rose, "FDA Regulation of Human Cloning: Usurpation or Statesmanship?," *Harvard Journal of Law and Technology* 15(1) (2001): 85–148, available at http://jolt.law.harvard.edu/articles/pdf/v15/15HarvJLTech085.pdf; Richard A. Merrill, "Human Tissues and Reproductive Cloning: New Technologies Challenge FDA," *Houston Journal of Health Law and Policy* 3 (2002): 1–86; Rebecca Dresser, "Human Cloning and the FDA," *Hastings Center Report* 33(3) (2003): 7–8; Louis M. Guenin, "Stem Cells, Cloning, and Regulation," *Mayo Clinic Proceedings* 80(2) (2005): 241–250, available at www.mayoclinicproceedings.com; Gail H. Javitt and Kathy Hudson, "Regulating for the Benefit of Future Persons: A Different Perspective on the FDA's Jurisdiction to Regulate Human Reproductive Cloning," *Utah Law Review* 2003: 1201–1229, Gregory J. Rokosz, "Human Cloning: Is the Reach of FDA Authority Too Far a Stretch?," *Seton Hall Law Review*

30 (1999–2000): 464–515. For a general discussion of human reproductive (and nonreproductive) cloning from this era, see California Advisory Committee on Human Cloning, "Cloning Californians: Report of the California Advisory Committee on Human Cloning," *Hastings Law Journal* 53 (2002): 1143–1203.

22. Steve Connor, "Three-Parent Baby Pioneer Jamie Grifo: 'The Brits Will Be Ahead of the World,'" *The Independent,* August 13, 2015, available at http://www.geneticsandsociety.org/article.php?id=8314; see also Kristina Fiore, "Rocky Road for Mitochondrial Transfer," *MedPage Today,* February 28, 2014, available at http://www.medpagetoday.com/Endocrinology/General Endocrinology/44530.

23. Dorothy R. Haskett, "Ooplasmic Transfer Technology," *Embryo Project Encyclopedia,* August 18, 2014, available at https://embryo.asu.edu/pages/ooplasmic-transfer-technology.

24. Ewen Callaway, "Scientists Cheer Vote to Allow Three-Person Embryo," *Nature News,* February 3, 2015, doi:10.1038/nature.2015.16843, also available at http://www.nature.com/news/scientists-cheer-vote-to-allow-three-person-embryos-1.16843.

25. Henry T. Greely, "Heather Has Three Parents," *Center for Law and the Biosciences* (Blog), March 2, 2014, available at https://blogs.law.stanford.edu/lawandbiosciences/2014/03/02/heather-has-three-parents/.

26. *U.S. v. Regenerative Sciences,* 741 F.3d 1314 (D.C. Cir. 2014).

11. Politics

1. Fr. Tad Pacholczyk, "Do Embryos Have Souls?," National Catholic Bioethics Center, available at http://www.ncbcenter.org/document.doc?id=846.

2. Louisiana Revised Statutes, 9, §123 states that an "in vitro fertilized human ovum is a juridical person until such time as the in vitro fertilized ovum is implanted in the womb; or at any other time when rights attach to an unborn child in accordance with law." Section 129 states, "A viable in vitro fertilized human ovum is a juridical person which shall not be intentionally destroyed by any natural or other juridical person or through the actions of any other such person." The penalties for violating the ban are not specified.

3. Proposition 71, which established the California Institute for Regenerative Medicine (CIRM), requires the governing board of that group to establish standards "prohibiting compensation to research donors or participants, while permitting reimbursement of expenses." See California Stem Cell Research and Cures Act, Section 125290.35. The California legislature added additional restrictions in SB 1260 in 2006, banning any compensation in California for human eggs used in research beyond reasonable reimbursements of direct expenses.

See California Health and Safety Code, §125355. In 2013, the legislature changed course and voted to allow research donors to be compensated up to $5,000 above and beyond reimbursement for their expenses, but Governor Brown vetoed the bill. Al Donner, "Jerry Brown Vetoes Sales of Eggs in California," *National Catholic Register,* September 13, 2013, available at http://www.ncregister.com/daily-news/jerry-brown-vetoes-sales-of-human-eggs-in-california/.

4. Seventeen states have statutes permitting surrogacy and another seven have court opinions approving it. Five states make surrogacy contracts unenforceable and the District of Columbia makes it a crime. Tamar Lewin, "Surrogates and Couples Face a Maze of Laws, State by State," *New York Times,* September 18, 2014, A1.

5. *Troxel v. Granville,* 530 U.S. 57 (2000).

6. *Obergefell v. Hodges,* 576 U.S., 135 S. Ct. 2584 (2015). See also *United States v. Windsor,* 570 U.S., 133 S.Ct. 2675 (2013), striking down the federal Defense of Marriage Act.

7. "A physician may not intentionally perform or attempt to perform an abortion with knowledge that the pregnant woman is seeking the abortion solely: a. On account of the sex of the unborn child; or b. Because the unborn child has been diagnosed with either a genetic abnormality or a potential for a genetic abnormality." North Dakota Century Code, Chapter 14-02.1. Violation is a Class A misdemeanor. The North Dakota law, part of a broad package of abortion restrictions, was initially challenged in federal district court but the challenge to the sex and genetic abnormality provisions was dropped by the plaintiffs. See the very useful discussion of North Dakota (and other states banning sex and, in one case, race as a grounds for abortion) in Part II of Carole J. Petersen, "Reproductive Justice, Public Policy, and Abortion on the Basis of Fetal Impairment: Lessons from International Human Rights Law and the Potential Impact of the Convention on the Rights of Persons with Disabilities," *Journal of Law and Health* 28 (2015): 121–163, available at http://engagedscholarship.csuohio.edu/jlh/vol28/iss1/7.

8. See Robert C. Green and Nita A. Farahany, "Regulation: The FDA is Overcautious on Consumer Genomics," *Nature* 505 (2014): 286–287.

9. Peterson, "Reproductive Justice," 128.

10. For an extended discussion of sex imbalances, see Mara Hvistendahl, *Unnatural Selection: Choosing Boys over Girls and the Consequences of a World Full of Men* (New York: PublicAffairs, 2011). We will return to this issue, and this book, in Chapter 15.

11. Ethics Committee of the American Society for Reproductive Medicine, "Use of Reproductive Technology for Sex Selection for Nonmedical Reasons," *Fertility and Sterility* 103 (2015): 418–422. This opinion expressly replaced the two earlier Ethics Committee opinions, published as Ethics Committee of the

American Society for Reproductive Medicine, "Sex Selection and Preimplantation Genetic Diagnosis," *Fertility and Sterility* 82 (2004): S245–248, and Ethics Committee of the American Society for Reproductive Medicine, "Preconception Gender Selection for Nonmedical Reasons," *Fertility and Sterility* 82 (2004): S232–235. The 2015 opinion usefully surveys the somewhat limited evidence about sex selection in the United States and less limited literature arguing about its ethics.

12. One 2004 survey found very little difference in American preferences for boys or girls. Edgar Dahl et al., "Preconception Sex Selection Demand and Preferences in the United States," *Fertility and Sterility* 85 (2006): 468–473. On the other hand, the Gallup Organization's polling shows a 40 to 28 percent preference for boys when people are asked would they prefer a boy or a girl if they could have only a boy or a girl. Frank Newport, "Americans Prefer Boys to Girls, Just as They Did in 1941," Gallup, June 23, 2011, available at http://www.gallup.com/poll/148187/americans-prefer-boys-girls-1941.aspx. On the other hand, there are many anecdotal reports that Americans considering sex selection actually prefer girls, at least when they are not recent immigrants from countries with strong male preferences.

13. Maybe the best-known example of deaf selection, although one using the old-fashioned technology of artificial insemination, was discussed in Liza Mundy, "A World of Their Own," *Washington Post Magazine,* March 31, 2000, W22.

14. See Holly C. Gooding et al., "Issues Surrounding Prenatal Genetic Testing for Achondroplasia," *Prenatal Diagnosis* 22 (2002): 933–940. This article notes that few would actually consider aborting a non-achondroplasic fetus. It is worth noting, as another example of the diversity of human genetics, that 70 percent of people with achondroplasia do not have parents with the condition or at particular risk for the condition, but are the result of new mutations in the relevant gene. (The same is even truer of the chromosomal condition in Down syndrome.)

15. *Rust v. Sullivan,* 500 U.S. 173 (1991).

16. As of 2008, thirteen U.S. states have banned human reproductive cloning: California (the first, less than six months after Dolly was announced), Arkansas, Connecticut, Indiana, Iowa, Maryland, Massachusetts, Michigan, Rhode Island, New Jersey, North Dakota, South Dakota, and Virginia. "Cloning Bans," National Conference of State Legislatures, available at http://www.ncsl.org/research/health/human-cloning-laws.aspx (updated through January 2008). Globally, the Center for Genetics and Society counts forty-six countries with express bans on human reproductive cloning. "Human Cloning Policies," Center for Genetics and Society, available at http://www.geneticsandsociety.org/article.php?id=325. This does not include countries like the United States, where the FDA takes the position that it would have to approve, in advance, any human reproductive cloning.

17. Ira Levin, *Boys from Brazil* (New York: Random House, 1976); *Boys from Brazil,* directed by Franklin J. Schaffner (London: ITC Entertainment, 1978), motion picture; Kazuo Ishiguro, *Never Let Me Go* (London: Faber and Faber, 2005); *Never Let Me Go,* directed by Mark Romanek (London: DNA Films, 2010), motion picture; *The Island,* directed by Michael Bay (Los Angeles: DreamWorks, 2005), motion picture; *Star Wars Episode II: Attack of the Clones,* directed by George Lucas (Los Angeles: Twentieth Century Fox, 2002), motion picture; Aldous Huxley, *Brave New World* (London: Chatto and Windus, 1970); *Gattaca,* directed by A. Niccol (Los Angeles: Columbia Pictures Corp., 1997), motion picture.

18. For Malta, El Salvador, and Uruguay, see Cora Fernandez Anderson, "The Politics of Abortion in Latin America," *RH Reality Check,* July 17, 2013, available at http://rhrealitycheck.org/article/2013/07/17/the-politics-of-abortion-in-latin-america/. In Spain abortion became widely legal in 2010; a more conservative government abandoned an attempt to retighten the laws in 2013. "Spanish Abortion: Rajoy Abandons Tightened Law," *BBC News,* September 23, 2014, available at http://www.bbc.com/news/world-europe-29322561. In France, abortion had been available for women "in a state of distress" during the first twelve weeks of pregnancy since 1975; the law was liberalized in 2014 to allow abortion on demand during that period; subsequent abortions require evidence of harm to the woman or a likely serious disease of any resulting child. "Abortion in France," *Wikipedia,* available at https://en.wikipedia.org/wiki/Abortion_in_France.

19. "Religious Response to Assisted Reproductive Technology," *Wikipedia,* available at https://en.wikipedia.org/wiki/Religious_response_to_assisted_reproductive_technology#Islam. For an exhaustive discussion of the much more complicated situation of Islamic views of assisted reproduction, see Marcia C. Inhorn and Soraya Tremayne, eds., *Islam and Assisted Reproductive Technologies: Sunni and Shia Perspectives* (New York: Berghahn Books, 2012).

20. Ruth Levush, "Israel: Reproduction and Abortion: Law and Policy," Law Library of Congress, February 2012, available at http://www.loc.gov/law/help/israel_reproduction_law_policy.php.

21. German law prohibits donating eggs and embryos as well as surrogacy. Only three eggs can be fertilized and transferred in one reproductive cycle. Freezing embryos is limited to cases where the prospective mother's medical condition would not allow an immediate pregnancy. "Germany," *Infertility Answers,* available at http://infertilityanswers.org/art_in_germany. PGD was legalized after a court decision in favor of parents who need to use PGD to avoid a serious genetic condition in their offspring. Annette Tuffs, "Germany Allows Restricted Access to Preimplantation Genetic Testing," *BMJ* 343 (July 12, 2011), doi:http://dx.doi.org/10.1136/bmj.d4425.

22. Hvistendahl, *Unnatural Selection.*

23. But he apparently was not the origin of the phrase. O'Neill first used it in 1935 but it appears in a newspaper column in 1932. See Barry Popik, "Origin

of 'All Politics Is Local' (Attention Rachel Maddow)," *RedState,* September 28, 2010, available at http://www.redstate.com/diary/barrypopik/2010/09/28/origin-of-all-politics-is-local-attention-rachel-maddow/; Fred R. Shapiro, "Quote . . . Misquote—A Commentary," *New York Times,* July 21, 2008.

12. Some Other Possible Uses of New Technologies in Reproduction

1. "List of Animals That Have Been Cloned," *Wikipedia,* available https://en.wikipedia.org/wiki/List_of_animals_that_have_been_cloned.

2. The FDA produced a wonderfully detailed study of health hazards to cloned animals (and their gestational mothers). It focused mainly on cattle, but also somewhat on pigs and goats. It concluded that, in cattle, both the young clones and their mothers faced increased health risks, but that clones that survived early problems were normal. Center for Veterinary Medicine, "Animal Health Risks," Chapter 8 of *Animal Cloning: A Risk Assessment* (Rockville, MD: FDA, 2008), available at http://www.fda.gov/AnimalVeterinary/SafetyHealth/AnimalCloning/UCM124756.pdf. On the other hand, cloned mammals have been a success in at least one area; see Haley Cohen, "How Champion-Pony Clones Have Transformed the Game of Polo," *Vanity Fair,* August 2015, available at http://www.vanityfair.com/news/2015/07/polo-horse-cloning-adolfo-cambiaso. On the other hand, an American appellate court rejected an antitrust lawsuit claiming the American Quarter Horse Association's ban on cloned horses is an illegal restraint of trade. *Abraham & Veneklasen Joint Venture v. American Quarter Horse Assoc.,* 776 F.3d 321 (Fifth Cir. 2015).

3. If cloning does become common, for humans or others, we will need a new word to describe the source of the nucleus that provides almost all the DNA for the clone. I think "progenitor" may be the best word, but the language has its reasons that reason does not know.

4. Both *Science* and *Nature,* the world's two leading general scientific journals, have created useful collections of their articles about CRISPR. See "The CRISPR Revolution," *Science,* available at http://www.sciencemag.org/site/extra/crispr/?intcmp=HP-COLLECTION-PROMO-crispr and "CRISPR: The Good, the Bad, and the Unknown," *Nature,* available at http://www.nature.com/news/crispr-1.17547.

5. A researcher at the Massachusetts Institute of Technology, Feng Zhang, was also using CRISPR in the early days. He (through MIT) applied for a patent on the technology and received it, even though Doudna (and the University of California) had applied for a patent sooner. How this is possible is hidden deep in the weeds of the U.S. patent system, but the fight is now in what is called an "interference" proceeding over whether one patent application, the other, both, or neither will stand. Hundreds of millions of dollars may be at stake. See the

excellent article by Antonio Regalado, "Who Owns the Biggest Biotech Discovery of the Century?," *MIT Technology Review,* December 4, 2014, available at http://www.technologyreview.com/featuredstory/532796/who-owns-the-biggest-biotech-discovery-of-the-century/. Stockholm is not bound by the U.S. Patent and Trademark Office; I suspect Doudna and Charpentier are shoe-ins for early Nobel Prizes; Zhang may also be included because the Nobel rules now allow three living recipients for each prize.

6. David Baltimore et al., "A Prudent Path Forward for Genomic Engineering and Germline Gene Modification," *Science* 348 (2015): 36–38.

7. Ralph J. Cicerone and Victor J. Dzau, "National Academy of Sciences and National Academy of Medicine Announce Initiative on Human Gene Editing," *News from the National Academies,* May 18, 2015, available at http://www8.nationalacademies.org/onpinews/newsitem.aspx?RecordID=05182015. The "Summit" is tentatively scheduled for December 2015 and will have been held by the time you read this. The committee report will probably not be done until late 2016 at the very earliest.

8. National Academies, Human Gene-Editing Initiative, available at http://nationalacademies.org/gene-editing/index.htm. For a discussion of the Summit results, see Henry T. Greely, "The International Summit on Human Gene Editing: A Successful Production", Center for Law and the Biosciences blog, available at https://law.stanford.edu/2015/12/05/the-international-summit-on-human-gene-editing-a-successful-production/ (December 2015).

9. Of course, if genome editing became *very* effective, through CRISPR/Cas9 or otherwise, germline editing would be even less necessary. Parents, or eventually the grown-up children themselves, could use the technique to change the somatic cells in their bodies. This would avoid any special risks of intervening in the embryo and, most likely, provide a better basis for an informed and consenting decision. Only when the alleles affect prenatal development would editing at or before the embryo stage be important.

10. Puping Liang et al., "CRISPR/Cas9-Mediated Gene Editing in Human Tripronuclear Zygotes," *Protein Cell* 6 (2015): 363–372. A different laboratory in China had earlier announced the successful birth of cynomolgus monkeys after CRISPR editing of their embryos. Yuyu Niu et al., "Generation of Gene-Modified Cynomolgus Monkey via Cas9/RNA-Mediated Gene Targeting in One-Cell Embryos," *Cell* 156 (2014): 836–843.

11. Thomas Wirth, Nigel Parker, and Seppo Ylä-Herttuala, "History of Gene Therapy," *Gene* 525 (2013): 162–169.

12. Sabrina Richards, "Gene Therapy Arrives in Europe," *The Scientist,* November 6, 2012, available at http://www.the-scientist.com/?articles.view/articleNo/33166/title/Gene-Therapy-Arrives-in-Europe/.

13. For explanations of this process, see "RNA Interference (RNAi)," National Center for Biotechnology Information, available at http://www.ncbi.nlm.

nih.gov/probe/docs/techrnai/, or a five-minute animation produced by Nature Reviews Genetics, *RNA Interference (RNAi),* available at http://www.nature.com/nrg/multimedia/rnai/animation/index.html.

14. Daniel G. Gibson et al., "Creation of a Bacterial Cell Controlled by a Chemically Synthesized Genome," *Science* 329 (2010): 52–56.

15. Narayana Annuluru et al., "Total Synthesis of a Functional Designer Eukaryotic Chromosome," *Science* 344 (2014): 55–58.

16. There seems to be some truth to the Japanese goat stories. Perri Klass, "The Artificial Womb Is Born," *New York Times,* September 29, 1997, available at http://www.nytimes.com/1996/09/29/magazine/the-artificial-womb-is-born.html?pagewanted=3. More generally, see the interesting discussion in David Warmflash, "Artificial Wombs: The Coming Era of Motherless Births?," *Genetic Literacy Project,* June 12, 2015, available at http://www.geneticliteracyproject.org/2015/06/12/artificial-wombs-the-coming-era-of-motherless-births/. There is a useful scientific review of prior work and prospects in Carlo Bulletti et al., "The Artificial Womb," *Annals of the New York Academy of Sciences* 1221 (2011): 124–128.

17. The first bladder so created was transplanted in 1999; the first trachea in 2011. Marissa Cevallos, "Transplanted Trachea, Born in Lab, Is One of Several Engineered-Organ Success Stories," *Los Angeles Times,* July 8, 2011, available at http://articles.latimes.com/2011/jul/08/news/la-heb-trachea-transplant-stem-cell-20110708.

18. One approach that might speed up this time table is not growing a new uterus for use in the artificial but transplanting a uterus from a woman, either living or dead. The safety and efficacy of such a "transplanted uterus in a box" is certainly unclear, as is the potential availability of such organs for transplant. This approach would, however, avoid one of the problems of the artificial womb suggested in the text—growing the uterus.

19. The "uterine replicator" plays a part in several of Lois McMaster Bujold's Vorkosigan series of novels but is particularly prominent in Bujold, *Barrayar* (Wake Forest, NC: Baen Books, 1991) (a well-deserved Hugo Award winner).

20. Sonia Suter: "*In Vitro* Gametogenesis: Just Another Way to Have a Baby?," *Journal of Law and the Biosciences* 2 (2015), doi: 10.1093/jlb/lsv047. Ironically I am one of three co-editors in chief of that journal but I had not seen this article, part of a symposium issue, or known its title or subject, during the editorial process.

Second Interlude

1. William D. Mosher, Jo Jones, and Joyce C. Abma, "Intended and Unintended Births in the United States: 1982–2010," *National Health Statistics,* No. 55, July 24, 2012, table 1. Interestingly, these results are very similar to those from the

earliest version of this survey, in 1982. On the other hand, a report from the respected Guttmacher Institute claims that 49 percent of pregnancies in the United States in 2006 were unintended. Lawrence B. Finer and Mia R. Zolna, "Unintended Pregnancy in the United States: Incidence and Disparities," *Contraception* 84 (2011): 478–485. The difference is probably because the first source looks at births (thus excluding abortions) while the second looks at pregnancies, many of which end in abortion. I think for our purpose the first measure is more relevant.

2. I could not find what percentage of American pregnancies involved an ultrasound, but they are clearly very common. For a discussion of the rising tide of ultrasounds, see Kevin Helliker, "Pregnant Women Get More Ultrasounds, without Clear Medical Need," *Wall Street Journal*, (July 17, 2015), available at http://www.wsj.com/articles/pregnant-women-get-more-ultrasounds-without-clear-medical-need-1437141219. The article states that in 2014 the average delivery in the United States involved 5.2 ultrasounds, based on insurance reimbursement (and so not counting "keepsake," non-medical ultrasounds). In its 2009 practice guideline on ultrasound, the American College of Obstetricians and Gynecologists stated that most women have at least one ultrasound examination in every pregnancy. The bulletin does not recommend ultrasound as a necessary part of prenatal care but does say it should be discussed with every pregnant woman and that it is reasonable for a doctor to honor a patient's request for ultrasound. ACOG Committee on Practice Bulletins–Obstetrics, "Ultrasonography in Pregnancy" ACOG Practice Bulletin, 101 (2009).

3. California Code of Regulations, Title 17, §6527(a) requires all clinicians to provide "to all pregnant women in their care before the 140th day of gestation" information about prenatal screening; §6527(c) requires them to provide an informed consent form to all women who choose screening.

4. Glenn E. Palomaki et al., "Screening for Down Syndrome in the United States: Results of Surveys in 2011 and 2012," *Archives of Pathology and Laboratory Medicine* 137 (2013): 921–926.

5. Early in my work on this book, I had a conversation with the late Dr. Carl Djerassi, one of the fathers of the oral contraceptive. He told me that eventually people would be sterilized at puberty, either in a way that is reversible or with the storage of eggs and sperm in order to start a pregnancy later, when, and only when, planned. Easy PGD could, of course, fit right into Djerassi's world—gametes from iPSCs would avoid the need for either reversible sterilization or storage of gametes.

Part III

1. Aldous Huxley, *Brave New World* (London: Chatto and Windus, 1970); H. G. Wells, *The Time Machine* (Eastford, CT: Martino Fine Books, 2011)

(reprint of the original edition, published in 1895); Mary Wollstonecraft Shelley, *Frankenstein: Or the Modern Prometheus* (Oxford: Oxford University Press, 2008).

2. The number of books on these topics is vast; the number of articles is, for all practical purposes (what a person might be able to read), infinite. Following are the titles of a dozen of what I consider the best accessible books around these issues (many focusing on human enhancement through genetics), listed in alphabetical order by author, with one forthcoming book as a bonus. Lori B. Andrews, *The Clone Age: Adventures in the New World of Reproductive Technology* (New York: Henry Holt, 1999); Allen E. Buchanan, *Better than Human: The Promise and Perils of Enhancing Ourselves*, (New York: Oxford University Press, 2011); Allen E. Buchanan, Dan W. Brock, Norman Daniels, and Daniel Wikler, *From Chance to Choice: Genetics and Justice* (Cambridge: Cambridge University Press, 2000); Francis Fukuyama, *Our Posthuman Future: Consequences of the Biotechnology Revolution* (New York: Farrar, Straus and Giroux, 2002); Ronald Michael Green, *Babies by Design: The Ethics of Genetic Choice*, (New Haven: Yale University Press, 2007); John Harris, *Enhancing Evolution: The Ethical Case for Making Better People* (Princeton: Princeton University Press 2007); Daniel J. Kevles, *In the Name of Eugenics: Genetics and the Uses of Human Heredity* (Cambridge, MA.: Harvard University. Press, 1995); Philip Kitcher, *The Lives to Come: The Genetic Revolution and Human Possibilities* (New York: Simon & Schuster, 1997); Maxwell J. Mehlman, *Wondergenes: Genetic Enhancement and the Future of Society* (Bloomington: Indiana University Press, 2003); John A. Robertson, *Children of Choice: Freedom and the New Reproductive Technologies* (Princeton: Princeton University Press, 1994); Michael J. Sandel, *The Case against Perfection: Ethics in the Age of Genetic Engineering* (Cambridge, MA: Belknap Press of Harvard University Press, 2007); and Lee Silver, *Remaking Eden: Cloning and beyond in a Brave New World* (New York: Avon, 1997). The forthcoming book is Judith Daar, *The New Eugenics: Selective Breeding in an Era of Reproductive Technologies*, (New Haven, Yale University Press, forthcoming 2016)

13. Safety

1. "Accidents Do Happen," Yosemite's Half Dome, available at http://hikehalfdome.com/accidents/.

2. Division of Reproductive Health, *Assisted Reproduction Technology: 2012 National Summary Report* (Atlanta, GA: National Center for Chronic Disease Prevention and Health Promotion, Centers for Disease Control and Prevention, 2014), available at http://www.cdc.gov/art/reports/2012/national-summary.html.

3. See the very detailed review by B. C. Fauser et al., "Health Outcomes of Children Born after IVF/ICSI: A Review of Current Expert Opinion and Literature," *Reproductive Biomedicine Online* 28 (2014): 162–182.

4. J. Reefhuis et al., "Assisted Reproductive Technology and Major Structural Birth Defects, United States," *Human Reproduction* 24 (2009): 360–366.

5. Urs Scherrer et al., "Cardiovascular Dysfunction in Children Conceived by Assisted Reproductive Technologies," *European Heart Journal* 18 (2015): 115–119.

6. Laura Ozer Kettner et al., "Assisted Reproduction Technology and Somatic Morbidity in Childhood: A Systematic Review," *Fertility and Sterility* 103 (2015): 707–719.

7. See the discussion of imprinting diseases, including Beckwith-Weideman and Angelman in Fauser, "Health Outcomes," 170–174.

8. Hayashi et al., "Reconstitution"; Hayashi et al., "Offspring."

9. See the discussion at Kenneth Bridges, "Malaria and the Sickle Hemoglobin Gene," Information Center for Sickle Cell and Thalassemic Disorders, Brigham and Women's Hospital, April 2, 2002, available at http://sickle.bwh.harvard.edu/malaria_sickle.html.

10. Malaria is, in some respects, a bad example as it might be wiped out through, for example, genetic editing of the mosquitoes that are an essential part of its life cycle. See the discussion in Sharon Begley, "Mosquito DNA Altered to Block Malaria, not Spread It," *Stat*, (2015), available at http://www.statnews.com/2015/11/23/malaria-mosquitoes-gene-drive-crispr/. On the other hand, it is the best example we know about the heterozygote advantage situation, where having two copies of an allele is a disease, but having one is a protection.

11. For example, Shoukhrat Mitalipov and his group at Oregon Health and Science University (the first to make SCNT work in primates) have worked extensively on macaque reproduction since 1998.

12. As I edited this chapter, Congress has just amended the appropriations bill for fiscal year 2016 to prevent, in effect, the FDA from allowing any clinical trials of genome editing in human embryos (even though no such trials are likely for many years). Kelly Servick, "FDA Gets 5% Bump and Ban on Gene Editing", *Science Insider* (December 18, 2015), available at http://news.sciencemag.org/funding/2015/12/budget-agreement-boosts-u-s-science#FDA.

13. For a discussion of the trial, see Ganesh Suntharalingam et al., "Cytokine Storm in a Phase 1 Trial of the Anti-CD28 Monoclonal Antibody TGN 1412," *New England Journal of Medicine* 355 (2006); 1018–1028. For a discussion of ethical issues in this trial, see Ezekiel J. Emanuel and Franklin G. Miller, "Money and Distorted Ethical Judgments about Research: Ethical Assessment of the TeGenero TGN1412 Trial," *American Journal of Bioethics* 7 (2007): 76–81.

14. For a discussion of several very questionable stem cell clinics, see Stephen Barrett, "The Shady Side of Embryonic Stem Cell Therapy," *QuackWatch*, available at http://www.quackwatch.org/06ResearchProjects/stemcell.html. A

good review of stem cell "tourism" is Eliza Barclay, "'Stem Cell Tourists' Go Abroad for Unproven Treatments," *National Geographic News,* December 3, 2008, available at http://news.nationalgeographic.com/news/2008/12/081203-stem-cell-tourism.html. Another study shows how stem cell clinics (misleadingly) portray themselves to possible patients; see Darren Lau et al., "Stem Cell Clinics Online: The Direct-to-Consumer Portrayal of Stem Cell Medicine," *Cell Stem Cell* 3 (2008): 591–594.

15. See I. Glenn Cohen, "How To Regulate Medical Tourism (and Why It Matters for Bioethics)," *Developing World Bioethics Journal* 12 (2012): 9–20. And see more generally Cohen's book on medical tourism, *Patients with Passports: Medical Tourism, Law, and Ethics* (New York: Oxford University Press, 2014).

14. Family Relationships

1. Leo Tolstoy, *Anna Karenina* (first published in book form in 1878 after serialization in 1873–1877).

2. Professor Susan Golombok at the University of Cambridge has spent many years studying the psychological effects of "new" families on their members. For a few examples of her IVF work, see Susan Golombok et al., "Children Conceived by Gamete Donation: Psychology Adjustment and Mother-Child Relationships at Age 7," *Journal of Family Psychology* 25 (2011): 230–239; P. Casey et al., "Families Created by Donor Insemination: Father-Child Relationships at Age 7," *Journal of Marriage and Family* 75 (2013): 858–870.

3. Only about 25 percent of cases of intellectual disability (formerly known as mental retardation) have a known cause. "Intellectual Disability," *Medline Plus,* available at http://www.nlm.nih.gov/medlineplus/ency/article/001523.htm.

4. Easy PGD has an advantage here over fetal genetic testing. In early discussions of genetics and reproduction, Barbara Katz Rothman argued that prenatal testing, with the possibility of abortion, prevents women from bonding early and fully with their children. Barbara Katz Rothman, *The Tentative Pregnancy: How Amniocentesis Changes the Experience of Motherhood* (New York, NY: W.W. Norton & Company 1993). A woman carrying a fetus after Easy PGD will have many reasons for uncertainty, but the fetus's genetic tests will not be among them. That aspect of the child she already knows.

5. Joel Feinberg, "The Child's Right to an Open Future," in William Aiken and Hugh LaFollette, eds., *Whose Child? Children's Rights, Parental Authority, and State Power* (Lanham, MD: Rowman & Littlefield, 1980).

6. *Wisconsin v. Yoder,* 406 U.S. 205 (1972).

7. See Donald M. Borchert, ed., *Macmillan Encyclopedia of Philosophy,* 2nd ed. (Detroit, MI: Macmillan Reference, 2005). Neither the *Internet Encyclopedia of Philosophy* nor the *Stanford Encyclopedia of Philosophy* have biographical entries for Feinberg.

8. Dena S. Davis, "Genetic Dilemmas and the Child's Right to an Open Future," *Rutgers Law Journal* 28 (1997): 549–592.

9. If one believed that sexual orientation was substantially influenced by genetic variations, one might expect gametes from two homosexual people to be more likely to produce homosexual children. The science is deeply unclear and the prejudice it expresses against people with same-sex orientation is, to say the least, out of fashion.

10. A survey of thirteen European countries showed that seven had upper age limitations for IVF with donor gametes for women, either under forty-five, under fifty, or during "naturally child bearing years." Only one, Sweden, had a limit for men, requiring "that they be sufficiently young to parent an infant through childhood" (whatever that means). K. Berg Brigham, B. Cadier, and K. Chevreul, "The Diversity of Regulation and Public Financing of IVF in Europe and Its Impact of Utilization," *Human Reproduction* 28 (2013): 666–675, 669. Many clinics even in countries that do not have restrictive legislation put their own age limits on women using IVF with donor eggs. The ASRM, for example, recommends an upper age limit of fifty-five. Ethics Committee of ASRM, "Oocyte or Embryo Donation to Women of Advanced Age, A Committee Opinion," *Fertility and Sterility* 100 (2013): 337–340.

11. *Hecht v. Superior Court (Kane),* 50 Cal. Rptr. 1289 (1996). Kane, the dead man, had committed suicide, but left a note for Hecht, expressing his hope that she would use sperm he was leaving to her in his will to make children. After eventually winning her case, Ms. Hecht, then 42, tried several times to become pregnant with Kane's sperm, but without success. Debran Rowland, *The Boundaries of Her Body: The Troubling History of Women's Rights in America,* 457, n. 223 (Naperville, Ill.: Sphinx Publishing, 2004).

12. See the discussion in Ruth Landau, "Posthumous Sperm Retrieval for the Purpose of Later Insemination or IVF in Israel: An Ethical and Psychosocial Critique," *Human Reproduction* 19 (2004): 1952–1956. The issue continues to be litigated, with a recent case involving the dead man's parents' wishes for the sperm over the widow's objection. Rahel Jaskow, "Dead Reservist's Parents May Use His Sperm, against Widow's Wishes," *Times of Israel,* March 26, 2015, available at http://www.timesofisrael.com/dead-reservists-parents-may-use-his-sperm-against-widows-wishes/.

13. See the interesting discussion of the historical roots of incest prohibitions in Martin Ottenheimer, *Forbidden Relatives: The American Myth of Cousin Marriage* (Chicago: University of Illinois Press, 1996).

14. Ibid. See the map of state laws in "Cousin Marriage Law in the United States by State," *Wikipedia,* available at https://en.wikipedia.org/wiki/Cousin_marriage_law_in_the_United_States_by_state.

15. David Plotz, "The Myths of the Nobel Sperm Bank," *Slate*, February 23, 2001, available at http://www.slate.com/articles/life/seed/2001/02/the_myths_of_the_nobel_sperm_bank.html.

15. Fairness, Justice, and Equality

1. H. G. Wells, *The Time Machine* (Eastford, CT: Martino Fine Books, 2011) (reprint of the original edition, published in 1895.

2. *Gattaca,* directed by A. Niccol (Los Angeles: Columbia Pictures Corp., 1997), motion picture.

3. Maxwell J. Mehlman, *Wondergenes: Genetic Enhancement and the Future of Society* (Bloomington: Indiana University Press, 2003).

4. See James R. Flynn, *What Is Intelligence? Beyond the Flynn Effect,* exp. ed. (New York: Cambridge University Press, 2009).

5. For some of the variations in European systems, see K. Berg Brigham, B. Cadier, and K. Chevreul, "The Diversity of Regulation and Public Financing of IVF in Europe and Its Impact of Utilization," *Human Reproduction* 28 (2013): 666–675.

6. I owe this insight to my Stanford colleague Darrell Duffie, who teaches, perhaps not coincidentally, at Stanford's Graduate School of Business.

7. Of American Catholic women, for example, 68 percent have used highly effective contraceptives, just slightly under the 73 to 74 percent rate of Protestant women. Rachel K. Jones and Joerg Dreweke, *Countering Conventional Wisdom: New Evidence on Religion and Conceptive Use* (New York: Guttmacher Institute, 2011). Very slightly more American Catholics than American Protestants see IVF as moral or "not a moral issue"—77 percent to 76 percent. "Abortion Viewed in Moral Terms: Fewer See Stem Cell Research and IVF as Moral Issues," Pew Research Center, Religion and Public Life, August 15, 2013, available at http://www.pewforum.org/2013/08/15/abortion-viewed-in-moral-terms/.

8. It is hard to find good evidence about the accuracy of this view, widely held by clinicians, but there is a little evidence that bears on it in related contexts. One study in Los Angeles found that Hispanic women for whom amniocentesis had been recommended were much less likely to accept it than the general population. Debra Baker, Senait Teklehaimanot, Rosetta Hasan, and Carol Guze, "A Look at a Hispanic and African American Population in an Urban Prenatal Diagnostic Center: Referral Reasons, Amniocentesis Acceptance, and Abnormalities Detected," *Genetics in Medicine* 6 (2004): 211–218. Another more recent study showed that Hispanic women were two to three times more likely to decline additional testing, whether by NIPT or invasive methods. Shilpa Chetty, Matthew J. Garabedian, and Mary E. Norton, "Uptake of Noninvasive Prenatal Testing (NIPT) in Women Following Positive Aneuploidy Screening," *Prenatal Diagnosis* 33 (2013): 542–546. A difference among California Hispanic women might be attributed to religion, but increasing numbers of Hispanics are Protestant and a large number of non-Hispanic women are Catholic.

9. Randy Newman, "Short People," in *Wasting Light*, Warner Brothers Records, http://itunes.com (1977).

10. Of course, if she never existed, she would never be in a position to complain that she never existed. This touches on a deep philosophical topic developed by Derek Parfit and called the "nonidentity problem." I encourage those of you with an interest in philosophy to explore it further; for me, although ultimately I think Parfit is right, it makes my head hurt. "The Nonidentity Problem," *Stanford Encyclopedia of Philosophy,* available at http://plato.stanford.edu/entries/nonidentity-problem/. Parfit's own full explanation can be found in Derek Parfit, *Reasons and Persons,* 2nd ed. (Oxford: Clarendon Press, 1987). (The first edition in 1984 seems to have been his first exploration of the problem, though a few other philosophers had touched on it earlier.)

11. Easy PGD might also be able to identify some cases of "intersexuality" with genetic causes, embryos that would be born with ambiguous genitalia, neither entirely male nor entirely female. Although this can be viewed as another kind of "sex," it may be more appropriately viewed culturally as a disability and be subject to the analysis above. For a discussion of issues around intersex people, see Katrina Karzakis, *Fixing Sex: Intersex, Medical Authority, and Lived Experience* (Durham, NC: Duke University Press, 2009).

12. Mara Hvistendahl, *Unnatural Selection: Choosing Boys over Girls and the Consequences of a World Full of Men* (New York: PublicAffairs, 2011). Hvistendahl's book provides almost all the facts for the following discussion.

13. See Chapter 11, footnote 12.

16. Coercion

1. *Buck v. Bell,* 274 U.S. 200, 207 (1927).

2. There are several histories of the eugenics movement. I quite like Daniel Kevles, *In the Name of Eugenics* (Cambridge, MA: Harvard University Press, 1998). For an even more critical overview, see Diane Paul, *Controlling Human Heredity: 1865 to the Present* (Amherst, NY: Humanity Books, 1995).

3. In Germany, which was about one-third Catholic, the first law on compulsory sterilization was passed by the Nazi government six days before the announcement of the treaty, or concordat, between Germany and the Vatican, by which the Vatican, in effect, agreed to step aside from German politics. See "Forced Sterilization," *United States Holocaust Memorial Museum,* available at http://www.ushmm.org/learn/students/learning-materials-and-resources/mentally-and-physically-handicapped-victims-of-the-nazi-era/forced-sterilization.

4. Paul Lombardo's book is an excellent discussion of the case and its background. Paul A. Lombardo, *Three Generations, No Imbeciles: Eugenics, the Supreme Court, and Buck v. Bell* (Baltimore, MD: Johns Hopkins University Press, 2010). Also notable for bringing early attention to the questionable

diagnoses involved in the case is an article by Stephen Jay Gould, "Carrie Buck's Daughter," *Natural History* 93 (July 1984): 14.

5. Victoria Nourse, "*Buck v. Bell*: A Constitutional Tragedy from a Lost World," *Pepperdine Law Review* 39 (2011): 101–117.

6. *Jacobson v. Massachusetts,* 197 U.S. 11 (1905).

7. For an excellent book on the background of these two cases, see William G. Ross, *Forging New Freedoms: Nativism, Education, and the Constitution, 1917–1927* (Lincoln: University of Nebraska Press, 1994).

8. *Meyer v. Nebraska,* 262 U.S. 390 (1923).

9. *Bartels v. Iowa,* 262 U.S. 404 (1923).

10. *Pierce v. Society of Sisters,* 268 U.S. 510 (1925).

11. *Skinner v. Oklahoma,* 316 U.S. 535 (1942).

12. For more information about this case see the excellent book by Victoria F. Nourse, *In Reckless Hands: Skinner v. Oklahoma and the Near Triumph of American Eugenics* (New York: W. W. Norton, 2008).

13. *Troxel v. Granville,* 530 U.S. 57 (2000).

14. *Stanley v. Illinois,* 405 U.S. 645, 92 S. Ct. 1208 (1972) (unwed father has a constitutional right to a hearing before his parental rights are severed by adoption); *Caban v. Mohammed,* 441 U.S. 380 (1979) (state could not, under the Equal Protection Clause, give more rights to unwed mothers to block adoptions than to unwed fathers); *Santosky v. Kramer,* 455 U.S. 745 (1982) (holding the Constitution requires at least clear and convincing evidence before severing parental rights).

15. *Griswold v. Connecticut,* 381 U.S. 479 (1965) (holds unconstitutional bans on use of contraceptives by married couples); *Eisenstadt v. Baird,* 405 U.S. 438 (1972) (extends constitutional holding of *Griswold* to unmarried people); *Roe v. Wade,* 410 U.S. 113 (1973) (holds unconstitutional some bans on abortion).

16. "Population Control in Singapore," *Wikipedia,* available at https://en.wikipedia.org/wiki/Population_control_in_Singapore. For a story on the "love boats," see Bryan Walsh, "Love, Exciting and New," *Time* (February 17, 2003).

17. For a detailed discussion of this law, see Sun-Wei Guo, "China: The Maternal and Infant Health Care Law," *eLS (Encyclopedia of the Life Sciences): Citable Reviews in the Life Sciences* (Hoboken, NJ: John Wiley & Sons, 2012).

18. Elisabeth Rosenthal, "Scientists Debate China's Law on Sterilizing the Carriers of Genetic Defects," *N.Y. Times* (Aug. 16, 1998):14.

19. Cass Sunstein and Richard Thaler, *Nudge* (New Haven, CT: Yale University Press, 2008).

20. "Nutrition Information," Wendy's fast food restaurant, available at https://www.wendys.com/en-us/hamburgers/daves-hot-n-juicy-triple#. Adding various toppings such as lettuce, tomato, mayonnaise, ketchup, mustard, and pickles, takes the toll to 1150 calories.

21. Even disease risks with low penetrance might be viewed in this light. An embryo with a 100 percent chance of eventually developing type 2 diabetes is a medical issue, but is choosing between embryos with a 5 percent or 15 percent chance of developing type 2 diabetes—or a 95 percent or 85 percent chance of avoiding it—a "disease" issue or an "enhancement"?,

22. Or, in the case of those abortion statutes, they may view the unenforceable but politically attractive statute as a stepping stone to changing the Supreme Court's doctrine on abortion more broadly.

23. *Rust v. Sullivan,* 500 U.S. 173 (1991).

17. Just Plain Wrong

1. Calif. Penal Code, §§598c, 598d. "California Proposition 6 (1998)," *Wikipedia,* available at https://en.wikipedia.org/wiki/California_Proposition_6_%281998%29.

2. My first year contracts professor, Art Leff, wrote an essay, one part of which I have always remembered. After concluding that there is no clear source for morality, he ends, "Nevertheless: Napalming babies is bad. Starving the poor is wicked. Buying and selling each other is depraved. Those who stood up to and died resisting Hitler, Stalin, Amin, and Pol Pot—and General Custer too—have earned salvation. Those who acquiesced deserve to be damned. There is in the world such a thing as evil. [All together now:] Sez who? God help us." Arthur Allen Leff, "Unspeakable Ethics, Unnatural Law," *Duke Law Journal* 1979: 1229–1249, 1249. (This essay was, in many respects, an extension of his earlier book review of Richard Posner's *Economic Analysis of Law.* Arthur Allen Leff, "Economic Analysis of Law: Some Realism about Nominalism," *Virginia Law Review* 60 (1974): 451–482.)

3. U.S. Catholic Church, *Catechism of the Catholic Church,* Section 2369 (Washington, DC: USCCB Publishing, 1995). The Catechism of the Catholic Church further discusses marital relations with particular relevance to reproduction in Section 2366, 2376, and 2377. Section 2366 stresses "the inseparable connection, established by God, which man on his own initiative may not break, between the unitive significance and the procreative significance which are both inherent to the marriage act." Section 2377 condemns artificial insemination and IVF (even without donor gametes) because "they dissociate the sexual act from the procreative act."

4. *Encyclical Letter Humanae Vitae of the Supreme Pontiff Paul VI,* Section 12, July 25, 1968, available at http://w2.vatican.va/content/paul-vi/en/encyclicals/documents/hf_p-vi_enc_25071968_humanae-vitae.html.

5. In another technology, gamete intra-fallopian transfer (GIFT), a surgically removed egg and sperm "left over" from marital sexual intercourse are injected together into the fallopian tube in the hopes fertilization will follow. (The need to

use sperm retrieved from marital sex comes both from the desire to rely on natural marital sexual intercourse and from a ban on masturbation.) The morality of GIFT is under continuing discussion and the Vatican has not pronounced a definitive position on it, so infertile married couples may use it or not as their consciences dictate. John M. Haas, *Begotten, Not Made: A Catholic View of Reproductive Technology* (Washington, DC: United States Catholic Conference, 1998), available at http://www.usccb.org/issues-and-action/human-life-and-dignity/reproductive-technology/begotten-not-made-a-catholic-view-of-reproductive-technology.cfm.

6. Fr. Tad Pacholczyk, "Do Embryos Have Souls?," National Catholic Bioethics Center, available at http://www.ncbcenter.org/document.doc?id=846.

7. The following publications are from the Committee on Jewish Law and Standards of the Conservative Rabbinical Assembly: Rabbi Elliot Dorff, "Artificial Insemination, Egg Donation, and Adoption," 1994, Rabbinical Assembly, available at https://www.rabbinicalassembly.org/sites/default/files/public/halakhah/teshuvot/19912000/dorff_artificial.pdf; Rabbi Aaron L. Mackler, "In Vitro Fertilization," 1995, Rabbinical Assembly, available at https://www.rabbinicalassembly.org/sites/default/files/public/halakhah/teshuvot/19912000/mackler_ivf.pdf; Rabbi Mark Popovsky, "Choosing Our Children's Genes: The Use of Preimplantation Genetic Diagnosis," 2000, Rabbinical Assembly, available at https://www.rabbinicalassembly.org/sites/default/files/public/halakhah/teshuvot/20052010/Popovsky_FINAL_preimplantation.pdf.

8. See the discussion in Marcia C. Inhorn, "Fatwas and ARTs: IVF and Gamete Donation in Sunni v. Shia Islam," *Journal of Gender, Race and Justice* 9 (2005): 291.

9. George Bernard Shaw, *Caesar and Cleopatra,* Act II, in *Three Plays for Puritans* (1901).

10. See the discussion in "Moral Non-Naturalism," *Stanford Encyclopedia of Philosophy,* available at http://plato.stanford.edu/entries/moral-non-naturalism/#NatFal.

11. Leon R. Kass, "The Wisdom of Repugnance," *The New Republic* (June 2, 1997): 17–26.

12. Ibid., 20.

13. Leon R. Kass, "Making Babies—The New Biology and the 'Old' Morality," *Public Interest,* No. 26 (Winter 1972): 18–56. Interestingly, to me, his first argument echoes my own concern about the safety of untried experiments on unconsenting embryos. He does not, however, stop there.

18. Enforcement and Implementation

1. "This Constitution, and the Laws of the United States which shall be made in Pursuance thereof; and all Treaties made, or which shall be made, under the Authority of the United States, shall be the supreme Law of the Land; and the

Judges in every State shall be bound thereby, any Thing in the Constitution or Laws of any State to the Contrary notwithstanding." U.S. Constitution, Art. VI, Cl. 2.

2. *United States v. Lopez,* 514 U.S. 549 (1995). This case struck down the Gun-Free School Zones Act of 1990.

3. *United States v. Morrison,* 529 U.S. 598 (2000). This case also ruled that the statute could not be upheld under the congressional power to enforce the Fourteenth Amendment.

4. *National Federation of Independent Business v. Sebelius,* 567 U.S. ___, 132 S.Ct 2566 (2012).

5. Partial-Birth Abortion Ban Act of 2003, 18 U.S.C. §1531.

6. *Gonzales v. Carhart,* 550 U.S. 124 (2007).

7. Ibid. at 169.

8. Similar objections can be made under many state constitutions, although state constitutions can also give rise to other claims. For example, California's constitution has a broad right of privacy that might be invoked in an Easy PGD case. State constitutional provisions are too numerous and varied to be discussed here; I just note that they may well be important.

9. *Lochner v. New York,* 198 U.S. 45 (1905).

10. *Griswold v. Connecticut,* 381 U.S. 479 (1965).

11. *Skinner v. Oklahoma,* 316 U.S. 535 (1942).

12. Professor John Robertson has long been the leading proponent of the reproductive liberty position. John A. Robertson, *Children of Choice: Freedom and the New Reproductive Technologies* (Princeton, NJ: Princeton University Press, 1996). Robertson himself has taken a somewhat more limited position on the ethical propriety of PGD for less serious traits and hence, implicitly, on some uses of Easy PGD. See, e.g., John A. Robertson, "Extending Preimplantation Genetic Diagnosis: Medical and Non-Medical Uses," *Journal of Medical Ethics* 29 (2002): 213–216. Professor Jamie King provides a somewhat different analysis in Jamie Staples King, "Predicting Probability: Regulating the Future of Preimplantation Genetic Screening," *Yale Journal of Health Policy, Law, and Ethics* 8 (2008): 283–358.

13. *Moore v. City of East Cleveland,* 431 U.S. 494 (1977). The power of the *Moore* decision is undercut by the fact that only four justices joined the main opinion striking down the ordinance. Justice Stevens concurred only in the judgment, not the reasoning, and four other justices dissented. As a result, the plurality opinion's reasoning is not a binding precedent, though it may have persuasive power.

14. *Troxel v. Granville,* 530 U.S. 57 (2000). The fact that there was no majority opinion is an indication that these issues are difficult for the justices.

15. *Wisconsin v.Yoder,* 406 U.S. 205 (1972).

16. Italian Constitutional Court (No. 151/09), available at http://www.cortecostituzionale.it/actionSchedaPronuncia.do?anno=2009&numero=151; German Federal Court of Justice decision of 6 July, 2010, 5 StR 306/09 (NJW 2010, 2672; NStZ 2010, 579). [Citation to German case].

17. This case, *Wollschlaeger v. Governor*, has a remarkably tangled history. On December 17 2015, the same three judge panel issued its third opinion on the case, each of them upholding the law by the same 2 to 1 vote. *Wollschlaeger v. Governor of Fla.* (*Wollschlaeger I*), 760 F.3d 1195 (11th Cir. 2014), *opinion vacated and superseded on reh'g, Wollschlaeger v. Governor of Fla.* (*Wollschlaeger II*), 797 F.3d 859 (11th Cir. 2015), *Wollschlaeger v. Governor of Fla.* (*Wollschlaeger III*), *opinion vacated sua sponte and superseded on reh'g.* __ F.3d __ (11th Cir. Dec. 14, 2015). The plaintiffs have sought a rehearing by the en banc Eleventh Circuit and the case might ultimately reach the Supreme Court. See the critical discussions by Eugene Volokh of both *Wollschlaeger II*, "Court Upholds Florida Law Restricting Doctor-Patient Speech about Guns," *Washington Post,* July 29, 2015, available at https://www.washingtonpost.com/news/volokh-conspiracy/wp/2015/07/29/court-upholds-restriction-on-doctor-patient-speech-about-guns/, and *Wollschlaeger III*, "'Docs v. Glocks' Decision Guts Free Speech,"*Anything Peaceful* (blog) Dec. 17, 2015), available at http://fee.org/anythingpeaceful/docs-vs-glocks-law-violates-of-free-speech/

18. *Pickup v. Brown,* 740 F.3d. 1208 (9th Cir. amended opin. Jan. 29, 2014), *cert. den.* The Ninth Circuit denied an en banc rehearing in the case over the dissent of several judges; the Supreme Court declined to review the case.

19. *Stuart v. Camnitz,* 774 F.3d 238 (4th Cir. 2014), *cert. denied,* __ U.S. ___, 135 S. Ct. 2838 (2015)

20. The *Stuart* decision appears to conflict directly with the Fifth Circuit's decision that forcing doctors to show ultrasounds of fetuses to patients did not violate the doctors' First Amendment rights. *Texas Medical Providers Performing Abortion Services v. Lakey,* 667 F.3d 570 (5th Cir. 2012). See Lyle Denniston, "Ultrasound Issue Headed to the Court," *SCOTUSblog,* December 28, 2014, available at http://www.scotusblog.com/2014/12/ultrasound-issue-headed-to-the-court/. Nonetheless, the Supreme Court in June 2015 declined to review *Stuart.*

21. *Rust v. Sullivan,* 500 U.S. 173 (1991).

22. I. Glenn Cohen, *Patients with Passports: Medical Tourism, Law, and Ethics* (New York: Oxford University Press, 2014).

23. "Additional Protocol to the Convention for the Protection of Human Rights and Dignity of the Human Being with Regard to the Application of Biology and Medicine, on the Prohibition of Cloning Human Beings," Council of Europe, January 12, 1998, available at http://conventions.coe.int/Treaty/en/Treaties/Html/168.htm. The protocol's likely effect is discussed broadly in N.

Somekh, "The European Total Ban on Human Cloning: An Analysis of the Council of Europe's Actions in Prohibiting Human Cloning," *Boston University International Law Journal* 17 (1999): 397–423.

24. "Child Sex Tourism," *Wikipedia,* available at https://en.wikipedia.org/wiki/Child_sex_tourism#Penalties_for_child_sex_tourists.

25. Guido Pennings, et al., "Cross Border Reproductive Care in Belgium," *Human Reproduction*, 24 (2009): 3108–3118.

26. Maxwell J. Mehlman, *Wondergenes: Genetic Enhancement and the Future of Society* (Bloomington: Indiana University Press, 2003, 191.

Conclusion

1. See, for example, James R. Ferguson, "Scientific Inquiry and the First Amendment," *Cornell Law Review* 64 (1979): 639; Steve Keane, "The Case against Blanket First Amendment Protection of Scientific Research: Articulating a More Limited Scope of Protection," *Stanford Law Review* 59 (2006): 505–550. (I cannot resist noting that this article started as a student paper in one of my classes.)

2. "Pre-Implantation Genetic Diagnosis (PGD)," Fertility Treatment Abroad, available at http://fertility.treatmentabroad.com/tests-and-investigations/preimplantation-genetic-diagnosis-pgd. The article lists Switzerland as banning PGD, but 60 percent of the voters said yes to a referendum to relax that ban in June 2015. "Switzerland Votes 'Yes' in PGD Embryo Tests Referendum," *DW*, available at http://www.dw.com/en/switzerland-votes-yes-in-pgd-embryo-tests-referendum/a-18516967. The article states that the amendment will allow PGD for carriers of severe hereditary diseases, as well as those unable to naturally conceive a child. The process should not be used, however, "to bring about certain characteristics in the child or to operate research."

3. European Society for Human Reproduction and Embryology, "European Court of Human Rights Denounces Italy's Ban on PGD as 'Disproportionate'," available at https://www.eshre.eu/Guidelines-and-Legal/Legislation-for-MAR-treatments/Specific-legislation-for-MAR-treatments/Italys-ban-on-PGD.aspx. A copy of the judgment in the case, *Costa and Pavan v. Italy*, is available at http://hudoc.echr.coe.int/eng?i=001-112993.

4. German Federal Court of Justice decision of 6 July, 2010, 5 StR 306/09 (NJW 2010, 2672; NStZ 2010, 579)See the discussion in Susanne Benöhr-Laqueur, "Fighting in the Legal Grey Area: An Analysis of the German Federal Court of Justice Decision in Case Preimplantation Genetic Diagnosis," *Poiesis and Praxis* 8 (2011): 3–8

5. Of course, if the permissible line for selection is set at sufficiently serious diseases—those invariably fatal at an early age or inevitably causing serious

intellectual deficits—no activist will have those diseases. Therefore, no individuals would be able to say that "under this system I would never have been born." Their family members would have to make that argument for them.

6. See Pier Jaarsma and Stellan Welin, "Autism as a Natural Human Variation: Reflections on the Claims of the Neurodiversity Movement," *Health Care Analysis* 20 (2011): 20–30.

GLOSSARY

aCGH Array comprehensive genomic hybridization, sometimes abbreviated ACGH or array CGH, is a laboratory technique that can detect aneuploidies (samples with the wrong number of chromosomes) as well as much smaller deletions, duplications, or movements of DNA within or between chromosomes.

Allele One variant of a gene. A gene may have many different alleles, each with a slightly different DNA sequence. Some may be harmful, some helpful; some may make neutral changes, and many of them will make no discernible change in the organism. The most common allele of a gene is sometimes called the "wild type" allele.

Alpha-fetoprotein A protein produced by a human fetus. Some of it crosses the placenta and can be found in the pregnant woman's blood serum, where its levels have some value in predicting Down syndrome or neural tube defects.

Aneuploid A cell that has the wrong number of chromosomes. Human cells should normally have 46 chromosomes (92 when preparing to divide). Trisomies (three copies of what should be a paired chromosome) are the most common aneuploidies in living humans.

ASRM The American Society for Reproductive Medicine is a nonprofit organization dedicated to reproductive medicine. It has an affiliate only for reproductive medicine clinics called SART, the Society for Assisted Reproductive Technologies.

Autosome One of the chromosomes that is not a sex chromosome. They come in pairs—in humans 22 pairs named chromosomes 1 through 22.

Blastocyst A human embryo from the fifth day after fertilization until implantation. It takes the form of a hollow sphere containing the inner cell mass.

Centromere The constricted part of a chromosome between its p (short) and q (long) arms. It is essential for the proper allocation of chromosomes to daughter cells when cells divide.

Chromosome A long DNA molecule wrapped around a protein backbone. Chromosomes carry the DNA in cells and, by their duplication and allocation to daughter cells, preserve that genetic information across cell divisions and generations.

CLIA The Clinical Laboratories Improvements Amendments of 1988 are the main way in which the federal government, acting through the FDA, CDC, and the Center for Medicare and Medicaid Services (CMS), regulates clinical laboratories.

Clone/cloning A clone is an exact copy of something and cloning is the process of copying. In biology it generally refers to nonsexual reproduction, which produces offspring that have the same DNA (except for occasional mutations) as their progenitors. Identical twins are clones created when one embryo separates into two. Other kinds of cloning in mammals are achieved through a process called somatic cell nuclear transfer (SCNT).

Codon Three "letters" or bases of DNA or RNA that stand for an amino acid, stop, or start.

CNVs Copy number variations are places in DNA where some individuals have more or fewer copies of a particular gene sequence than normal. Some of the genes may have been duplicated or some existing copies may have been deleted.

Diploid A cell with two copies of each of the chromosomes, a pair of each autosome and two sex chromosomes. This is the normal state of cells except when they are preparing to divide, when their chromosomes are doubled.

DNA Deoxyribonucleic acid, the molecule that conveys information about protein sequence and expression within one cell and across generations of a cell and of an organism. DNA was discovered in 1869 in pus-soaked bandages, but its significance was not widely known or accepted until the 1950s.

Dominant A Mendelian trait is called dominant if it only takes one allele of a gene to produce the associated trait (except for traits in males where the relevant gene is on the X chromosome, called X-linked, or the Y chromosome). Hence, these traits will appear even if only one of the parents contributes the allele.

Easy PGD A name I made up for inexpensive whole genome analysis of an embryo before implantation when the egg that contributed to the embryo was artificially derived from stem cells.

Embryo The earliest developmental stage of a multicellular eukaryote. In humans, embryos are often (but not universally) seen as coming into being with fertilization of the egg and lasting until (by convention) the end of the eighth week after fertilization, when it is renamed a fetus. Different early stages of embryonic development are sometimes referred to by different names, such as blastocyst and morula.

Eukaryote An organism whose cells contain a nucleus and other organelles protected within the cell by their own membranes. Eukaryotes are one of the three types of life, along with the prokaryotes, bacteria and archaea (single-celled microbes that are, outwardly at least, very similar to bacteria).

Euploid A cell with the "right" number of chromosomes. In humans that is usually 46 chromosomes, in 22 pairs from 1 to 22 and with two sex chromosomes, though it also includes cells ready to divide with four sets of chromosomes and gametes with only one set of chromosomes.

Exome The part of a genome made up of exons, the parts of a DNA sequence that specify the order and identity of amino acids in a protein. The exome is less than 2 percent of the human genome.

Exon The part of a gene that specifies the order and identity of the amino acids in a protein. Exons in humans and other eukaryotes are separated by introns.

FDA The Food and Drug Administration, which in the United States regulates drugs, medical devices, and biological products, among other things. It is particularly important where its approval is required before a product can be marketed.

FDCA The Federal Food, Drug, and Cosmetic Act, passed in 1938 and frequently amended. This statute gives the FDA its power to regulate drugs and medical devices.

FISH Fluorescence in situ hybridization, a technique that attaches a fluorescent marker to pieces of DNA or RNA, which then attach to sites containing sequences that are complementary to theirs (CGGTAT's complementary sequence is GCCATA in DNA; the equivalents in RNA would be CGGUAU and GCCAUA).

Gamete A mature germ cell (egg or sperm) that is able to unite with a gamete from the other sex to form a zygote.

Gene A unit of heredity that is now known to be physically embodied in DNA. The exact definition of gene is disputed but it generally means a stretch of DNA that specifies the structure for a protein or an RNA molecule that is not messenger or transfer RNA, two kinds intimately associated with the "reading" of DNA.

Gene family A set of very similar genes, usually produced by past duplication of an original gene. The functional members of a gene family will often do similar or identical jobs.

Genetics The study of genes and heredity. (See **Genome.**)

Genome The sequence of all the DNA in an organism or individual.

Genomics A term to describe the study not of individual genes but of large parts of a genome. Supposedly coined in 1986 it has become increasingly trendy as a fancy-sounding term since 2000. (See **Genetics**).

Genotype The genetic makeup of a cell or organism. It is like the genome but "genotype" usually is used in the context of which particular DNA variations are found in a cell or organism.

GWAS A genome-wide association study is an effort to look at many different DNA variations in order to determine which if any of the variations are associated with a given trait. GWAS originally mainly referred to research with SNP chips but can now include studies using broader sequencing techniques, including whole exome and whole genome sequencing.

Haploid A cell with only half of its normal number of chromosomes as a result of meiosis. In humans haploid cells have 23 chromosomes with only one each of chromosomes 1 through 22 and one sex chromosome. The only human cells that are normally haploid are sperm and, for a brief time just after fertilization when they have completed meiosis but before their pronuclei merge with that of the sperm, eggs.

hESCs Human embryonic stem cells are cells derived from the inner cell masses of blastocysts and kept alive, and undifferentiated, in laboratory equipment containing culture medium.

Heterochromatin Stretches of chromosomes where the DNA is very tightly packed. Heterochromatin is rarely expressed (or read for its information content). It is also often difficult to sequence accurately. It is contrasted with "euchromatin," the remainder of the chromosomes, which contain most active genes.

HTC/P Human cells, tissues, and cellular or tissue-based products, a classification used by the FDA in regulation. It includes not only tissues like skin, bone, and corneas, but also cell replacement therapies and gene therapies.

ICSI Intracytoplasmic sperm injection is an assisted reproduction technique where a single sperm will be injected directly into an egg, initiating fertilization.

Indels Insertions and deletions in DNA that are fairly common variations in DNA sequences and can have serious negative effects, especially if the insertion or deletion is not a number of bases divisible evenly by three, because those indels can cause "frame shifts," causing the codons after them to be read differently.

Inner cell mass The cluster of cells in the inside of a blastocyst that later form the embryo, fetus, baby, and book reader. Inner cell mass cells are the source of hESCs.

iPSCs Induced pluripotent stem cells are cells that have been modified by the injection of certain genes or proteins so that they revert to something similar to an embryonic state, where they appear to be capable of forming many different, and perhaps all, cell types.

IVF In vitro fertilization is an assisted reproduction technique where egg and sperm are united outside a woman's body to form a zygote and embryo, which is intended to be transferred into a women's uterus for implantation and subsequent pregnancy.

Karyotype The number and appearance of chromosomes in a cell, used in genetic testing to find aneuploidies, large deletions or insertions, and large translocations of DNA from one place in the chromosomes to another.

LINES Long interspersed elements are pieces of DNA that have been copied and put into new locations by transposons or retrotransposons. They contrast with SINES.

Mitochondria/Mitochondrion Organelles found in the cytoplasm (the fluid outside the nucleus) of eukaryotic cells that are involved in the production of energy for the cell. They are invariably described as shaped like kidney beans.

Mitochondrial DNA Mitochondria have their own DNA found in one circular chromosome inside the mitochondrion. In humans the mitochondrial DNA has about 16,600 base pairs and codes for about thirty-seven genes. It is thought that all eukaryotes are descended from a merger of two prokaryotes, one of which became the mitochondrion.

Mosaicism An organism that has more than one version of a genome, all derived from the same zygote. (If the different cell populations descend from different zygotes, the organism is a chimera.)

MSAFP Maternal serum alpha-fetoprotein screening is the process of looking for alpha-fetoprotein in a pregnant woman's blood serum as a screening test for Down syndrome or neural tube defects. It is now one part of a large screening test and may be replaced, for some purposes, by NIPT.

Mutation A mutation is a permanent change in DNA sequences in a cell or organism. Mutations can take many forms and may be harmful, beneficial, or neutral. The mutations may occur in germ cells and thus affect subsequent generations or may appear only in the body's cells and thus die with the organism (if not before).

NIPT Noninvasive prenatal testing is a technique for sequencing and assessing the DNA of a fetus without physically invading the uterus. It relies on sequencing very large numbers of small pieces of DNA in the blood that are created when cells, from the fetus or the pregnant woman, break down. In pregnant women around 10 percent of this cell-free DNA is from the fetus.

Nuclear DNA The DNA found in a cell's nucleus, contrasted with the mitochondrial DNA (or, in photosynthesizing plants, also the DNA in the chloroplast, organelles involved in photosynthesis.

Nucleus The part of a eukaryotic cell partitioned off from the rest of the cell by the nuclear membrane that contains the chromosomes. Like the nucleus of an atom, it comes from the classical Latin word for "kernel" or "core," a diminutive of the word for "nut."

Oligonucleotide A short stretch of DNA or RNA. Oligonucleotides (often referred to as oligos) are often used in various DNA analysis techniques. The term usually refers to twenty-five or fewer bases, but long pieces of DNA with 200 bases have been referred to as oligos—there seems to be no formal size limit.

ORF An open reading frame is the part of a DNA sequence that can be transcribed into messenger RNA and hence can give rise to proteins or functional RNAs. ORFs start with the start codon and end with one of the three stop codons.

Organelle One of several kinds of small functional bodies inside a cell. Organelles are separated from the rest of the cell by their own membranes. The word is derived from a term meaning "little organ." Only eukaryotes have organelles.

Pathogenic Causing disease. In the context of genetics, it refers to a pathogenic variation or allele, though it can be pathogenic without certainly causing the disease (i.e., it may have less than perfect penetrance).

Penetrance The percentage of the time a particular DNA variation will cause an associated disease, trait, or other "phenotype."

PGD Preimplantation genetic diagnosis is the process of doing genetic testing on one or more cells from an embryo for determining the likely traits of those embryos in order to decide which embryos to transfer for possible implantation and ultimately birth.

Phenotype An individual's observable traits (including traits not observable by the eye, such as ABO blood group). Genetics is mainly about the associations between genotypes and phenotypes.

Polymorphism Polymorphism comes from the Greek words for "many forms." In genetics it means a genomic variation that is (fairly) commonly found with different sequences in a population. Polymorphisms may be different alleles of a gene but they are more often differences in DNA that are not part of genes.

Pseudogene A nonfunctional stretch of DNA that looks like a functional gene, to which it is probably related by descent.

Recessive A Mendelian trait is called recessive if it takes two copies of an allele to produce the associated trait. Hence, these traits will only appear if each of the two parents contributes the allele (except for traits in males where the relevant gene is on the X chromosome, called X-linked, or the Y chromosome).

REI Reproductive endocrinology and infertility, the medical subspecialty (of obstetrics/gynecology) that deals with IVF and most other forms of assisted reproduction.

Retrotransposon A retrotransposon is a transposon that moves from one place to another by first being copied into RNA, which later is copied back into DNA that is inserted in a different location.

RNA Ribonucleic acid is a cousin of DNA. It differs by using sugar molecules called ribose in the sides of its molecules instead of deoxyribose. RNA has many important functions in cells and particularly in turning DNA instructions into cells and organisms.

SCNT Somatic cell nuclear transfer, or the Dolly process, is a form of cloning where the nucleus of one egg is removed and another cell's nucleus is inserted in its place (or through fusing another cell entirely into the much larger egg). SCNT is the technique involved in what people refer to as "cloning."

Sex chromosomes Two chromosomes that determine (usually) whether an organism will be male or female. In mammals these are the X and Y chromosomes. An embryo with two X chromosomes will become female; one with one X and one Y chromosome will become male. Organisms cannot normally have two Y chromosomes as their only sex chromosomes because their mothers normally would have only X chromosomes to give them. An embryo with only two Y chromosomes would not be viable. Some sex chromosome aneuploidies, such as X alone, XYY, XXY, XXX, and so on are viable and not rare.

SINES Short interspersed elements are short pieces of DNA copied or inserted into the genome by transposons or retrotransposons. Along with LINES, they make up much of what some people call "junk" DNA.

SNP Single nucleotide polymorphisms are common variations in DNA that occur at one base pair. For example, if 40 percent of people had a G at a particular position in their melanin gene and 60 percent had a T there, that location would be a SNP. SNPs have been very useful as signposts in DNA research, indicating the region in which a DNA variation with some phenotypic effect might be found.

SNP chip A chip or an array that contains hundreds of thousands or millions of different SNPs, allowing a DNA sample to be probed for its SNPs in a fast and inexpensive way.

Telomere Each chromosome has a telomere at each end, a long stretch of highly repetitive DNA that appears to convey no information but to serve to protect the informative parts of the chromosome near its ends during cell division.

Transposon A transposable element is a small piece of DNA that can move around inside a cell's genome. The changes in location of a transposon, a kind of mutation, can have effects on the genes and the organism.

Trophectoderm The outer part of a blastocyst, which forms the hollow ball in which the inner cell mass sits. The trophectoderm will eventually become the placenta and other supporting tissues of a pregnancy but not the fetus or baby.

Zygote A fertilized egg. The DNA from the sperm and the egg do not come together, though, until the zygote is in the process of dividing to become a two-cell embryo.

ACKNOWLEDGMENTS

I want to start by thanking the Big Bang, without which this book would be neither possible nor necessary. That's the problem with acknowledgments. How far back do you go, when the writing of a book was influenced by many different people and things over so many years?

The idea that became this book took shape after a conference in October 2010 in Munster, Germany, where I heard Professor Laurie Zoloth talk about some ramifications of making eggs and sperm from stem cells. (I saw Laurie in early December 2015, just before my final edit of this book—she remembered the talk and told me that she hadn't written it up. Yet.) In January 2011 I talked with Elizabeth Knoll, an editor (soon my editor) at Harvard University Press, about the idea for this book. By February 2011 it had its title, *The End of Sex*, suggested by my friend and colleague Buzz Thompson at a dinner with our wives. (The four of us, though, have at least three different recollections of the location of that dinner, but we all agree that as soon as Buzz said the title, we were all sold.) By April I had written a book proposal; by July I had a contract; and by November I had a working outline, which has changed only slightly. After that, things slowed down; it took over four and half years from that first conversation for me to deliver the final manuscript and over five years to actual publication.

During that time, many people helped. Several research assistants toiled on this tome. The latest, Kristin Liska, got stuck with the bulk of the drudgery, which she did, along with (I hope) some nondrudgery, extraordinarily well, but her predecessors from 2010 and onward—Mark Hernandez, Amanda Rubin, and Ben Chagnon–were also invaluable. The fellows during these years at the Center for Law and the Biosciences and the Stanford Program in Neuroscience and Society were good sources of comments on the book, as well as of constant stimulation:

Matt Lamkin, Jake Sherkow, Patti Zettler, Dmitry Karshtedt, Stephanie Bair, Roland Nadler, Andrea Wang, Natalie Salmanowitz, and Vera Blau-McCandliss. In a special category, Emily Murphy and Amy Knight provided valuable help before, during, and after their days as Stanford Law students.

Colleagues around Stanford University were enormously helpful, particularly Barry Behr, Mildred Cho, Kelly Ormond, and Chris Scott, who read draft chapters for me and corrected at least some of my errors, and Renée Reijo Pera, whose discussions with me about stem cells and gametes kept me believing in my argument. Darrell Duffie provided great insights and support, and, of course, Buzz Thompson gave me my title. I also benefited greatly from two conversations with the late Carl Djerassi, one of the fathers of the oral contraceptive and a deep and original thinker about human reproduction (among many other things).

I owe a huge debt to the twenty or thirty audiences with whom I have talked about the ideas in this book. They ranged from academic seminars and lectures at Stanford (many), Harvard, Duke, Davidson, and elsewhere to Stanford alumni clubs or local Rotary clubs. My Stanford Law School colleagues heard me give talks on this book at least three times and made many useful suggestions (as well as posed many difficult questions), as did the other academic audiences. In some ways, though, the nonacademic audiences were the best—they helped me work out how to explain this material to bright, interested, nonexperts. I wish I could remember every person who gave me an idea or a turn of phrase that made it into this book; I cannot, so a collective "thank you" will have to suffice.

Throughout the process, the Stanford's extraordinary Robert Crown Law Library, under the direction of both the late Paul Lomio and his successor Beth Williams, provided everything I needed, often before I knew I needed it. And I want to extend special thanks to some of my Stanford Law School deans: the late John Hart Ely for hiring me, Kathleen Sullivan for creating the Center for Law and the Biosciences and naming me director, and the very supportive deans during my years of writing this book, Larry Kramer and Elizabeth Magill. Stanford Law School and Stanford University are extraordinary institutions; this book would not exist without them.

Harvard University Press also proved a good home for this book. Elizabeth Knoll and Thomas LeBien, at different stages of the project and in different ways, provided much needed, and appreciated, guidance.

Which leads me a step farther back in my thanks, to before this book was even a twinkle in my eye. In 1990 my Stanford colleagues Paul Berg, David Botstein, and Lucy Shapiro added me to the planning committee for a two-and-a-half-day Stanford Centennial Symposium on the then-new Human Genome Project, held in January 1991. That launched me on the professional path of my past twenty-five years. Shortly thereafter, Luca Cavalli-Sforza, Marc Feldman, and (non-Stanford) Mary-Claire King cemented my connection to genetics through the Human Genome Diversity Project—and through their continuing help and friendship.

Who's left before we get back to the Big Bang? Too many to name, but I do want to single out one teacher and one writer. Ken Turknette, my speech and debate coach over forty-five years ago at Tustin High School, taught me many lessons that continue to help me, every day. And Lois McMaster Bujold has played with bioscience in her Vorkosigan universe in ways that helped inspire this book.

And, of course, for a book about human reproduction, my family. My parents, Mary Lou and the late Zett Greely, shaped me in manifold ways, of which their DNA was certainly not the most important. If not for my children, John and Eleanor, and the experience of parenthood they gave me, the deep meaning of this book for me would not exist. And, last but first, my beloved wife, Laura Butcher. Without her and her unstinting somewhat qualified belief in me as a spur, this book would have stayed just another dream. As would most of the other many good things in my life.

INDEX